INTERPRETING EVOLUTION

DARWIN & TEILHARD DE CHARDIN

H. James Birx

Prometheus Books
Buffalo, New York

Published 1991 by Prometheus Books

Editorial offices located at 700 East Amherst Street, Buffalo, New York, 14215, and distribution facilities at 59 John Glenn Drive, Amherst, New York 14228

Library of Congress Cataloging-in-Publication Data

Birx, H. James.
    Interpreting evolution: Darwin & Teilhard de Chardin / by H. James Birx.
       p.   cm.
    Includes bibliographical references and index.
    ISBN 0-87975-636-5
    1. Evolution. 2. Darwin, Charles, 1809-1882. 3. Teilhard de Chardin, Pierre, 1881-1955. 4. Evolution—Religious aspects—Christianity—History of doctrines—20th century. I. Title.
B818.B63   1990
113'.8—dc20   90-44530
                                                        CIP

Printed on acid-free paper in the United States of America

To the immortal memory of
Marvin Farber
(1901–1980)

In his honest and lucid critique of those philosophical themes of serious inquiry, especially the idealist movement from Plato to Husserl, Farber steadfastly sided with the advances in science and a rigorous use of reason. An uncompromising spokesman for both naturalism and humanism, he stressed a cosmic perspective and the evolutionary framework. His own view of reality is grounded in pervasive materialism, methodological pluralism, and a sober compassion for the human condition within sociocultural development.

Acknowledgments

I am especially thankful to Jack E. Kind for his friendship during the writing of this manuscript.

Also, I wish to express my sincere gratitude and deep appreciation to the following individuals who have been particularly supportive and helpful during the preparation of this project: Lorraine Baranski, Barbara Bergstrom, Marie C. Coffey, Terry L. DiDomenico, Valerie Ferenti-Cognetto, Todd B. Hilligas, Clifford G. Holland, Richard W. Hough, Tim Madigan, Ashley Montagu, Karl Schmitz-Moormann, Paul S. Sikorskyj, Victor A. Tomovich, and Tom Wolf.

Furthermore, I am profoundly thankful for those gracious accommodations given to me by Paul Kurtz, Steven L. Mitchell, and Eugene M. O'Connor of Prometheus Books throughout the production of this volume.

I also extend my heartfelt compliments to Paulette B. Kirsch for her expert, conscientious, and invaluable secretarial assistance during the writing of this manuscript.

Lastly, chapter six first appeared as the article "Darwin and Teilhard: Some Final Thoughts" in *Proteus: A Journal of Ideas* 6, No. 2 (Fall 1989): 38–46, published by Shippensburg University of Pennsylvania.

Darwin

"These huge reptiles, surrounded by black lava, the leafless shrubs, and large cacti, seemed to my fancy like some antediluvian animals. . . . Hence, both in space and time, we seem to be brought somewhere near to that great fact—that mystery of mysteries—the first appearance of new beings on this earth. . . . How great would be the desire in every admirer of nature to behold, if such were possible, the scenery of another planet!"

The Voyage of the Beagle (1839)
Chapters 17, 21

"This preservation of favorable individual differences and variations, and the destruction of those which are injurious, I have called Natural Selection, or the Survival of the Fittest. . . . There is a grandeur in this view of life, with its several powers, having been originally breathed into a few forms or into one; and that, while this planet has gone cycling on according to the fixed law of gravity, from so simple a beginning endless forms most beautiful and most wonderful have been, and are being, evolved."

On the Origin of Species (1859)
Chapters 4, 15

"In a series of forms graduating insensibly from some apelike creature to man as he now exists, it would be impossible to fix on any definite point when the term 'man' ought to be used. But this is a matter of very little

9

importance. . . . Man still bears in his bodily frame the indelible stamp of his lowly origin.")

The Descent of Man (1871)
Chapters 8, 21

"Considering how fiercely I have been attacked by the orthodox it seems ludicrous that I once intended to be a clergyman. . . . I gradually came to disbelieve in Christianity as a divine revelation. . . . Thus disbelief crept over me at a very slow rate, but was at last complete. . . . The voyage of the *Beagle* has been by far the most important event in my life and has determined my whole career. . . . Looking backwards, I can now perceive how my love for science gradually preponderated over every other taste. . . . With such moderate abilities as I possess, it is truly surprising that thus I should have influenced to a considerable extent the beliefs of scientific men on some important points."

Autobiography (1876)
Headings 2, 4, 3, 8

Teilhard de Chardin

"I, your priest, will make the whole earth my altar and on it will offer you all the labors and sufferings of the world. . . . You know how your creatures can come into being only, like shoot from stem, as part of an endlessly renewed process of evolution."

The Mass on the World (1923)
The Offering, Fire Over the Earth

"We may, perhaps, imagine that the creation was finished long ago. But that would be quite wrong. It continues still more magnificently, and at the highest levels of the world."

The Divine Milieu (1927)
Part One, 3., c.

"Is evolution a theory, a system or a hypothesis? It is much more: it is a general condition to which all theories, all hypotheses, all systems must bow and which they must satisfy henceforth if they are to be thinkable and true. Evolution is a light illuminating all facts, a curve that all lines must follow. . . . The consciousness of each of us is evolution looking at itself and reflecting upon itself. . . . Man is not the center of the universe as once we thought in our simplicity, but something much more wonderful—the arrow pointing the way to the final unification of the world in terms of life. Man alone constitutes the last-born, the freshest, the most

11

complicated, the most subtle of all the successive layers of life. . . . The universe has always been in motion and at this moment continues to be in motion. But will it still be in motion *tomorrow?* . . . What makes the world in which we live specifically modern is our discovery in it and around it of evolution. . . . Thus in all probability, between our modern earth and the ultimate earth, there stretches an immense period, characterized not by a slowing-down but a speeding up and by the definitive florescence of the forces of evolution along the line of the human shoot."

The Phenomenon of Man (1940)
Books 3, 4

"Man is, in appearance, a 'species,' no more than a twig, an offshoot from the branch of the primates—but one that we find to be endowed with absolutely prodigious biological properties. . . . Without the earth could there be man?"

Man's Place in Nature (1949)
Foreword, Chapter 1

Contents

Prologue

Facing Reality

Evolution is an established fact supported by both science and reason; our own species is linked to life, earth history, this galaxy, and the universe itself. Nevertheless, serious literature presents a wide range of interpretations of evolution, from mechanistic materialism through creative vitalism and finalistic spiritualism to cosmic mysticism.

About twenty billion years ago, according to modern astronomy and cosmology, an explosion of unimaginable magnitude brought this cosmos into existence: space-time and matter-energy were suddenly born! Indeed, there may even be other universes beyond this particular physical reality.

This awesome cosmos of objects, forces, events, and relationships has been changing and expanding ever since that first moment of its actual formation. The darkness of emptiness became a birthplace of stars, comets, meteors, planets, moons, quasars, pulsars, and black holes.

Countless stellar clusters have appeared, developed, and vanished throughout sidereal history. Literally, everything that exists owes its birth and death to the stars.

There are astonishing potentialities in those particles of matter, units of energy, and pervasive cosmic forces of this dynamic universe (including gravity, magnetism, and electricity). Actually, the possibilities within this physical reality seem limitless.

Those past sightings of comets and supernovas (especially in 1054, 1552, and 1604) helped to shatter the rigid Aristotelian worldview. As modern astronomers probe deeper and deeper into this universe, they actually gaze further and further back to that cosmos just this side of the emergence of space-time and matter-energy from that first singularity. Likewise, our present cosmology has dispelled the biblical time scale forever. Human inquiry now faces the principle of ignorance when it attempts to fathom what once existed before that first singularity suddenly gave birth to this evolving universe.

The nebular hypothesis maintains that our sun was born about five billion years ago as a result of the condensation of cosmic dust and gas; it is now halfway through its stellar cycle. Our planet is merely a material fragment of this solar system that, in turn, is just an infinitesimal part of the Milky Way galaxy, which consists of billions of stars. In fact, there are billions of such galaxies in this universe! From this overwhelming perspective, the human animal is simply a fleeting incident within evolving nature.

The earth is nearly 4.6 billion years old. Life has probably been evolving on its surface for over 4 billion years; the primate order emerged about 70 million years ago; and hominid forms have stood erect on land and walked upright with a bipedal gait for at least 4 million years.

The earth is a fascinating museum with an incomplete collection of rocks, fossils, and artifacts. Scientific evidence now documents the historical continuity and essential unity of all life on this planet. Our own species is within the web, pyramid, and complex flux of flora and fauna on this evolving globe.

Indeed, the emergence of our zoological group is a recent event within cosmic history. The human animal is a frail, finite, and fallible species on this precarious planet. It is a product of organic evolution as well as sociocultural development and psychological conditioning; being totally within nature, humankind comes under the rigorous purview of abductive inquiry as the continuous interworking of facts from induction and concepts from deduction at the same time.

One must not forget that nature is full of suffering and disaster. In fact, most of those species that have inhabited this planet are now extinct; death rules everything. Nevertheless, humankind in whatever future form may still have millions, if not billions, of years to explore and enjoy this dynamic universe of law and chance. In those ages to come, there may even be the cosmic study of comparative exoevolution.

Science is an ongoing voyage of discovery toward new horizons within this changing cosmos: the universe is its laboratory. As a method, it must always remain open to empirical facts and rational concepts as well as new perspectives. During the past three centuries, those triumphs of scientific inquiry have been staggering. One striking result is that vacuous legends and effete myths are superseded by documented evidence and rational explanation. Yet, it is truly incredible what a person will choose to believe when there is simply no scientific evidence or rational cause whatever to justify such blind commitments, while at the same time this person may vehemently deny scientific theories (such as the paradigm of evolution) despite those overwhelming and incontestable facts as well as ruthless logic. Likewise, the inadequacy of any particular science or theory at a stated time is always subject to correction in light of more information and critical reflection.

Science is the most powerful method now known to human inquiry. It has disclosed a reality of matter, energy, and forces; a dynamic universe rich in temporal objects, passing events, and complex relationships. Often, entrenched beliefs and mere opinions are actually deceptive flights from this concrete cosmos. However, there is no escape from evolving nature.

Is there any religion as such that encourages doubting, questioning, scientific inquiry, and rational deliberation? It smacks of anthropocentrism in the extreme to claim that this whole evolving universe was created by a personal God for the sole purpose of making human beings.

Is there a modern theology that embraces both a cosmic perspective and the evolutionary framework within an openended and myth-free worldview? Eventually, the blind dogma of religious faith with its outmoded stories and myopic values wears very thin, indeed; it is simply wishful thinking to believe in human immortality and a spiritual cosmos.

Today, arrogant assertions by biblical fundamentalist creationists represent a fatuous obviation of science and reason; such claims are completely nugatory so far as the search for truth and wisdom is concerned.

Actually, it is the brute fact of cosmic evolution that provides a sound scientific research program for comprehending human existence within this dynamic universe (but not that objective idealism of some metaphysical systems or those supernatural beliefs of theistic religions, no matter how tempting they may be). One may speak of creative materialism for, in the last analysis, it is matter that matters.

In the search for truth and wisdom, no dogmatic limits should be placed on empirical inquiry and the use of reason; one must overcome sentiment and ignorance and superstition. Both the stars in the sky and the fossils on earth have crucial things to tell us about the true place of our own species within cosmic history. Surely, despite both the microscope and telescope, the awesome possibilities of scientific inquiry through advancing technology have only begun to be realized. No doubt it is science that will more and more shape our human world of the distant future, yielding even adequate explanations for the origin of energy and life itself as well as human thought.

I advocate scientific naturalism and rational humanism within a cosmic perspective, the evolutionary framework, and a holistic view of our species within earth history. Such a materialist standpoint embraces documented facts and rational hypotheses of the special sciences (from astronomy and biology to anthropology and psychology) and adheres to the canons of logic.

An outlook grounded in naturalism and humanism upholds free inquiry, open expression, the use of reason, scientific investigation, and those values dedicated to increasing tolerance, eliminating misery, and inspiring both creative thought and prudent action (all necessary conditions for the continued development and ongoing success of human civilization).

There can be no doubt that evolution is a fact, but interpretations of this cosmic process range from crude materialism to absurd spiritualism. Over many years, I have enjoyed spending countless hours with the thoughts of Darwin and the visions of Teilhard de Chardin; their ideas and perspectives have been both illuminating and inspiring. Surely, other voices are yet to be heard. Yet, it does seem to me that the human species is a strange animal in an elusive universe.

I wish to emphasize that this book deliberately focuses on the life, thought, and lasting influence of both Darwin and Teilhard de Chardin*; their contrasting interpretations of evolution stress materialism and spiritualism, respectively.

In those ages to come and with that advantage of hindsight, all earthbound and human-centered interpretations of reality will be viewed as merely egocentric aberrations in natural philosophy. At the end of this universe, even space and time will vanish. However, there may be an eternal series, an infinite number, of other universes. Concerning existence itself, why there is something rather than nothing may be the one question that will always stupefy the human intellect. In short, a complete cosmology may remain forever beyond human comprehension.

From a terrestrial perspective, one essential question still remains to be answered by biologists: Is the modern neo-Darwinian synthesis sufficient to explain the whole history of life on this planet in terms of organic evolution?

Furthermore, within a cosmic framework, one important question needs to be taken seriously by scientists: Is organic evolution occurring on other planets elsewhere in this dynamic material universe?

---

*Hereafter and throughout this volume, Pierre Teilhard de Chardin will be referred to as Teilhard.

FURTHER READINGS

Appleman, Philip, ed. *Darwin*. 3d ed. New York: W. W. Norton, 1990.

Barrow, John D., and Joseph Silk. *The Left Hand of Creation: The Origin and Evolution of the Expanding Universe*. New York: Basic Books, 1983.

Bergson, Henri. *Creative Evolution* (1907). New York: Modern Library, 1944.

Birx, H. James. "Darwin and Teilhard: Some Final Thoughts." *Proteus: A Journal of Ideas* 6(2):38–46.

Boslough, John. *Stephen Hawking's Universe: An Introduction to the Most Remarkable Scientist of Our Time*. New York: Quill/William Morrow, 1985.

Bronowski, Jacob. *Science and Human Values*. New York: Harper Colophon, 1975.

Chaisson, Eric. *Cosmic Dawn: The Origins of Matter and Life*. Boston: Little/Brown, 1981.

———. *The Life Era: Cosmic Selection and Conscious Evolution*. New York: Atlantic Monthly Press, 1987.

———. *Universe: An Evolutionary Approach to Astronomy;*. Englewood Cliffs, N.J.: Prentice-Hall, 1988.

Darwin, Charles. *Autobiography* (1876). Edited by Nora Barlow. New York: W. W. Norton, 1969.

———. *On the Origin of Species by Natural Selection, or the Preservation of Favored Races in the Struggle for Life* (1859), and *The Descent of Man, and Selection in Relation to Sex* (1871). New York: Modern Library, 1936.

———. *The Voyage of the Beagle* (1839). Edited by Leonard Engel. Garden City, N.Y.: Anchor Books, 1962.

Einstein, Albert: *Relativity: Special and General*. New York: Random House, 1961.

Farber, Marvin. *Naturalism and Subjectivism* (1959). Albany, N.Y.: State University of New York Press, 1968.

Gould, Stephen Jay. "In Praise of Charles Darwin." *Proteus: A Journal of Ideas* 6(2):1–4.

Gribbin, John. *In Search of Schrödinger's Cat: Quantum Physics and Reality*. New York: Bantam Books, 1984.

———. *In Search of the Big Bang: Quantum Physics and Cosmology*. New York: Bantam Books, 1986.

Hawking, Stephen W. *A Brief History of Time: From the Big Bang to Black Holes*. New York: Bantam Books, 1988.

Hempel, Carl. *Philosophy of Natural Science*. Englewood Cliffs, N. J.: Prentice-Hall, 1966.

Kaku, Michio, and Jennifer Trainer. *Beyond Einstein: The Cosmic Quest for the Theory of the Universe*. New York: Bantam Books, 1987.

Kuhn, Thomas S. *The Structure of Scientific Revolutions*. 2d ed. Chicago: University of Chicago Press, 1970.

Lamont, Corliss. *The Philosophy of Humanism*. Rev. 2d ed. New York: Frederick Ungar, 1982.

Medawar, Peter B. *The Limits of Science*. Scranton, Penn.: Harper & Row, 1984.

Michalsky, Walt. "The Vatican Heresy." *The Humanist* 49(6):27, 48.

Montagu, Ashley, ed. *Science and Creationism*. Oxford, England: Oxford University Press, 1984.

Morris, Richard. *Time's Arrows: Scientific Attitudes toward Time*. New York: Touchstone Book, 1986.

Munits, Milton K. *Cosmic Understanding: Philosophy and Science of the Universe*. Princeton, N.J.: Princeton University Press, 1986.

Nagel, Thomas. *What Does It All Mean?* Oxford, England: Oxford University Press, 1987.

Naroll, Raoul. *The Moral Order: An Introduction to the Human Situation*. Beverly Hills, Calif.: Sage, 1983.

Nordlander, Robert E. "Charles Darwin: Agnostic/Atheist." *The American Rationalist* 28(5):69–71.

Organ, Troy. "Dignifying Humanity: The Humor of Stephen W. Hawking." *The Humanist* 49(4):29–30.

Pagels, Heinz R. *Perfect Symmetry: The Search for the Beginning of Time*. New York: Bantam Books, 1986.

Parker, Barry. *Einstein's Dream: The Search for a Unified Theory of the Universe*. New York: Plenum Press, 1986.

———. *Search for a Supertheory: From Atoms to Superstrings*. New York: Plenum Press, 1987.

Popper, Karl. "Natural Selection and the Emergence of Mind." *Dialectica* 32 (1978): 339-355.

———. *The Logic of Scientific Discovery*. London: Hutchinson, 1959.

Preston, Richard. *First Light: the Search for the Edge of the Universe*. New York: Atlantic Monthly Press, 1987.

Redondi, Pietro. *Galileo: Heretic*. Princeton, N.J.: Princeton University Press, 1987.

Reeves, Hubert. *Atoms of Silence: An Exploration of Cosmic Evolution*. Cambridge, Mass.: MIT Press, 1984.

Rolston, Holmes. *Science and Religion*. New York: Random House, 1987.

Sagan, Carl. *Cosmos*. New York: Random House, 1980.

Shapiro, Robert. *Origins: A Skeptic's Guide to the Creation of Life on Earth*. New York: Bantam Books, 1987.

Stenger, Victor J. *Not by Design: The Origin of the Universe*. Buffalo, N.Y.: Prometheus Books, 1988.

Szamosi, Géza. *The Twin Dimensions: Inventing Time and Space*. New York: McGraw-Hill, 1986.

Teilhard de Chardin, Pierre. *Man's Place in Nature: The Human Zoological Group*. New York: Harper & Row, 1966.

———. *The Divine Milieu*. Rev. ed. New York: Harper Torchbooks, 1968.

———. *The Phenomenon of Man*. 2d ed. New York: Harper Torchbooks, 1965.

Thaxton, Charles B., Walter L. Bradley, and Roger L. Olsen. *The Mystery of Life's Origin: Reassessing Current Theories*. New York: Philosophical Library, 1984.

Weinberg, Steven. *The First Three Minutes: A Modern View of the Origin of the Universe*. New York: Bantam Books, 1977.

Weiner, Jonathan. *Planet Earth*. New York: Bantam Books, 1986.

1

Introduction

On the Origin of Life

How, where, and when did life begin? Questions concerning the origin of living things on our planet and elsewhere in this cosmos have intrigued scientists as well as philosophers and theologians. Ideas about life itself are challenging, especially in light of those implications stemming from the theory of biological evolution and the possibilities inherent throughout this material universe. From rational speculations among the ancient thinkers to recent discoveries in the earth sciences (particularly geology and paleontology) along with astrochemistry and astrophysics in modern cosmology, naturalists have sought for an explanation that accounts for the existence of organic matter without referring to theology or metaphysics.

The origin of life is no longer such an elusive problem in modern biology and chemistry. Scientific experimentation and philosophical scrutiny in this century have shed considerable doubt on the credibility of all earlier attempts to explain organic existence on earth as due to divine intervention, a series of special creations, pervasive vitalism, spontaneous generation, or the panspermia hypothesis. The latter position maintains that life exists elsewhere in this universe and that microorganisms drifted to our globe from deep space (subsequently, biological evolution has been taking place).

Mechanistic materialists claim that living things first emerged on our planet from inorganic matter according to physical laws. Subsequently, organisms have evolved for about 4 billion years. Some philosophers and religionists still argue that life itself is the result of a metaphysical force or a theistic creation following the divine origin of this physical world. The naturalist approach to this problem of the origin of life is, in fact, diametrically opposed to the supernaturalist attitude.

Within the evolutionary continuum, however, the distinction between inert matter and life itself is not always clear. Likewise, there are problems that involve raw materials, monomers, polymers, isolation, replication, photosynthesis, and sexual reproduction (among other problems).

Speculations about the origin and nature of life as we know it probably go back to prehistoric times. Our early ancestors could not as yet account for birth, diseases, growth and development, sexual reproduction, natural catastrophes, unique cosmic phenomena, and the inevitability of death in terms of empirical evidence and logical deliberation. With the passage of time and as a result of increased awareness, people recognized the distinction between inorganic objects and living things. To account for the organic world, some thinkers referred to supernatural entities or forces that are immanent and/or transcendent. Later, philosophers resorted to dynamic concepts. Finally, naturalists presented explanations derived from critical observation and rational reflection.

In general, four major accounts have been given to explain the origin of life on earth: the willed creation of all living things by a supreme spiritual being, the arrival of living things to our planet from elsewhere in this universe, the sudden appearance of plant and animal forms from inorganic matter through spontaneous generation, or the gradual emergence of life from matter. Facts and relationships in the special sciences (e.g., geology, paleontology, embryology, morphology, taxonomy, biochemistry, and genetics) clearly support the biological theory of organic evolution.

Various philosophical positions from hylozoism and vitalism to the panspermia hypothesis and an organismic view of our earth or this cosmos have been offered as possible explanations for the seemingly elusive phenomenon of life in this material universe. Hylozoism maintains that all matter is in fact alive (life and matter are therefore inseparable), whereas vitalism holds that life is due to a unique force radically different from matter.

Scientific findings and rational speculations continue to shed light on the origin of life from a planetary framework and more recently, due to the developing science of exobiology, within a cosmic perspective. Exobiology is the search for the existence of life, including intelligence and civilization, elsewhere in this universe. Seriously considering the empirical evidence of those special sciences, a critical thinker may interpret life within a naturalistic viewpoint without recourse to theology or metaphysics. In fact, it still needs to be demonstrated that the supernatural realm is even necessary for any sound understanding of and proper appreciation for life within physical reality.

A precise definition of life is not possible at this time, especially if it is to include all living things both past and present as well as exclude all nonliving things. For example, it may be debated whether a virus (which cannot replicate on its own) is a living or nonliving entity since it exists at the borderline between complex megamolecules and simple organisms.

Living things as we know them on earth generally exhibit the following properties or characteristics: mobility, metabolism, self-replication or sexual reproduction, growth and development, irritability, mutability, and heterocatalysis (i.e., living things exhibit complex and different chemical reactions). Through time and change, organisms may even evolve into new forms of life.

That elemental foundation of all life on earth is based on the carbon atom, although organic phenomena elsewhere in the cosmos may be structured around silicon, ammonia, or some other molecule. Likewise, all living things on this planet consist of some naturally occurring twenty amino acids as those building blocks of structural proteins and enzymes. The materialistic viewpoint accounts for the historical continuity and essential unity of life in terms of biochemistry and organic evolution.

Before philosophy and science, explanations for organic phenomena relied on appeals to myth, legend, magic, religion, and personal opinion. Evidence suggests that, even in prehistoric times, our remote ancestors recognized the significance of life. Neanderthal people deliberately buried their dead with ritual. Later, Cro-Magnon people painted magnificent cave murals. Even today, technologically primitive and nonliterate peoples believe in various forms of animism and animatism. In anthropology the term animism refers to the magicoreligious belief in the existence of spiritual beings in natural objects or events, whereas animatism is the belief in an impersonal supernatural force (e.g., mana) or forces which may reside in both living and nonliving things. The emergence of institutionalized religion and, later, science and reason followed these early beliefs.

During the Presocratic period (600–400 B.C.E.), several naturalist cosmologists offered rational speculations to account for the origin of living things.[1] Thales wrote that life originated in water, the basic element of all reality; Anaxagoras claimed that cosmic seeds pervade the entire universe; and Empedocles gave a bizarre explanation for the origin of the first organisms from the haphazard synthesis of free-floating organs of various sizes and shapes. This view implied the evolutionary principles of multiplicity, variation, adaptation, survival, reproduction (heredity), and extinction; monstrosities perished and only those plants and animals compatible with their environments lived long enough to reproduce themselves.

In their steadfast appeal to the critical observation of dynamic nature itself and the rigorous use of reason (logic and mathematics), these early philosophers found no need to rely upon supernatural beliefs. In fact, these first serious thinkers in Western intellectual history paved the way for the coming of science itself.

Later, Aristotle (384–322 B.C.E.), as the father of biology, held that every plant and animal form is eternally fixed in its natural place within the great chain of being or static hierarchy of organic nature.[2] This Greek thinker did not concern himself with the creation, history, or extinction of things and, consequently, never took fossils seriously or anticipated the idea of biological evolution. He did, however, resort to spontaneous generation to account for the sudden appearance of some simple organisms from inanimate matter. Aristotle claimed that teleological (purposeful) development from potentiality to actuality is ultimately caused by the mere existence of an eternal and perfect unmoved mover as pure thought thinking only about itself beyond the fixed ceiling of stars. For him, man as a sociopolitical and rational animal occupies the terrestrial apex of this ascending ladder of increasing complexity and sensitivity.

Interestingly enough, two millennia later, Charles Darwin himself did not speculate on the origin of life (although his paternal grandfather, Erasmus Darwin, had held that all living things on earth had evolved from a single living filament that existed in the primordial ocean millions of years ago). Charles apparently thought this question to be beyond both the comprehension of the human mind and the domain of empirical sciences. Even today, the origin of life remains one of the central problems in evolutionary science and natural philosophy.

Some past thinkers did claim that life actually had a material origin, having first arisen and then evolved naturally as a result of physicochemical processes in nature. Following the successes of science and reason during the Italian Renaissance (particularly in astronomy and physics), some biologists and philosophers attempted to understand and appreciate life by comparing organisms to complex machines. Mechanistic materialism was supported by Julien Offroy de La Mettrie in his major work *Man the Machine* (1747)[3] and later advocated by Friedrich Engels in his posthumously published speculative volume *Dialectics of Nature* (1898).[4]

However, some believers still hold that at a specific moment in earth history the divine action of a supernatural creator caused the sudden origin of life and, as a result, this unique event sharply separates inorganic nature from the organic world. In the Judeo-Christian tradition, the living God may even be interpreted as an intelligent and personal creator who also sustains and nourishes organic evolution.[5] In sharp contrast to this liberal religious view, fundamentalist creationists continue to preach that God created each plant and animal species once and only once less than 10,000 years ago.[6] Although some are willing to accept microevolution below the species level, they all reject macroevolution in their dogmatic belief in the eternal fixity of all floral and faunal types.

Although there are numerous religious accounts for the creation of life, a rigorously naturalist explanation for the emergence of this organic world rejects any appeal to supernatural interpretations and the assumed existence of a spiritual realm transcending material nature. An appeal to a supernatural being in order to explain the origin of living things is, in principle, outside the realm of scientific investigation. There is no evidence to support the belief that life is contingent upon the existence of a divine will and its sudden or ongoing intervention into the natural order of material reality. Such a religious view is grounded in blind faith and personal hope rather than science and reason.

Through the extensive use of analogy, both Leonardo da Vinci and Alexander von Humboldt interpreted our earth as a living organism. They did not separate life from matter and, therefore, perceived our planet in terms of a pervasive metabolism. The organismic philosopher Alfred North Whitehead even envisioned all of reality as a process field of interacting feelings, an idealist position giving preference to rational speculations rather than the empirical sciences.[7] Recently, Lewis Thomas has written that our earth resembles nothing so much as a single living cell.[8] As a result of their animated conception of dynamic nature, these thinkers did not concern themselves with the origin and evolution of planetary life.

It has also been held that, in this physical universe, life is actually coextensive and coeternal with nonliving matter; therefore, life had no origin as such. From this viewpoint, even though both inorganic and organic structures and functions are interrelated, neither preceded the other in cosmic history. This curious position was supported by those cosmologists who appealed to the steady-state theory, which claims that the universe presents the same characteristics in all places and at all times throughout cosmic reality. However, according to the present big-bang theory, which does account for an evolving and expanding material universe, modern scientific evidence strongly argues that this dynamic cosmos did have an origin about twenty billion years ago.[9] One hypothesis even suggests that all cosmic matter is in fact merely the residue of that universal life that preceded it!

Philosophers advocating forms of vitalism or idealism assert that life originated as a result of the creative act of a personal God, spiritual source, higher intelligence, or perhaps even cosmic intelligences.

From Aristotle and the Greek anatomist-physician Galen (130–201 C.E.) to the German biologist-philosopher Hans Adolf Eduard Driesch (1867–1941) and Henri Bergson, some biologists and natural philosophers have explained life in terms of a hypothesis that supposes the necessary existence of a metaphysical force autonomous from inanimate matter.[10] According to this standpoint, living organisms are held to be something more than merely the systematic organization of nonliving parts. It is claimed that only a vital force is able to infuse matter with life. This metaphysical force is held to account for the increasing complexity and sensitivity, diversity and continuity, and essential unity of all life on earth. However, there is no scientific evidence whatsoever to support any vitalist account of life on this planet.

Some natural philosophers from Jean Baptiste de Lamarck to Teilhard have appealed to an active principle or vital force to explain the origin and evolution of life as we know it. Early in this century, Lamarckism pervaded Russian biology and stymied its progress.[11] Teilhard, a geopaleontologist and Jesuit priest, even believed that a personal God plays a key role in directing the organismic history of this alleged spiritual universe. In truth, however, vitalism remains a pseudoexplanation to account for the origin and evolution of life; it is grounded in an intellectual obscurantism that may take the form of unwarranted religious (if not even mystical) beliefs.

Whereas hylozoism is the doctrine that all matter is active with life, and that life is therefore inseparable from matter, panpsychism maintains that mind or spirit or consciousness pervades all of nature. Of course, both positions are unacceptable to the evolutionist as materialist.

Like Aristotle, even later biologists thought that worms, eels, fish, toads, rats, mice, flies, and fireflies (among other organisms) suddenly arise from substances such as mud, rags, slime, morning dew, fallen leaves, decaying wheat, and rotting meat. Clearly misunderstanding the relationship between matter and life, these early naturalists offered spontaneous generation as an explanation for that assumed sudden emergence of complex living things from the inorganic realm.[12]

Following the Italian Renaissance, the physician Francesco Redi (1626–1697) offered experimental evidence to discredit the prevailing Aristotelian concept of the spontaneous generation of visible organisms. Redi correctly concluded that fly larvae are not suddenly formed from the inanimate realm. Instead, adult flies lay their eggs on unprotected decaying meat. Later, these eggs give rise to maggots.

The Dutch lens-grinder and naturalist Anton van Leeuwenhoek (1632–1723) used his single-lens microscope to see microorganisms, which were held by some to be spontaneously generated from inorganic matter. This discovery revived the idea of spontaneous generation; this idea was criticized by Lazzaro Spallanzani (1729–1799) although supported by John Needham (1713–1781).

In 1864, the French chemist Louis Pasteur (1822–1895) performed several ingenious experiments to demonstrate that organisms are not spontaneously generated from inanimate matter but, on the contrary, are produced from other existing organic forms. For a few decades following Pasteur's work, the origin of life was no longer considered to be a legitimate scientific question worthy of empirical inquiry. Of course, discrediting the spontaneous generation of complex life forms does not mean that those first organic entities had to be generated only from other living forms. Briefly, Pasteur's findings are compatible with both a naturalistic explanation for the origin of life and the fact of organic evolution.

In modern science, spontaneous generation has been superseded by the theory of abiogenesis, which maintains that, under primitive terrestrial conditions very different from those now prevailing on the surface of our planet and over vast periods of time, the chemical evolution of matter slowly led to the emergence of those first primordial organic systems from complex, preliving protein macromolecules. Surely, there is a crucial distinction between the spontaneous generation of complex forms of flora and fauna on the planet today from nonliving matter (as was believed by early biologists) and the gradual abiogenetic emergence of those first entities of life under the environmental conditions of a primitive world as a result of chemical evolution. Again, although he contributed to the refutation of spontaneous generation, Pasteur did not discredit the origin of life itself from matter or the biological theory of organic evolution.

Entropy is the second law of thermodynamics. It maintains that closed systems increase in disorder as a result of the dissipation of energy. However, there is no conflict between entropy and evolution.[13] Since our planet is an open system with energy provided by the sun, life itself exists and has been evolving for about 4 billion years. Of course, individual organisms do die and all life on earth will become extinct when our exploding sun of the future eventually burns out.

Earth is an unusual planet because of its vast bodies of water; the geosphere consists of core, mantle, ocean, and atmosphere. In fact, water was necessary for the formation of organic compounds and is that dominant component of living tissue. Today, scientists use the most advanced technologies to explore those earliest ages of earth history and to find those first traces of life on the planet.

Life as we know it requires the following six essential elements: carbon, sulfur, oxygen, hydrogen, nitrogen, and phosphorus. The five major chemical constituents of all living systems are fats, water, proteins, carbohydrates, and nucleic acids (DNA and RNA). The natural synthesis of amino acids into complex proteins may have occurred first on beaches or in tide pools, sites of periodically dried environments.

Some maintain that a naturalist argument for the origin of life in terms of random events in chemical evolution is not plausible because of the assumed extremely improbable occurrence of protolife substances (eobionts) when considering statistical calculations and the age of earth. However, life does exist on this planet. Likewise, one must distinguish between empirical evidence and formal reasoning. For the scientist and natural philosopher, the seemingly very improbable event of life arising from matter may be certain to occur given enough time as well as those required chemical substances in a suitable environment. Actually, considering the vast age and material uniformity of this cosmos, the presence of life may even be a natural imperative on numerous other worlds throughout this dynamic universe. Rational speculations on the existence of life elsewhere have given rise to the science of exobiology, which infers exoevolution.

Early in this century, both the English biologist J. B. S. Haldane (1892–1964) and the Russian biochemist A. I. Oparin (1894–1980) developed the idea of a "dilute organic soup" to describe those surface waters of our primitive earth.

A major breakthrough in that systematic attempt to account scientifically for the emergence of living things on earth was first offered by Oparin in his classic volume *The Origin of Life* (1936).[14] In 1922, he had proposed the idea that the first organic compounds emerged from inorganic matter on earth in the lifeless Precambrian environment with its generally oxygen-free atmosphere (this unique event followed a long period of prebiological synthesis). Oparin argued that the atmosphere of our primitive planet contained almost no free oxygen and had no upper ozone layer to act as a shield against ultraviolet radiation. At that time in earth history, the surface of our early planet contained both seas and pools of so-called probiotic soup. They were rich in chemical elements that reacted to the intense radiation from outer space as well as terrestrial sources of energy. As a result of prolonged photochemical activity, these inorganic mixtures gave rise to organic compounds (including amino acids). Through time and chemical selection, these eobionts or organic systems increased in complexity and stability, becoming the immediate precursors of living things.

Oparin conceived of coacervate molecules (colloidal gels) as precursors of those molecules that could reproduce themselves. He was successful in separating prebiotic coacervate liquid molecules from the aqueous solutions of several organic polymers. These prebiotic molecules were able to absorb energy from the external environment and, consequently, grow slowly or rapidly depending upon the surrounding conditions. In fact, microspheres may have first formed in ice or mud or the primitive atmosphere instead of emerging as surface films on the primordial waters.

On the basis of empirical facts accumulated in this century, one may consider the origin of life in this material universe to be a natural process of the evolution of carbonaceous compounds, which are present not only on our planet but also throughout interstellar space (including the chemical compositions of comets, meteors, and cosmic dust/gas clouds).

Thus, the beginning of life is an event that may have occurred at a single time or at different times, and at only one location or at different locations. The presence of carbonaceous compounds throughout sidereal regions of the universe suggests that organic substances originated from inanimate matter before the appearance of life on earth, and no doubt even before the formation of our solar system or this particular galaxy. Such pervasive compounds give credence to the abiogenetic hypothesis for the origin of organic substances from inorganic matter. This scientific hypothesis is empirically strengthened by recent discoveries in astronomy, biochemistry, and geopaleontology.

The American chemist Harold C. Urey theorized that the primitive reducing atmosphere of earth must have contained water vapor and gaseous hydrogen as well as carbonaceous compounds of methane and cyanogen (along with nitrogen in the form of ammonia as well as carbon monoxide and carbon dioxide). This atmosphere slowly changed through time. The evolution of simple anaerobic forms contributed their waste product, free oxygen, to the environment. As a result, the atmosphere gradually acquired oxidative properties that eventually made possible the emergence of first plant and then animal life on this planet.

Urged by Urey, in 1953, Stanley L. Miller conducted a now classic simple experiment to scientifically support the theory of abiogenetic synthesis.[15] Under laboratory conditions, he succeeded in producing several organic compounds (including twenty-five amino acids) by introducing a high-energy electrical spark through a steam-gas mixture of water, methane, hydrogen, and ammonia presumed to simulate the assumed atmosphere of primeval earth (note that no free oxygen is present). His subsequent experiments produced glycine, adanine, uracil, porphyrins, and adenosine triphosphate (among other things).

In 1956, the Russian scientists A. G. Pasynskii and T. E. Pavlovskais demonstrated that it is also possible to create some amino acids by irradiating with ultraviolet light a gaseous mixture containing ammonia salts and formaldehyde. These experiments gave further credibility to the mechanistic materialist theory of protobiogenesis (the origin of life from nonlife).

Sidney W. Fox, in 1957, obtained complex organic molecules similar to proteins from a heated dry mixture of various amino acids.[16] Twelve years later, through abiogenetic synthesis, he formed proteinlike substances and polypeptides.

In 1960, the Spanish scientist J. Oro had even abiogenetically synthesized the components of the DNA and RNA molecules (purines, pyrimidines, ribose, and deoxyribose).

Then, in 1970, Cyril Ponnamperuma produced ATP (adenosine tri-phosphate), which is the immediate source of metabolic energy in living organisms.[17] His experiments further supported the idea of an origin of organic compounds from inorganic substances in the universe through the stimulation of various cosmic energies. Probionts may have been synthesized on the subvital surface of our primitive planet through the natural transition from chemical development to biological evolution as an inevitable occurrence. If so, then the transition from matter to life was the necessary consequence of ongoing environmental circumstances rather than due to an accidental change producing a chance event.

In sharp contrast to those other hypotheses, modern science continues to support the theory that the chemical evolution of matter paved the way for the slow emergence of life in the hydrosphere of this world. The earth formed about 5 billion years ago and, extrapolating from recent fossil evidence, the earliest forms of life probably appeared less than a billion years later: chemical transformations formed first simple and then later complex organic substances (from prebionts through probionts to protocells). Still a puzzle is whether proteins or nucleic acids emerged first, or if they appeared jointly. Furthermore, did DNA or RNA emerge first? Being reproductively parasitic, the viruses have shed no light on the origin of life as such. The emergence of genetic information imperative for the synthesis of amino acids into structural proteins and enzymes remains a crucial area for future scientific investigation.

Briefly, organic molecules may be synthesized from inorganic matter: amino acids are the essential constituents of life as we know it (the DNA and RNA molecules as the codes of life are nucleic acids very distinct from free-floating amino acids).

Although viruses are simple, composed of only a protein sheath and a nucleic acid core, their parasitic nature prevents them from being the earliest form of life since they require a living host cell for mutation and reproduction.

The most primitive yet complex microorganisms on earth are the anaerobic bacteria, found where there is little or no free molecular oxygen (e.g., in the depths of the oceans and in specific areas of living organisms). In evolution, these anaerobic bacteria preceded both the autotrophic and heterotrophic bacteria as well as other simple organisms.

Those natural events that occurred throughout earth history and gave rise to life itself and organic evolution have probably happened (are happening or may happen) on other worlds elsewhere in this universe. It is the primary assumption underlying that emerging science of exobiology that infers exo-evolution. The existence of life on other planets among those countless galaxies is inferred by statistical probability and the fact that organic molecules are found in outer space. Even so, no concrete empirical evidence has yet confirmed the presence of life or intelligence in those different regions of our own solar system or even in those sidereal depths of this cosmos.

In the last century, both Lord Kelvin (William Thompson) and Hermann Helmholtz hypothesized that those first germs of life came to our earth from deep space through the arrival of meteorites carrying organic particles (or bits of cosmic dust with living germs that are propelled through this material universe by such forces as light quanta or the pressure of radiation, with some of them visiting this planet). Early in this century, S. A. Arrhenius and J. B. S. Haldane defended the panspermia hypothesis.

Recently, both astronomer Fred Hoyle and mathematician Chandra Wickramasinghe hypothesized that life was brought to our earth by comets or meteors, which allegedly carry simple organic forms throughout an assumed eternal steady-state universe.[18] At present, there is no empirical evidence to suggest that cosmic seeds or living spores pervade outer space. Likewise, Hoyle and Wickramasinghe do not account for the actual origin of life but merely place its existence elsewhere in this material universe. Finally, no modern cosmologist takes a steady-state model of this universe seriously.

Francis H. C. Crick and Leslie E. Orgel have suggested that our sterile earth may have been deliberately seeded by intelligent beings living in those other solar systems of this cosmos whose planets are billions of years ahead of our own in their stages of evolutionary development. Crick argues for the directed panspermia hypothesis in his book *Life Itself* (1981).[19] He suggests that a bacteria-carrying rocket was sent to this planet by a civilization elsewhere in this universe!

However, there are difficulties with this approach to explaining the existence of life on our earth. To begin with, it does not account for the appearance of those first living things presumed to inhabit those depths of outer space. Likewise, no irrefutable empirical evidence of such organic entities as cosmic seeds or biospores has yet been found in regions beyond this planet. In fact, the intense cold and harmful radiation of this universe pose serious threats to any life as we know it (not to mention the vast emptiness of this cosmos). These facts make it extremely improbable that viable organic germs could ever reach our planet from those remote abysses of this material universe.

At this time in scientific inquiry, any theory of how life first came into being must include rational speculation without disregarding the established facts and logical consistency. One needs to distinguish among inorganic chemistry, organic chemistry, and biological evolution (although there is no clear separation between plants and animals, or between matter and life).

The atmosphere of our primitive earth was an open system containing methane, ammonia, and water vapor (carbon monoxide, carbon dioxide, and free oxygen were added later). Free energy came from volcanic heat, electrical discharge, and both cosmic and terrestrial radiation. The origin of life was a natural process that occurred at least four billion years ago, with those first living things on this planet appearing gradually after a long period of inorganic evolution. Controversy still surrounds the biochemical explanation for the origin of life on earth, particularly whether something analogous to the DNA or RNA molecule arose first or, instead, those basic amino acids necessary for protein synthesis. Living things emerged when organic systems became capable of metabolism and reproduction. The development of inorganic syntheses in chemical evolution had paved the way for biological evolution and subsequently the adaptive radiation of more and more complex and diversified forms of life on this planet.

Life may have first appeared in the atmosphere and/or in slightly acidic aqueous environments such as the seas, if not in pools or clays or ice crystals. The development of eucaryotic cells and subsequent multicelled organisms from procaryotic cells took longer than the relatively short time necessary for the emergence of monomers (e.g., sugars, organic bases, and amino acids) and, later, polymers (e.g., proteins, nucleic acids, lipids, and ATP as the primary molecule of energy storage and exchange in all living organisms on earth).

Today, according to some taxonomists, there may be as many as seven separate organic kingdoms.[20] Each kingdom may have had a distinct origin. Nevertheless, diverse anaerobic bacteria are those earliest known organic inhabitants of this planet, as evidenced by the discovery of Precambrian microfossils and stromatolites.

Since 1954, Precambrian microfossils of procaryotes (bacteria or blue-green algae) have been found at over twenty-four localities. Such primitive organisms from the Bitter Springs chert deposits of central Australia are nearly 1 billion years old. However, of particular significance are those microfossils from the Gunflint Chert formation along those shores of Lake Superior in Ontario, Canada; these ancient organisms are about 2 billion years old. An even younger site is the exposed Fig Tree group of sedimentary cherts in those Barberton Mountains of the Transvaal Region of South Africa; it contains procaryotes that are over 3 billion years old. Presently, the oldest fossil evidence of diversified life on this planet now known to science is about 3.5 billion years in age and is found in the Hamlin Pool at Shark Bay of the Warawoona Group in West Australia.

Stromatolites are Precambrian microfossil remains of complex forms that lived in tide pool communities. They are direct chemical traces of early life as layered hemispherial masses of limestone or chert that were deposited by those metabolic activities of ancient procaryotes. Such stromatolite remains also occur in Phanerozoic sedimentary rocks and are even being formed in some shallow waters on our earth today.

Briefly, the emergence of an oxygenic atmosphere in the Precambrian and subsequent photosynthesis were later followed by sexuality in the Phanerozoic time with the resultant Cambrian explosion of complex forms of advanced life. This so-called explosion of life marks the beginning of abundant fossil remains in the earth's sedimentary rock strata.

Although various explanations have been offered to depict how life could have begun, not one of these has yet been sufficiently documented and experimentally tested. However, the law of parsimony dictates that neither religious beliefs nor metaphysical assumptions are necessary for a sound understanding of and proper appreciation for the origin of life as we know it. Furthermore, it is even possible that organic phenomena elsewhere in this universe are beyond technological detection and human cognition. At present, the naturalist theory of biochemical evolution to account for the emergence of life from prelife is the best explanation in terms of science and reason. Indeed, the problem of the origin of life on earth continues to both excite and challenge scientists as well as natural philosophers and enlightened theologians.

It is debated whether life emerged once or several times, and if this process could occur again in light of those changes in our atmosphere and on the surface of the planet. Also, it is still questioned whether chance or necessity was the predominant factor in the origin of life on earth. Of course, the emergence of life on this planet and the idea that it arrived on earth from deep space are not mutually exclusive hypotheses.

Although some religionists as theists still argue that divine intervention caused the transition from inanimate matter to living things, most naturalists as materialists now explain that first appearance of life in terms of statistical probabilities and the results of chemical evolution.

Modern biologists are still far from explaining all those awesome complexities surrounding the terrestrial origin of living things. Even so, the chemical emergence of all life on earth is now the most widely accepted theory to account for the beginning of organic history. Today, an interdisciplinary team of scientific specialists is needed to continue the research required to add to the empirical evidence supporting a materialist explanation for the origin of life on our planet, if not also in deep space. In the distant future, one may even anticipate scientists synthesizing complex organisms under natural conditions in test tubes where neither theology nor metaphysics need be required.

ENDNOTES AND SELECTED REFERENCES

1. Cf. G. S. Kirk and J. E. Raven, *The Presocratic Philosophers: A Critical History with a Selection of Texts* (Cambridge, England: Cambridge University Press, 1964).

2. Cf. Richard McKeon, ed., *The Basic Works of Aristotle* (New York: Random House, 1941).

3. Cf. Julien Offroy de La Mettrie, *Man the Machine* (LaSalle, Ill.: Open Court, 1961).

4. Cf. Friedrich Engels, *Dialectics of Nature* (New York: International Publishers, 1963).

5. Cf. Pierre Teilhard de Chardin, *The Phenomenon of Man*, 2d ed. (New York: Harper Torchbooks, 1965).

6. Cf. Ashley Montagu, ed., *Science and Creationism* (Oxford, England: Oxford University Press, 1984).

7. Cf. Alfred North Whitehead, *Process and Reality: An Essay in Cosmology* (New York: Free Press, 1969).

8. Cf. Lewis Thomas, *The Lives of a Cell: Notes of a Biology Watcher* (New York: Bantam Books, 1975), p. 4.

9. Cf. Joseph Silk, *The Big Bang: The Creation and Evolution of the Universe* (San Francisco: W. H. Freeman, 1980).

10. Cf. Rainer Schubert-Soldern, *Mechanism and Vitalism: Philosophical Aspects of Biology* (Notre Dame, Ind.: University of Notre Dame Press, 1962).

11. Cf. Loren R. Graham, *Science, Philosophy, and Human Behavior in the Soviet Union* (New York: Columbia University Press, 1987).

12. Cf. John Farley, *The Spontaneous Generation Controversy From Descartes to Oparin* (Baltimore, Md.: Johns Hopkins University Press, 1977).

13. Cf. Daniel R. Brooks and E. O. Wiley, *Evolution as Entropy: Toward a Unified Theory of Biology* (Chicago: University of Chicago Press, 1986). Also refer to Jeremy Rifkin, *Entropy: A New World View* (New York: Bantam Books, 1981).

14. Cf. A. I. Oparin, *The Origin of Life*, 2d ed. (New York: Dover, 1965).

15. Cf. Stanley L. Miller, "The Origins of Life," in *This is Life: Essays in Modern Biology*, ed. Willis H. Johnson and William C. Steere (New York: Holt, Rinehart and Winston, 1962), pp. 316–41.

16. Cf. Sidney W. Fox and Klaus Dose, *Molecular Evolution and the Origin of Life* (San Francisco: W. H. Freeman, 1972), esp. pp. 1–15, 316–44.

17. Cf. R. Buvet and C. Ponnamperuma, eds., *Chemical Evolution and the Origin of Life* (New York: Elsevier, 1971), esp. pp. 3–9, 420–24, 432–45, 495–547.

18. Cf. Fred Hoyle and Chandra Wickramasinghe, *Lifecloud: The Origin of Life in the Universe* (New York: Harper & Row, 1978).

19. Cf. Francis Crick, *Life Itself: Its Origin and Nature* (New York: Touchstone Book, 1982).

20. Cf. Lynn Margulis and Karlene V. Schwartz, *Five Kingdoms: An Illustrated Guide to the Phyla of Life on Earth*, 2d ed. (New York: W. H. Freeman, 1988).

FURTHER READINGS

Blum, Harold F. *Time's Arrow and Evolution.* Princeton, N.J.: Princeton University Press, 1968.

Buvet, R., and C. Ponnamperuma, eds. *Chemical Evolution and the Origin of Life.* New York: Elsevier, 1971.

Conant, James B. *Pasteur's and Tyndall's Study of Spontaneous Generation.* Cambridge, Mass.: Harvard University Press, 1953.

Crick, Francis. *Life Itself: Its Origin and Nature.* New York: Touchstone Books, 1982.

———. *What Mad Pursuit: A Personal View of Scientific Discovery.* New York: Basic Books, 1988.

Dauvillier, A. *The Photochemical Origin of Life.* New York: Academic Press, 1965.

Deamer, David W., and Peter B. Armstrong. "The Edge of Life." *Natural History* 92(2):40–42.

Dickerson, Richard E. "Chemical Evolution and the Origin of Life." *Scientific American* 239(3):70–86.

Ditfurth, Hoimer V. *The Origins of Life: Evolution as Creation.* San Francisco: Harper & Row, 1982.

Ehrensvärd, Gösta. *Life: Origin and Development.* Chicago: Phoenix Books, 1962.

Folsome, Clair Edwin. *The Origin of Life: A Warm Little Pond.* San Francisco: W. H. Freeman, 1979.

Fox, Sidney W., and Klaus Dose. *Molecular Evolution and the Origin of Life.* San Francisco: W. H. Freeman, 1972.

Gurin, Joel. "In the Beginning." *Science 80,* 1(5):44–51.

Hoyle, Fred, and Chandra Wickramasinghe. *Lifecloud: The Origin of Life and the Universe.* New York: Harper & Row, 1978.

Keosian, John. *The Origin of Life.* New York: Reinhold, 1964.

Margulis, Lynn. *Early Life.* Boston: Science Books International, 1982.

Margulis, Lynn, and Mark McMenamin. "Marriage of Convenience: The Motility of the Modern Cell May Reflect an Ancient Symbiotic Union." *The Sciences* 30 (5) (September/October 1990): 31–36.

Miller, Stanley L. "The Origin of Life." In *This Is Life: Essays in Modern Biology*, edited by Willis H. Johnson and William C. Steere, pp. 316–41. New York: Holt, Rinehart and Winston, 1962.

Miller, Stanley L., and Leslie E. Orgel. *The Origins of Life on the Earth.* Englewood Cliffs, N. J.: Prentice-Hall, 1974.

Oparin, A. I.: *Genesis and Evolutionary Development of Life.* New York: Academic Press, 1968.

———. *The Chemical Origin of Life.* Springfield, Ill.: Charles C. Thomas, 1964.

———. *The Origin of Life* (1936). 2d ed. New York: Dover, 1965.

———. *The Origin of Life on the Earth.* New York: Academic Press, 1957.

Orgel, Leslie E. *The Origins of Life.* New York: John Wiley and Sons, 1973.

Osterbrock, Donald E., and Peter H. Raven, eds. *Origins and Extinctions.* New Haven, Conn.: Yale University Press, 1988.

Sagan, Carl. *Cosmos,* esp. pp. 22–43. New York: Random House, 1980.

Schopf, J. William. "The Evolution of the Earliest Cells." *Scientific American* 239(3):110-12, 114, 116-20, 126, 128-34, 137-38.

———., ed. *Earth's Earliest Biosphere: Its Origin and Evolution.* Princeton, N.J.: Princeton University Press, 1983.

Shapiro, Robert. *Origins: A Skeptic's Guide to the Creation of Life on Earth.* New York: Bantam Books, 1987.

Stansfield, William D. *The Science of Evolution,* esp. pp. 50–66. New York: Macmillan, 1977.

Thaxton, Charles B., Walter L. Bradley, and Roger L. Olsen. *The Mystery of Life's Origin: Reassessing Current Theories.* New York: Philosophical Library, 1984.

Vallery-Radot, René. *The Life of Pasteur.* Garden City, N.Y.: Doubleday, Page, 1923.

Wald, George. "The Origin of Life." *Scientific American* 191(2):44-53.

Watson, James D. *The Double Helix: A Personal Account of the Discovery of the Structure of DNA.* New York: Mentor Books, 1977.

2

Interpreting Evolution: An Overview

In Western intellectual thought, the scientific theory of organic evolution represents a major challenge to religious beliefs and traditional values within the Judeo-Christian interpretation of the meaning and purpose of our own species immersed in this material universe. A serious acceptance of the evolutionary framework, with its emphasis on geological time and biological change, clearly threatens all previous geostatic and zoostatic views of global history that have supported an earth-bound and human-centered conception of the cosmos.

In sharp contrast to all those philosophies and religions that have emphasized pervasive fixity, the fact of evolution acknowledges the impermanence of geological structures along with the mutability of plant and animal species. As such, an evolutionary worldview grounded in science and reason argues for the historical continuity and fundamental unity of all life on this planet. The earth is but a cosmic speck within a physical reality of matter and energy that seems to be eternal in time, infinite in space, and endlessly changing. Likewise, humankind itself is merely a product of and totally within organic history.

One must distinguish among the fact, theory, evidence, hypotheses, mechanisms, rates, models, levels, distances, relationships, perspectives, speculations, and interpretations of evolution. The idea of evolution has developed from rational speculations in the naturalist cosmologies of antiquity, through

39

theoretical writings of Darwin and Wallace (among others), to its present status as a powerful fact in the modern synthesis of the special sciences, along with its obvious materialist consequences for both dynamic philosophy and process theology.

In particular, the theory of evolution provides modern biology with a comprehensive and intelligible framework that is applicable to all living beings. There is no area of human inquiry that has not been influenced to some degree by the fact of evolution and its often-devastating implications.

There is no certain means available in modern science for determining how our prehistoric ancestors perceived this physical world. No doubt, they, too, wondered about those stars above and death below. While struggling to survive, these early humans probably expressed their fears and joys in sounds and gestures long lost to the flow of time. Today, their own thoughts and behavior patterns can only be inferred from such evidence as stone tools, cave murals, and burial sites.

The earliest recorded accounts of the origin or existence of our planet and its life forms, including the human animal, were culture-bound worldviews. They expressed the myopic perspectives and vested interests of particular societies that lacked any useful body of empirical knowledge and were devoid of the scientific method for rigorous inquiry. These ancient cosmologies are a mixture of wishes and desires and assumptions, usually grounded in geocentrism, zoocentrism, or anthropocentrism and ethnocentrism. Such models of nature are the products of human beings preoccupied with magic, religion, opinions, and beliefs as well as the sociopolitical problems and values of their age.

In ancient India, among the orthodox texts of Hinduism, the Upanishads (ca. 800 B.C.E.) are the most indigenous of the Vedic writings.[1] Unfortunately, they are open to several conflicting interpretations. In general, however, these religious and philosophical speculations attempted to deal with the alleged sacred power implicit in sacrificial rites and ritual performances. One tradition maintains that this immutable and neutral force is, in fact, the single principle that pervades this entire cosmos and underlies its multiplicity of phenomena. The human individual as a microcosm mirrors this whole world process as the macrocosm (there is also the later-added belief in moral progress through reincarnation or metempsychosis, i.e., the transmigration of souls). According to the Upanishads, an eternal element within finite man is claimed to be identical with that divine power sustaining this infinite universe. Acknowledging the limitations of the human perspective, this statement of an illusive process metaphysics clearly regards man and life and material objects as being merely those temporal products of an endless series of finite transformations of that same unknowable substance in space grounded in a transcendent reality. It is no surprise, but of serious results, that this worldview did not engender the scientific attitude in India. Although the Upanishads did recognize both time and change, they nevertheless held this seemingly material cosmos to be ultimately an illusion. As such, human inquiry into the nature of things is forever inconsequential and therefore of no value.

In ancient China, an individual or a group of wise elders referred to as Lao Tzu (ca. sixth century B.C.E.) is said to have founded the Taoist school of thought. This philosophy, known as Taoism and recorded in the classic works entitled Lao tzu or Tao te ching, takes both this material world and the human situation seriously.[2] Tao is that central concept identical to the one or the totality of nature itself: it is the simple, eternal, absolute, mystical, spontaneous, indescribable, and forever unknowable beginning and way of all things. Tao is that single and independent creative source and sustaining support of the myriad things in this inexhaustible real universe as it goes through endless cycles of slow growth and rapid decay. The supreme goal of all life, especially human existence, is survival (unlike Darwinism, in Taoism the submissive and weak survive and thrive in this changing world). In the first century B.C.E., Chuang Tzu interpreted the Tao as that dynamic and equalizing principle in the universal flux of all particular things. As such, this neutral Tao has now taken on a moral significance favoring the good. In an obviously crude form, Chuang Tzu seems to have anticipated natural evolution in his conception of the transformation of species in a cyclical reality that is ever-changing and developing from the simple to the complex. All things come from the originative process of nature and return to the originative process of nature.

Later, in the cryptic I Ching or Book of Changes, which was influenced in part by Taoism, one finds a more mechanical statement of this process law of cyclic change through the interaction of a passive yin and an active yang, those two natural forces engendered by the great ultimate.[3] Be that as it may, these neo-Taoists devoted themselves primarily to the chemical sciences rather than to biological research (except for descriptive, but not comparative, anatomy).

In Western antiquity (600–400 B.C.E.), several astute Presocratic natural cosmologists presented concepts in their speculations on life and this universe that directly or indirectly anticipated the idea of evolution.[4] They rejected earlier myths, legends, personal opinions, and religious beliefs. Instead, their thoughts on this dynamic cosmos were grounded in the observation of nature itself as well as a critical reflection on human experience and the rigorous use of reason (logic and mathematics).

Thales, the first philosopher, claimed that life had first originated in water (the basic substance of this cosmos). Then, through time and change, some living forms appeared that were able to remain on dry land. Consequently, this materialist cosmologist grasped that essential unity of process reality as well as the ongoing creativity of organic history.

Thales' student Anaximander went further. He claimed that, in the organic history of life from primordial water to dry land, the line leading to the human animal had once passed through a fishlike stage of development. As such, Anaximander may be considered the founder of comparative anatomy.

Xenophanes was the first philosopher to recognize correctly both the biological and historical significance of fossils in rock strata as those mineralized remains of once-living organisms (even today, the impressive paleonotological record is the single most convincing body of evidence to support the fact of evolution). Furthermore, Xenophanes boldy rejected outright all earth-bound and human-centered worldviews as well as all religious interpretations of material reality that support forms of anthropocentrism and anthropomorphism.

Heraclitus dealt with both the being and becoming of nature, teaching that cyclical change is the fundamental characteristic of reality. He symbolically likened dynamic nature to the burning phoenix of mythology or a flowing river that always endures (being) but is never the same (becoming).

Yet, it was Empedocles who came closest to anticipating the modern theory of organic evolution. In his bizarre but insightful attempt to explain the origins of plants and then animals, he envisioned the primordial surface of this planet covered with free-floating organs of different sizes and shapes that haphazardly came together by sheer chance, forming the first organisms. Most of these resulting animals (including plants) were monstrosities that perished in the struggle for existence; nevertheless, some organisms did have the necessary structures and functions that enabled them to adapt successfully to their environments and thereby survive and reproduce. In this account of the origin of life on an early earth, Empedocles had introduced those ideas of multiplicity, variation, adaptation, survival, reproduction, and extinction. In doing so, he actually glimpsed in a general way the Darwin/Wallace explanatory mechanism of natural selection that Herbert Spencer extended to the human realm in his own biosocial concept of the survival of the fittest.

Suffice it to say that over two thousand years ago there were a few bold thinkers who, in their philosophical speculations on the structures and functions of nature, recognized both the historical continuity and essential unity of all living things on this planet. Unfortunately, this creative advance in human thought, with its emphasis on free inquiry and a naturalist attitude, was to end with the emergence of the encyclopedic but dogmatic Aristotelian worldview.

In retrospect, it is ironic that Aristotle (384–322 B.C.E.), the father of biology and several other scientific areas, should have come so close to discovering the principle of evolution because he preferred to believe in the eternal fixity of all plant and animal forms. However, Aristotle did pioneering studies in taxonomy, morphology, and embryology. His biological writings include *On Motion in Animals, On the Generation of Animals, On the Parts of Animals,* and *The History of Animals.*[5]

Despite his concern with reproductive methods as well as the growth and development of organisms, Aristotle grounded his view of life in that general concept of a latent and static "Great Chain of Being" or terrestrial continuum. He held to the immutability of biological types, arguing that each organic kind has its own eternally fixed natural place in this hierarchical order of the living world depending on its degree of complexity and sensitivity or intelligence. Briefly, Aristotle referred to the human animal as that featherless sociopolitical biped with a rational faculty placed at the apex of this ladder of nature; therefore, our own species falls midway between an ape and the moon.

Aristotle's natural philosophy rejected creation, extinction, the significance of fossils, and therefore organic evolution. However, it did maintain that spontaneous generation accounts for the sudden appearance of at least some lower animal forms on earth (e.g., worms, frogs, mice, and flies).

Aristotle ignored those implications of the paleontological record by claiming fossils to be merely chance aberrations in rock strata. He even claimed that no form of life had ever become extinct. Likewise, Aristotle ignored the implications of comparative morphology that suggest a common ancestry for similar forms. As such, he never contributed even an anticipatory foreshadowing of the theory of evolution. It is unfortunate that the followers of this Aristotelian worldview dogmatically accepted its rigid ideas rather than encouraging the inductive method of scientific inquiry. Actually, Aristotle's views dominated Western science, philosophy, and theology for almost two thousand years!

Although Darwin did refer to Aristotle as "the philosopher" and saw himself as only a schoolboy when compared to this Greek's comprehensive and profound genius, it is nevertheless obvious that (in terms of evolution) this English naturalist is far closer to a Presocratic outlook than to the subsequent Aristotelian worldview. For the Stagirite, nature is merely the monotonously eternal recurrence of the same (there is no awareness of a creative history). So, in the final analysis, Darwinian science and Aristotelian philosophy do not mix.

Among other early philosophers, Lucretius (99–55 B.C.E.) presented a mechanistic and materialistic model of this universe free from a static metaphysics and religious beliefs. In his poetic masterpiece *On the Nature of Things,* inspired by Epicurus, Lucretius outlined the origin of this cosmos and life on earth as well as the sociocultural development of humankind from a prehistoric-caveman stage, through hunting/gathering and agriculture, to the emergence of civilization with language and metallurgy.[6]

Lucretius maintained that this material universe is godless, eternal, and endless; it is composed of an infinite number of atoms moving throughout an infinite void with an infinite number of different worlds (each created by chance). He also claimed that life, including intelligent animals and perhaps even immortal godlike beings, exists elsewhere on other inhabited planets. This wealthy aristocrat held that the earth itself gives birth to life, e.g., the spontaneous generation of worms. First our planet produced plants and birds, then came larger animals, and finally humankind. Organic history is filled with both creativity and extinction.

According to Lucretius, our robust and naked primitive ancestors wandered the woodlands and forests. They ate acorns and berries and pears, searched for wild beasts (e.g., lions, panthers, and boars) with clubs and stones, and lived in caves. Later they wore animal skins, learned the use of fire, lived in social units, and even waged war. Eventually, the human animal developed the use of symbolic language as articulate speech. Agriculture brought the cultivation of plants and the domestication of animals; people could now live in cities where law, art, religion, and the use of metals (copper, bronze, and iron as well as gold, silver, and lead) flourished.

The thoughts of Lucretius are remarkably sound, considering that he recorded them about two thousand years ago. Unfortunately, they were lost in the pervasive umbra cast by dogmatic Aristotelianism. In fact, overcoming Aristotle took almost twenty centuries!

In the *Enneads*, the dynamic idealist and pantheist mystic Plotinus (205–270 C.E.), the founder of Neoplatonism, presented a process metaphysics grounded in a cyclical view of reality: the three immanent spheres of matter or void, life or soul, and mind or intelligence necessarily fulgurate from and return to the divine one as the transcendent good itself.[7] As such, this dynamic worldview is an early form of panentheism, i.e., the belief that God is simultaneously both immanent as the world and transcendent as its distinct source.

In medieval Islam, Avicenna of Afghanistan (980–1037) was the leading naturalist and humanist.[8] Integrating his own work as a physician and logician, he advocated a holistic view of humankind within nature and presented a worldview grounded in a synthesis of science, philosophy, and religion. Besides a devotion to medicine, his encyclopedic interests included astronomy, physics, geology, paleontology, zoology, anatomy, sociology, and geometry. Avicenna even introduced the idea that the origin of our own species was in an earlier animal form. His pantheistic and mystical interpretation of this universe argued for the emanation of multiplicity from unity.

Following the Greco-Roman era, the so-called Dark Ages and medieval period saw natural philosophy and rational inquiry replaced by blind faith in religious beliefs. Thinkers were more concerned with theology than with free inquiry into the nature of this universe and the history of life on earth. There was very little scientific research during this millennium; priority was given both to metaphysics and mysticism rather than empirical investigation into the material cosmos or the use of reason to understand and appreciate nature itself and its history.

The writings of Augustine and Aquinas were attempts to reconcile Christianity with the thoughts of Plato and Aristotle, respectively. To maintain the integrity of creation as a single unique event, the theologian Augustine (354-430 C.E.) may have intuited organic evolution when he wrote that in the beginning a personal God had created only latent causes or germs of those forms of life on earth which were afterwards to be gradually expressed in the development of life throughout the course of planetary time. However, he was not interested in cosmology, astronomy, or a philosophy of history. Furthermore, the theologian Aquinas (1225-1274 C.E.) grounded his comprehensive philosophy in the Aristotelian worldview.

In general, the planet earth was held to be the physical center of the universe, while our own species allegedly represented the pinnacle of a divine creation, both static and complete. In this scheme of things, the human being is held to be the special and final product of divine creation: a link falling midway between all the lower animals and the higher angels.

The Renaissance and later Enlightenment represented a rebirth of learning. In these two periods, knowledge and wisdom necessitated the overcoming of the outmoded ideas of Aristotle and the religious beliefs of Aquinas (among other dogmatic thinkers).

It is not generally known that Leonardo da Vinci (1452–1519), the greatest genius of the Italian Renaissance, was interested in geology and paleontology besides his devotion to both art and anatomy (he rejected superstition and authority, never taking theology seriously).[9] His insatiable curiosity led him to boldly question and bravely explore the very workings of a material world: What is the age of this planet? Why are birds able to fly? How does the human eye work?

Leonardo's intellectual strength lay in an exceptionally acute perception of concrete material reality. His universal mind wanted to understand the human animal as well as the forces of the physical world; it could analyze bones and muscles, synthesize art and science, and even envision some of the technological marvels to come, e.g., the airplane, the armored tank, and the submarine.

A true natural philosopher, Leonardo was always open to new facts and experiences. He even made investigations into earth history. While wandering through the Swiss Alps and the mountains of Tuscany, Leonardo became fascinated with the beauty and detail of the geological and paleontological evidence about him and its implications. To him, the rock strata containing marine fossils suggested a story considerably different from that interpretation given by most thinkers in antiquity. Since fossils of many sizes and various species were found in disparate layers far above sea level, he knew that the surface of this planet had not been eternally fixed throughout earth history.

Leonardo sought a scientific explanation for this phenomenon, rejecting all religious interpretations of creation and destruction. He saw time as the evil destroyer of all things and became obsessed with catastrophic events involving wind, water, and fire; he thought water to be the basic modifier of the crust of this earth. Consequently, Leonardo envisioned the end of this planet as a result of the disappearance of all water into its interior, followed by fire destroying all terrestrial life (including the human species). Clearly, this view of things is a far cry from the static Aristotelian model of a safe and secure universe as an eternally fixed home for human existence.

In these layers of mountain stone, Leonardo studied the fossilized marine evidence: crabs, seashells, sea snails, oysters, corals, scallops, cockles, cuttlefish, traces of worms, and the bones and teeth of fishes. To account for these objects at the tops of high mountains far removed from the sea, he took time and change seriously. His penetrating intellect leaped ahead of contemporary thought to embrace conceptions of both geological and biological history (ideas that are still hard for some thinkers to accept today). Recognizing the plasticity of this earth, Leonardo claimed that there had been periodic upheavals of mud layers from the bottom of the salt waters; as such, mud layers from those ancient floors of a sea had emerged to form the mountains, and erosion by rivers later uncovered these rock strata with their marine fossils.

Leonardo appreciated both the biological and historical significance of those marine fossils he found imbedded in the mountain rocks at high elevations (recall the insight of Xenophanes, in sharp contrast to the view of Aristotle). He rightly held them to be the remains of creatures that had once lived on beaches or in the seas, habitats that had been later lifted up by the forces of nature.

Leonardo claimed our earth to be at least 200,000 years old. He remained fascinated by the destructive force of water on geological structures and its consequences over vast periods of time, and may have even intuited the fact of evolution (if only in a most general way). These were incredible discernments on his part, especially since most thinkers at that time held fast to the biblical account of a relatively recent divine creation and therefore did not take change and history seriously. Yet, Leonardo had speculated on the processes of sedimentation, fossilization (mineralization), and natural erosion. In doing so, he pushed back the horizon of time.

Leonardo may have even glimpsed the mutability of plant and animal species. He at least acknowledged similarities among the monkeys, apes, and man over two centuries before the works of Carolus Linnaeus and Jean Baptiste de Lamarck.

In his dynamic cosmology, Leonardo rejected an earth-centered model of the cosmos and even anticipated both the universality of gravitation and a scientific use of the telescope. In his mature vision of things, he vitalized all of reality: planet earth was viewed as a living organism. He wrote that animals are the image of this world; man is a microcosm while this universe is the macrocosm, these two objects being identical in their nature. Like Heraclitus, Leonardo eventually wrote that reality is in continuous flux (the ultimate basis of all phenomena is not mathematics and mechanics but rather life, movement, and change). In fact, his final overview supported a plurality of worlds within the eternality and infinity of this universe. He even shed theology for a natural philosophy grounded in pervasive necessity, simplicity, monistic change, and pantheism.

In short, these ideas of Leonardo da Vinci clearly represent a major break from all those classic and medieval interpretations of man and nature that stressed fixity and essentialism as well as a separation of the terrestrial and celestial realms. Even today, his visions and accomplishments remain an inspiration to the soaring intellect and creative imagination of each free seeker for truth and wisdom.

Giordano Bruno (1548–1600), the greatest philosopher of the Italian Renaissance, rejected both Aristotelian philosophy and medieval theology.[10] The most impressive cosmologist of his age, he courageously taught the eternity and infinity and ongoing creativity of this unfolding universe, and boldly claimed the existence of a plurality of inhabited worlds scattered throughout an unbounded space of spirited matter in endless time. Bruno at least glimpsed the fact of evolution. In maintaining the essential unity of nature and implying the development of lower or simple life forms into higher or complex organisms, he apparently recognized the historical transformation of all life on earth. His interpretation of the place of humankind within a dynamic cosmos pioneered modern science and rational philosophy free from religious dogmatism and an outmoded metaphysics; it even anticipated exobiology, the search for life (including intelligent beings) elsewhere in this universe.

Following the Italian Renaissance, the emergence of both modern science, particularly astronomy and physics (Kepler, Galileo, and Newton), and rational philosophy (Bacon, Descartes, and Spinoza) took place. A dynamic but mechanistic worldview grounded in mathematics, especially geometry and calculus, next dominated Western thought. This material universe was now represented as a vast machine running on divinely given celestial laws by a deity independent of, and indifferent to, physical reality as well as human existence. Astronomers openly acknowledged material changes in this imperfect cosmos, e.g., the sudden appearance of comets, meteors, and supernovas along with irregularities in seemingly circular orbits of the wandering stars or planets.

The pantheist Spinoza emphasized the unity of a reality utterly indifferent to the fleeting occurrence of our species.[11] In sharp contrast to the deist Descartes, who even held to a mechanistic interpretation of animals as merely complex machines, it was Leibniz who gave a process view of the history of life as a slowly emerging monadic continuum.[12] In doing so, Leibniz paved the way for the idea of evolution.

In the *Protogaea* (1691, 1749) and elsewhere, after having observed rock strata and the fossils of extinct ammonites, Leibniz speculated on the appearance and history of life on earth as well as the mutability of species. He even hazarded the thought that there are living species intermediate between apes and our species, but cautiously relegated the existence of such "missing links" in the unbroken cosmic continuum of all life to some other planet beyond this one!

In the *Monadology* (1714), Leibniz finally presented his rich and profound interpretation of the order, harmony, and teleology he saw in the unfolding universe. This Leibnizian system incorporates several important principles: monadic plenitude, identity of indiscernibles, dynamic panpsychism, process continuum, pervasive causality (necessity or determinism), ongoing fulgurations, preestablished harmony, and sufficient reason. Briefly, this process metaphysics argued for a psychic reality of endless progress composed of an infinite number of independently unique monads (nonspatial, indivisible psychic centers) eternally developing by degree in an unbroken hierarchy of ever-increasing awareness toward God's transcendent perfection. Leibniz's monadic continuum is a natural chain of ceaseless becoming that advances never by leaps but always in linear gradation by degrees (*natura non fecit saltus*). Like space, time is merely a relational concept abstracted from within the subjective manifold of each monad. There are preconscious (inorganic), conscious (organic), self-conscious (human), and superconscious (angelic) levels of monadic development. Thus spirits are linked with humans, humans with the animals, animals with the plants, and plants with the fossils, which in turn merge with those material bodies that our senses and our imagination represent to us as being absolutely inanimate. All these levels of beings actually form a single chain of dynamic continuity throughout the universe. God, the hypercategormatic absolute monad, transcreates this cosmos and (at each instant) holistically apperceives all things within all space and throughout all time.

In the final analysis, Leibniz is not an evolutionist because each isolated monad merely unfolds in a fixed sequence throughout the unending chain of existence (there is no mutability of species as such and no extinction of any monad). Nevertheless, his process metaphysics did pave the way for an evolutionary view of the world. One may argue that Leibniz is a link between the antievolutionary Aristotelian philosophy and the evolutionary Darwinian worldview.

It is an incredible example of parallel development that those advances of scientific knowledge and philosophical understanding in various fields have more or less repeated the actual order of cosmic history as seen from our planetary perspective, e.g., from astronomy and physics through biology to anthropology and psychology. In general, man first studied motion in those inorganic objects most remote from the earth and then only gradually became interested in those organic beings (plants and animals) nearer to him until, finally, he critically examined himself. Nevertheless, it still remained for a "Newton of biology" to synthesize the terrestrial sciences within a historical framework that took both planetary time and organic change seriously.

With due respect to Galen and Vesalius, Carolus Linnaeus (1707–1778) was the first significant biologist after Aristotle.[13] Linnaeus believed himself to be divinely chosen and inspired to devote his career to taxonomy to help fill in those numerous gaps in the assumed hierarchical continuum of the Aristotelian "Great Chain of Being" and thereby prove the existence of God by empirically demonstrating the assumed order, design, and intelligence pervasive throughout static nature.

This Swedish botanist and physician was particularly interested in the reproductive methods and sexual characteristics of the flowering plants in Lapland. This interest resulted in his devising a new system of artificial classification by means of binary designation called binomial nomenclature (each plant and animal type or kind is assigned its own Latinized scientific name consisting of two terms: a general first term indicating the genus, and a specific second term indicating the species). As such, he is considered the founder of modern taxonomy.

Linnaeus revived the science of biology, systematized floral and faunal taxonomy, and greatly expanded the descriptive classification of organisms. He presented his ideas on and scheme of this world in a book entitled *The System of Nature* (1735). As a naturalist, he did recognize obvious biological similarities among the primates: man, apes, monkeys, and lemurs; modern science excludes the bats from this order. He even referred to the human animal as *Homo sapiens* (the designation still used today), claimed man's closest living relative to be the orangutan, and also discerned that adaptive harmony between different animals and their various environments established through adequate senses and organs.

Having accepted the Aristotelian doctrine of the eternal fixity of all species, Linnaeus was understandably bewildered when he found varieties among those alleged immutable plant forms he studied so carefully; plant hybridization among species could produce a new species that is successful in terms of reproduction. With an ironic twist, modern biology uses a natural taxonomy as one of those several major areas of comparative evidence to support empirically the scientific truth of organic evolution (with a common ancestry for similar organisms).

One hundred years separate the appearance of Linnaeus's *The System of Nature* and Darwin's visit to the Galapagos Islands. For ten decades, the scientific pendulum was swinging away from that rigid idea of the external fixity of organic forms and toward a growing awareness of the ongoing evolution and pervasive extinction of species over vast periods of geological time and biological change on this planet.

During the Age of Enlightenment, French philosophers in the eighteenth century took both a historical perspective and the study of nature seriously.[14] They emphasized the use of reason, criticized traditional religious beliefs, encouraged the growth of science and technology, and called for educational reforms to improve the human condition. Their optimistic views upheld process as progress and outlined sociocultural development. The insights of Charles Bonnet, Denis Diderot, Etienne Geoffroy-Saint-Hilaire, and Baron de la Brede et de Montesquieu even glimpsed the theory of evolution.

In France, the scientist and philosopher Pierre Louis Moreau de Maupertuis (1698–1759) rejected in favor of epigenesis the hereditary theory of embryonic preformation popular at that time. Unable to give a mechanistic account for the origin and nature of life, he developed a metabiological stance grounded in corpuscular psychism or atomistic dualism; all matter has sensitivity as well as intelligence and memory to some degree. Maupertuis applied mathematical probability to his own particulate theory of heredity in a serious empirical investigation of albinism and polydactyly. Contributing to speculations on evolution, he maintained that a process of random chance mutations along with the survival of the fittest organisms in changing environments (especially geographically isolated areas) could explain structural and functional transformations of various species emerging throughout time on earth.

Although he died one hundred years before the publication of Darwin's *On the Origin of Species,* Maupertuis had actually suggested those two major explanatory mechanisms of neo-Darwinism or the synthetic theory of organic evolution in modern science: genetic variation and natural selection.

Another Frenchman, the mechanistic materialist Julien Offroy de La Mettrie, wrote *Man the Machine* (1747).[15] This book pointed out the close relationship between humankind and the apes. In doing so, it challenged those many popular beliefs on the assumed special nature of our own species.

The French naturalist Georges Louis Leclearc, Comte de Buffon (1707–1788), asserted early in his monumental forty-four volume work *Natural History: General and Particular* (1749–1804) that plant and animal species may undergo modifications and, consequently, produce varieties if not even new species.[16] If true, then all those existing forms of life now on earth could be the descendants of common ancestors in the remote past through biological generations from different organic forms; fossils are the remains of plants and animals in those distant ages of geological time. He also suggested that man and ape are related, having once shared a common ancestry.

Although drawing attention to vestigial organs, Buffon did not offer an explanation for the transformation of species. Even so, in 1751, his early ideas on evolution were both condemned and censured by those ecclesiastical authorities at the Sorbonne. As a result, Buffon's later opinions on this then very controversial subject of organic history fluctuated and (perhaps reluctantly but intentionally) became obscure; he all but withdrew his support for the heretical hypothesis of biological evolution.

Of course, Buffon the scientist had placed great emphasis on rigorous experimentation as well as the critical observation of nature. He wisely insisted that science and religion should be strictly separated. However, his own views often illustrate not only the power of rational speculation but also the unfortunate need for compromise.

In cosmogony, Buffon boldly proposed a nebular hypothesis to account for the origin of the earth: this planetary system had resulted when the glancing blow of a comet against the molten surface of our sun freed a mass of matter that eventually separated into various solar objects. In geology, he recognized seven distinct epochs of natural history, each spanning a vast period of time involving millions of years; thereby arguing for the formation and destruction of mountains by the flux and reflux of tides and currents, and explaining fossil evidence as those remains from the organic development of plants and animals in remote ages.

Within the framework of continuity and plenitude, Buffon clearly distinguished between true species as entities in nature and the limited mutability of varieties. He wrote about the sterility of hybrids, the heritability of acquired characters, the unity of type demonstrated in comparative anatomy studies, and the origin and natural history of birds within an implied evolutionary overview. Buffon is also remembered for his biological theory of organic molecules that, since they are assumed to be pervasive throughout nature, are a source of living organisms (perhaps through spontaneous generation).

It is deeply regrettable that Buffon's pioneering insights, as offered in his sweeping vision of cosmic development and terrestrial history, were silenced by some of those myopic and dogmatic religionists who were so influential during his lifetime. Ironically, like Aristotle, Buffon was both an asset and a liability to the emerging theory of organic evolution.

Speculating on the direction of sociocultural development in his major work *Sketch for a Historical Picture of the Progress of the Human Mind* (1795), the extraordinarily optimistic French philosopher M. J. A. de Caritat de Condorcet (1743–1794) envisioned a future tenth human epoch in which everyone on the planet will enjoy the awesome achievement of an indeterminate life span or practical immortality as a result of the ongoing advances in science, medicine, technology, and mathematics.[17]

Auguste Comte (1798–1857), the father of both sociology and positivism, extended this developmental framework to include human thought, the special sciences, and sociocultural history.[18] He called for a science of humankind free from theology and mataphysics.

In Germany, the early writings of Immanuel Kant (1724–1804) presented a process cosmology that upheld the nebular hypothesis in astronomy as well as the existence of life and even intelligent beings throughout an expanding universe of increasing perfection.[19] Although interested in both natural science and philosophy (with awe and admiration for the heavens above and the moral law within), this great thinker in ethics held to the fixity of species. He dogmatically claimed that science would never understand the growth and development of even a blade of grass, much less its origin in organic history.

Kant did suggest that planet earth could already be millions of years old. Yet, he maintained that the "first parents" of humankind suddenly began to exist as full-grown adults with no natural cause explaining their abrupt appearance on earth!

However, recognizing those similarities between man and the apes, Kant speculated that in a future epoch (after the occasion of some great revolution in nature) the orangutan or the chimpanzee could perfect those organs that serve for walking and touching and speaking into the biological structures of an articulate human being with a central organ for the use of understanding; thereby, this ape would gradually develop itself through sociocultural advances into a civilized man!

Influenced by Linnaeus and Buffon, Johann Friedrich Blumenbach (1752–1840) systematically applied the comparative study of human anatomy to derive a racial classification of our own species based on the morphological differences among geographically separated populations. This German naturalist published his original findings in anthropometry* as a volume titled *On the Natural Variety of Mankind* (1755) and is therefore credited as the founder of physical anthropology.

---

*The classification of human populations based upon the comparative study of body measurements

In England, the botanist and physician Erasmus Darwin (1731–1802) speculated that the origin of all life on earth may be traced back to a single organic filament that existed in the primordial oceans millions of ages ago.[20] In doing so, this poetic naturalist anticipated the current "primeval soup" hypothesis associated with the Russian biochemist A. I. Oparin and other scientists of the twentieth century.

Erasmus Darwin held that species gradually evolve and improve over vast periods of time. He emphasized the factors of mutation, adaptation, struggle for existence, reproduction of the strongest, and even those roles of both sexual selection and artificial selection in the ongoing transformation of plant and animal forms throughout organic history. These ideas were presented in his major work, the two-volume *Zoönomia: or, the Laws of Organic Life* (1794, 1796). Apparently these books did not impress his grandson Charles Darwin, even though their natural speculations had anticipated the scientific theory of biological evolution.

Erasmus Darwin's long poem *The Temple of Nature* (1802) was published posthumously. This same year also saw Jean Baptiste de Lamarck, G. R. Treviranus, and Lorenz Oken independently outline their principles of biology and views on evolution.[21]

Lamarck was the first naturalist to write a book solely for the purpose of presenting a theory of animal evolution. Treviranus claimed that, in the flux of nature, man has not as yet reached the highest term of his existence but will progress to still higher regions and thereby produce a nobler type of being. Oken presented and defended his conception of an evolving God whose development is manifested in the increasing complexity and progressive diversity of living things.[22] In this unique pantheistic interpretation of the gradually evolving ascent of all life from some primeval form through plants to animals, man is becoming a godlike being in a dynamic universe created by a God that actually metamorphosed itself into this unfolding cosmos! As such, mind is merely a mirror in which time or the absolute and human consciousness as the goal and end of creative nature can behold themselves.

Although important conceptual advances were being made in the emerging science of biology before the nineteenth century, most early biologists accounted for the origin of life as a result of a divine creation or a series of special creations or the spontaneous generation of the organic from the inorganic.[23] Likewise, one does find the following representative views in the literature of that time: fallen leaves on water slowly transform into fishes and those on land slowly transform into birds; rats emerge from decaying wheat, mice from rotting rags, and flies from decaying meat; orchids touching the ground give birth to both birds and very small men; birds are derived from flying fishes, lions from sea lions, and man from a metamorphosis of the husband of a mermaid. In general, these naturalists also taught the immutability of plant and animal species, thereby attesting to the staying power of both Thomism and Aristotelianism.

Nevertheless, the thoughts and perspectives of some natural philosophers in the Age of Enlightenment provided an intellectual atmosphere that paved the way for the emergence of several earth sciences (particularly geology, paleontology, and archeology) and subsequently the empirical theory of biological evolution in the nineteenth century.

Early in the last century, the French natural philosopher Jean Baptiste de Lamarck (1774–1829) wrote the first comprehensive theory of animal evolution.[24] He had a Jesuit education for the priesthood but then spent time doing military service as a lieutenant in the army and, after studying medicine, finally turned to the natural sciences. He was devoted to forming theoretical principles rather than collecting meticulous facts. His interests included chemistry, geology, meteorology, invertebrate paleontology and zoology, taxonomy, and biology (especially botany).

As a deist, Lamarck argued for a mechanistic and materialistic view of life and earth history; nature consists of the mineral, plant, and animal kingdoms. He devised a natural system of classification based on comparative structural relationships and distinguished between invertebrate and vertebrate animals. In 1800, he rejected the fixity of organisms and gave his previous theory of degradation an evolutionary interpretation. In 1802, he coined the term biology, thereby separating physics from the scientific study of the living world as well as pointing out the similarities between plants and animals.

Lamarck's ideas on the mutability of fauna are presented in his major book *Zoological Philosophy* (1809), as well as an introduction to his seven-volume work *The Natural History of Invertebrate Animals* (1815–1822). The former title appeared in the year of Darwin's birth and fifty years before the publication of *On the Origin of Species* (1859). Whereas Lamarck's views on organic history are primarily speculative but nonetheless important, Darwin's theory of evolution gives preference to science and reason.

Lamarck held that plant and animal classes have slowly changed over the immense span of earth history, although he claimed that there have been no extinctions of such groups in the living world. Dynamic reality is ultimately grounded in a supreme author independent of this material world. Likewise, he assumed that a pervasive power in evolving nature has been propelling life toward increasing complexity and greater perfection, resulting in a branching hierarchy of zoological forms ranging from a simple protozoan up to the human species.

A quasi-vitalist, Lamarck offered four basic mechanisms to account for animal evolution: chemical spontaneous generation to explain the appearance of only those first primitive organisms, the two natural laws of use/disuse and the inheritance of acquired characteristics, and an inner physical feeling found only in complex fauna that can respond to natural challenges and thereby (as a directing force to satisfy those needs and desires that emerge) bring about those alterations of structures and functions necessary so that these higher animals are able to adapt, survive, and reproduce in changing environments. Higher animals develop involuntary or voluntary motions as adaptive habits, which in turn may bring about those required structural or functional alterations necessary so that these complex fauna can continue to survive and reproduce (even new organs and behavior patterns would eventually result). It must be added that the ability of these mechanisms to account for animal evolution has never been supported by conclusive scientific evidence. Lamarck held to separate ongoing acts of spontaneous generation at different points in time, and also separate lines of parallel organic progression (but he refused to accept the possibility of extinction).

The famous example given to demonstrate Lamarckian evolution is the history of the giraffe. According to Lamarck, giraffes were once short-necked animals eating the leaves of short trees. Yet in order to reach and feed off the ever-higher trees, the giraffes had to constantly stretch their necks from generation to generation. This acquired characteristic of an extra length of neck, as a direct result of ongoing use, was inherited by the offspring. Over many generations, the continuous accumulation of this specific characteristic has produced the long-necked giraffe of today. In sharp contrast, modern Darwinian evolution emphasizes the results of the struggle for existence throughout organic history in terms of chance genetic variation and necessary natural selection.

Lamarck boldly suggested that the human animal originated from an orangutanlike form somewhere in the vastness of Asia and, through slow evolution, gradually acquired its present physical features and mental faculties. Furthermore, he argued that our own species differs merely in degree from the apes and monkeys; its higher powers like reason are grounded in a far more complex central nervous system and brain.

Lamarck died blind and poor without giving a scientific explanation for organic history. However, he did openly champion the evolutionary framework at a time when most naturalists advocated divine special creation and the fixity of all species. Nevertheless, the influence of Lamarck's ideas on Darwin's theory is still debated; clearly, there is a significant difference between the speculations of the French naturalist and that empirical orientation of the English scientist.

In Germany, the idealist philosopher and pervasive pessimist Arthur Schopenhauer (1788–1860) presented a monistic metaphysics in his major two-volume work, *The World as Will and Representation* (1818, 1844).[25] His dynamic overview outlined both cosmic and planetary evolution; the essence of everything in the universe is a blind, eternal, irrational, and always evil Will manifesting itself as restless movement and creative diversification throughout changing nature.

Schopenhauer distinguished among three major temporal divisions of global development: the inorganic (chemical), the organic (plants and all nonhuman animals), and humanity itself. Earth history represents a successive hierarchy of struggling plant and animal forms, with our own species being the final stage of a pyramidally structured terrestrial evolution.

For Schopenhauer, humankind is a product of this timelessly unfolding but never satisfied Will. In essence, *Homo homini lupus* (man is a wolf to man) is an animal manifesting both a will-to-survive and a will-to-sex. Furthermore, Schopenhauer wrote that our own species had a simian origin in those tropics of the Eastern Hemisphere; the first human beings were suddenly born from the chimpanzee in Africa and the orangutan in Asia.

Schopenhauer referred to pantheism as being only a polite form of atheism; his own worldview was influenced by Oriental philosophy rather than Christian theology. Unfortunately, he is now remembered more for his doctrine of the evil Will than for his glimpse into evolutionary history.

Dissatisfied with both traditional theology and idealist thought, particularly Hegelianism, the unorthodox Ludwig Feuerbach (1804–1872) strongly advocated a scientific, philosophical anthropology grounded in the cosmic perspective and a materialistic interpretation of planetary evolution.[26] This German thinker stressed that the true place of humankind within natural history can only be correctly understood and properly appreciated in light of those special sciences (e.g., astronomy, biology, physiology, and psychology) and the critical use of reason.

As a naturalist humanist, Feuerbach pointed out that to think one must first eat, emphasizing that this physical universe both preceded and is independent of human experience. He boldly argued that all theological beliefs, objects, and events are merely the objectification of man's creative subjectivity; the wishful ideas of religion are ultimately the products of human feelings and emotions and the imagination of the heart rather than referents to alleged metaphysical correlates of faith.

Feuerbach's own starting point is the dynamic physical universe itself: he clearly taught the great geological age of this planet, was aware of the vast environmental changes on the earth throughout time, held to the scientific origin of life from matter, acknowledged the extinction of various plants and animals in the remote past, presented a strictly evolutionary account of the emergence of new flora and fauna (including our own species), and always emphasized the importance of a cosmic perspective and its awesome consequences. He saw humankind as a newcomer totally within the organic evolution of this material planet.

In his anonymously published book titled *Vestiges of the Natural History of Creation* (1844), Robert Chambers (1802–1871), a successful Edinburgh printer and prolific writer, presented an evolutionary worldview grounded in a cosmic perspective and a planetary framework.[27] His scandalous but popular volume was the singularly most important work on organic evolution to appear between the writings of Lamarck and Darwin.

Although a deist, the unorthodox Chambers rejected both the divine creation of life and the miracles of strict biblical literalism. However, he did believe in a First Cause or Supreme Being as the ultimate source of the two dynamic and essential laws of nature: gravitation and evolution. He further claimed that the grand plan of organic nature is the direct result of a divine preconception unfolding itself according to these two laws of a dynamic world.

Chambers's numerous interests included astronomy, physics, chemistry, and meteorology as well as geology, geography, paleontology, and biology (especially taxonomy, embryology, and ecology). In scientific cosmogony, this gifted naturalist upheld the nebular hypothesis: the emergence and formation of all galactic objects of material immensity (e.g., stars, comets, planets, and moons) from an original diffused mass of universal firemist as a direct result of those natural laws operating throughout the sidereal depths of atomic reality. His sweeping framework of stellar evolution embraced vast astral clusters with countless solar systems. Chambers even maintained that there exists a plurality of other remote worlds more or less similar to this planet. An early exobiologist, he speculated that life and intelligent beings do or will exist on those innumerable spheres developing elsewhere throughout space and time.

Chambers argued that this mechanical and material universe manifests a cosmic design governed by a single process of advancing evolution, forever unfolding from inorganic dust first through primordial protoplasm and then organic development to rational man and perhaps even beyond (thus the historical continuity and essential unity of dynamic existence). After studying both the rock strata and the fossil record, he was convinced of the adaptive and progressive organic transformation of all life on this globe: plant and animal species have evolved from common ancestors over the vast eras of earth history. In this ongoing succession of living things, the appearance of simple marine plants and sea animals preceded the emergence of complex flora and fauna types on dry land.

Chambers maintained that our own species had a single origin from an Asian ape and now it is the paradigm of all terrestrial life. Concerning the emergence of the central nervous system and brain, he concluded that the human mind with language and reason differs merely in degree rather than in kind from the source of those mental activities in the earlier or lower animals. His speculations even foresaw the human animal evolving into a far more superior type that would thereby complete the assumed geozoological circle of this developing sphere.

Chambers argued for the accidental electrochemical origin of life from inorganic matter and acknowledged the role that monstrosities (sudden mutants) play in producing major leaps within the graded scale of organic development that advances analogously in both the plant and animal kingdoms from the simple to the more complex. He was also aware of the evolutionary significance of "missing links" as well as the crucial role that extinction plays throughout the history of life and in the subsequent appearance of new species (e.g., the demise of the trilobites and dinosaurs, and the emergence of reptiles and mammals). Yet, with all these insights, no scientific mechanism was offered to explain the natural process of biological evolution.

Unfortunately, despite its serious acceptance of the mutability of species and fascinating speculations, *Vestiges* is filled with errors of both fact and interpretation. For example, Chambers wrote that those crystalline designs of frosted vapor on winter windowpanes reminiscent of fern leaves illustrate the identity of inorganic and organic matter. He also claimed that lime laid on waste ground will spontaneously spring up as clover, and that oat plants cropped before maturity can metamorphose into rye while in the ground during a single winter season. In fact, he even maintained that plant forms are determined by the laws of electricity and that a fish might suddenly develop a reptile heart; this latter idea is a mutation theory with a vengeance!

With all these shortcomings, it is not surprising that *Vestiges* had little influence on the acceptance of an evolutionary framework by serious scientific and philosophical communities at that time. In short, Chambers had presented neither a rigorous nor a convincing process cosmology or scientific theory of organic evolution. He died in the year Darwin published *The Descent of Man* (1871).

The inquisitive Charles Robert Darwin (1809–1882) had neither the first word nor the last word on the scientific theory of organic evolution.* Yet, he remains the most important thinker in the history of evolutionary thought. In fact, his innovative writings are still a major source of useful ideas for ongoing research in the evolutionary sciences.

Darwin was a human being of flesh and blood whose curiosity and creativity sought to grapple with that "mystery of mysteries" (the appearance of new species on earth). He studied actual objects in the real world, and

---

*For a more detailed treatment of Darwin's life and thought, refer to chapter 3.

the immediate effect of this research was a major shift in biology from merely philosophical speculation to crucial empirical inquiry. However, his analytic and synthetic mind did not develop the theory of evolution in a vacuum. In geology, he was particularly influenced by Charles Lyell as well as James Hutton and John Playfair. It was Lyell's three-volume work *Principles of Geology* (1831–1833) that gave the young naturalist that sweeping geological framework of space, time and change within which he could later conceive of the slow and continuous evolution of plant and animal species throughout earth history due to natural forces.

A five-year global voyage on HMS *Beagle* (1831–1836) was the single most important event in Darwin's life, as it gave the young scientist the time he needed to both critically observe and rigorously reflect upon nature itself. In retrospect, Darwin focused his attention upon those oceanic archipelagoes (especially the Galapagos Islands) that offered crucial data against the dogma of divine special creation while at the same time providing scientific evidence for descent with modification. The natural invasion, dispersal, and differentiation of flora and fauna populations on such oceanic islands over vast periods of time and change argued for speciation and, in turn, organic evolution.

Through an ingenious use of the hypothetico-deductive method with its preconceptions and anticipations, Darwin looked beyond unique events and particular species in order to formulate generalizations about organic history (including the origin and development of humankind). The geological column, paleontological record, geographical distribution of life, comparative studies in embryology and morphology, and the emergence of varieties or incipient species among cultivated plants and domesticated animals as well as in wild flora and fauna populations gave sufficient evidence to argue in favor of descent with modification and against the fixity of life.

Convinced that species are mutable, Darwin still needed an explanatory mechanism to account for organic evolution in terms of science and reason. In October 1838, he happened by chance to read casually but critically Thomas Robert Malthus's *An Essay on the Principle of Population* (1798, 1803). Briefly, the Malthusian treatise on population growth claims that animal numbers, including man, have a potential to increase at a geometrical rate, whereas food resources from plants have a potential to increase at best only at an arithmetical rate. Since geometrical progression outstrips arithmetic growth despite checks in nature, then inevitably there will be a ruthless struggle for existence pervasive throughout the living world.

Upon reading Malthus's monograph, Darwin suddenly intuited the explanatory mechanism of natural selection as that primary driving force throughout organic evolution. To Darwin, any beneficial-because-it-is-useful slight variation or major modification enhancing the adaptation of an organism to its own changing niche would at the same time provide this plant or animal with a survival advantage that would (more likely than not) also allow for its producing offspring to insure future generations. As a result of blind competition among the many, only a few favored organisms are destined to survive and reproduce, thereby perpetuating their variations through inheritance. Given enough time and change, higher taxonomic categories are able to emerge throughout the vast history of life on earth.

In short, Malthus's principle of population was the theoretical catalyst that gave Darwin (and later Wallace) the explanatory mechanism of natural selection, which his contemporary, Herbert Spencer, referred to as the "survival of the fittest," to adapt and reproduce. For Darwin, the awesome task was now to synthesize all those facts and concepts and relationships in such a way as to demonstrate clearly the scientific truth of organic evolution in a comprehensive and intelligible manner.

Darwin's two major books are *On the Origin of Species* (1859) and *The Descent of Man* (1871). Taken together, all his volumes represent a single persuasive argument for the mutability of plant and animal species throughout earth history, a unity of thought that represents a major turning point in the history and philosophy of natural science.

On 24 November 1859, Darwin published his *On the Origin of Species;* it became the most important book of the nineteenth century. Although this volume made only one reference to our own species, an inescapable implication was that the human animal is also a product of organic evolution and had descended from some earlier apelike form. Predictably, most readers were outraged; there was acrimonious debate and incredulous opposition.

In *On the Origin of Species,* Darwin merely wrote that light would be shed on the history of man and his nature. However, in *The Descent of Man,* he did make several controversial generalizations about human evolution: our own species and two great apes (chimpanzee and gorilla) share a common ancestry in the remote past; fossils representing this common ancestry will be found in Africa; the difference between those three living pongids or great apes (including the orangutan) and the human animal is merely one of degree rather than kind; and sexual selection has influenced organic evolution, including the emergence of our own species.

There is a crucial distinction between the fact of evolution and those differing interpretations of this process, which range from mechanistic materialism through creative vitalism and finalistic spiritualism to cosmic mysticism. However incomplete, Darwin's own worldview is essentially mechanistic and materialistic. He explained organic evolution in terms of three selective mechanisms (natural, sexual, and artificial) as well as his provisional hypothesis of pangenesis (1869) and those assumed environmental influences that cause the appearance of those needed variations. Pangenesis was the erroneous Darwinian hypothesis that all the organs of an animal contribute hereditary particles that are influenced directly by the use or disuse of the organs; consequently, those slight variations or major modifications in the parents were held to be inherited by and manifested in the offspring. In modern biology, chance genetic variation and necessary natural selection or differential reproduction are the two basic mechanisms of organic evolution.

Darwin's theory of evolution was staunchly supported by Thomas Henry Huxley in England, Ernst Haeckel in Germany, Asa Gray in the United States, William Dawson LeSueur in Canada, Marquis Gastonde Saporta and Albert Gaudry Quenot in France, Cesare Lombroso in Italy, Bozidar Knezević in Yugoslavia, and Vladimir O. Kovalevskii in Russia (among others). It has profound significance for interpreting the history of all life on this planet. Grounded in facts and logic along with those necessary modifications to keep it in step with the advancements of relevant special sciences (e.g., biochemistry, genetics, embryology, anatomy, physiology, immunology, paleontology, and ecology), the theory of evolution continues to withstand onslaughts from ignorance. Its triumph is due to a final emancipation from vitalism, teleology, and idealism in favor of evidence, experience, experimentation, and rigorous reflection now free from traditional theology and peripatetic metaphysics, although there is always a need for rational speculation. Indeed, the Darwinian revolution caused a collapse of our intellectual tradition, including the demise of Aristotle and Aquinas and Kant (among others). This is due to its geological perspective, the mutability of species, and the fact that humankind has emerged from an apelike ancestor (as well as its rejection of cosmic purpose and biblical literalism in favor of a mechanistic and materialistic worldview). In fact, some modern evolutionists, e.g., Stephen Jay Gould, feel liberated by its philosophical materialism.

It must be remembered that Darwin's writings were published when there was no molecular biology, population genetics, paleoanthropology, and primate ethology. Radiometric dating techniques and the global science of plate tectonics (continental drift theory) did not appear until this century. During Darwin's time, no scientist took seriously the thought that celestial influences may have contributed to both the evolution and extinction of species on earth. Likewise, exobiology was only a subject for speculation in rational philosophy and debate in natural theology.

There were remarkable parallels between Darwin and Alfred Russel Wallace (1823–1913):[28] each was a young English naturalist with interests in both geology and biology who had explored South America; each visited a unique archipelago; each studied orchids and collected beetles; each revered life while doing original research on the barnacle and the butterfly, respectively; each read Paley, Humboldt, Lyell, and Malthus; each accepted the scientific theory of organic evolution as a true framework of biological history; and each independently discovered that key explanatory mechanism of natural selection to account for the origin, adaptation, survival, divergence, evolution, and extinction of species on earth.

As an avid naturalist, Wallace devoted his life to entomology as well as to the study of tropical plants and animals. His extensive research took him to the Amazon and the Malay Archipelago. "Wallace's Line" (1863) refers to a zoogeographical boundary that separates Australian and Indian flora and fauna regions; it passes through the middle of Malaysia. He saw four distinct but successive stages in planetary history: the inorganic, organic, conscious, and spiritual levels of evolution. His interest in general anthropology focused on the human brain and human intelligence.

In 1858, Wallace sent an unpublished manuscript to Darwin at Down House. In it Wallace presented the mechanism of natural selection to explain in part the organic evolution of species. Darwin was startled to see his own scientific and rational interpretation of organic evolution independently formulated by another naturalist on the other side of the globe. To receive credit for his many years of thought and research, Darwin quickly published the one-volume On the Origin of Species (1859).

In later years, Wallace turned from naturalism to spiritualism. He claimed that materialistic Darwinism was insufficient to account for both the evolution of life in general and the emergence of humankind in particular. His final position argued for the uniqueness of our own species within a cosmic scheme of things. If Darwin had not gone on the voyage of the Beagle, then perhaps Wallace may have received the credit for fathering the scientific theory of organic evolution. In terms of its superior brain, Wallace argued that the human animal clearly has mental abilities far beyond those needed merely for survival. (He viewed this fact as a discrepancy in nature requiring a spiritualistic explanation.)

Wallace maintained that man's numerous faculties (artistic, musical, mathematical, and metaphysical) as well as wit, speech, humor, morality, intelligence, perfected brain, specialized hands and feet, and naked sensitive skin could not have slowly but progressively evolved from the lower primates only as a result of chance variations and selective mechanisms (natural and sexual). Therefore, Wallace eventually held that our own species differs not merely in degree but significantly in kind from the three great apes. Furthermore, he believed in the personal immortality of the human soul.

Wallace even saw the end of organic evolution on earth in terms of human spirit. He dogmatically claimed that our own species is absolutely alone in the great diversity of a finite universe. He held that an infinite deity designed this entire cosmos and is the acting mental power within the purposive evolution of all life; human development is guided to an enduring spiritual existence by higher intelligences or a supreme will. Certainly his final views of humankind, organic evolution, and the dynamic universe were very far removed from any ideas or speculations that Darwin himself would ever take seriously.

As "Darwin's bulldog" in England, the celebrated agnostic scientist Thomas Henry Huxley (1825–1895) gladly and successfully defended the Darwinian theory of evolution and, in fact, even remarked that it was very stupid of him not to have discovered the explanatory mechanism of natural selection as Darwin and Wallace had done. Although advocating cosmic evolution, Huxley sharply separated evolving and competitive nature "red in tooth and claw" with its "ape and tiger methods" from the human world of ethical conduct. However, Peter A. Kropotkin rejected Huxley's portrayal of the nonhuman world as being separated from the human realm.

Huxley held that "question of questions" to be the origin of our own species. In fact, he was the first naturalist to write a serious book solely for the purpose of extending the scientific theory of organic evolution to account for the emergence of humankind and its relationship to the apes. *Evidence as to Man's Place in Nature* (1863) includes the results of his own comparative research in anatomy, embryology, primatology, and paleontology.[29] In it, as a direct outgrowth of those studies on the central nervous system and brain of higher primates, Huxley presented his pithecometra hypothesis: in terms of biology and evolution, man is closer to the three great apes (orangutan, chimpanzee, and gorilla) than they are to the two lesser apes (gibbon and siamang). Briefly, there is no new structure or function peculiar to the human animal alone that does not already exist to some degree in the pongids. As a consequence, Huxley maintained that our own species differs only slightly from the three great apes; there are only quantitative rather than qualitative differences between man and these pongids.

In Germany, the great zoologist and naturalist philosopher Ernst Heinrich Haeckel (1834–1919) rigorously defended both the Darwinian theory of evolution and the pithecometra hypothesis, especially through his own research in comparative embryology that demonstrated that all living primates share a common ancestry. The pithecometra hypothesis maintains that the human animal is closer to the pongids (orangutan, chimpanzee, and gorilla) than they are to the hylobates (gibbon and siamang). This generalization was the outgrowth of comparative studies in primate morphology. Haeckel emphasized this close anatomical affinity between man and the three great apes, these four hominoids being the latest products of primate evolution.

Speculating on the origin and evolution of the human animal, Haeckel wrote and lectured that the "missing link" *Pithecanthropus alalus* (apeman without speech) would be found in Asia. In fact, his hypothesis inspired the student Eugene Dubois to search for such a fossil hominid form with subsequent success and controversy. However, Darwin was correct in maintaining that Africa was the birthplace of our species.

An outspoken pantheist, Haeckel called for scientific inquiry free from idealist philosophy and theistic religion. His most popular book remains *The Riddle of the Universe at the Close of the Nineteenth Century* (1900), in which he boldly presents a philosophy of monism grounded in evolution, materialism, and the cosmic perspective.[30]

In Germany, Karl Marx (1818–1883) emerged as a revolutionary socialist and economic theorist.[31] This atheist and humanist as rigorous critic emphasized the crucial need for a historical perspective. In his *Economic and Philosophical Manuscripts* (1844), Marx presented a most controversial view of man and society within history. It was a dramatic break from philosophical idealism in favor of dialectical materialism.

With his friend Friedrich Engels (1820–1895), Marx wrote the *Communist Manifesto* (1848). It is an interpretation of social history, an analysis of capitalism, and a call to revolutionary action to establish a world free from alienation and exploitation. Financially and morally supported by Engels, Marx lived in London where he spent most of his years in the British Museum gathering material for his monumental study of economic history. The result was *Das Kapital*, a significant contribution from his passionate quest for comprehending the human condition in terms of progressive sociocultural development.

Today, this Marxian synthesis is held by some to be scientific; its philosophy advocates historico-dialectical materialism and its ideology calls for a new egalitarian civilization (actually, it was Engels who developed dialectical materialism as a metaphysical justification for Marx's own economic and historical materialism).

Marx had been greatly influenced by the major ideas of Hegel and Feuerbach; he later found inspiration in the evolutionary perspectives of Darwin and Lewis Henry Morgan. Marx himself advocated a conflict theory that focused on the alienation and exploitation pervasive throughout the industrialized human situation. He envisioned that, through enlightened awareness and collective action, the proletariat would bring about an inevitable revolution followed by the establishment of global communism. In the near future, humankind would then enjoy ongoing fulfillment in a society without money, classes, religion, state government, monogamy, private property, and the profit motive. Of course, all political ideologies and social systems have their own advantages and disadvantages; who would really choose to live in Plato's ideal republic? If humankind survives, one can only speculate on what its sociopolitical milieu will be like hundreds, thousands, or millions of years from now.

Karl Marx taught that history progresses by struggle with opposition;

qualitative changes occur in revolutionary leaps (contrast with the quantitative changes of gradual development). Whether one endorses revolution or evolution, it remains a fact that few social theories have had such a profound influence on actual man in the concrete world as has Marxism in this century.

In June 1876, Friedrich Engels (1820–1895) wrote an article entitled "The Part Played by Labour in the Transition from Ape to Man"; it was not published until 1896.[32] Greatly influenced by Marx and Darwin, Engels held that the theory of evolution is central to comprehending the natural rather than the divine origin of humankind from the animal kingdom. Furthermore, he argued that the explanatory mechanism of natural selection shed light not only on that ongoing active relationship between an organism and its environment, but also on that pervasive struggle between progress and reaction throughout the history of our own species.

Engels grounded his interpretation of the emergence of humankind in atheistic dialectical materialism, in sharp contrast to theistic metaphysical idealism. His labor theory of hominid evolution holds that our own species emerged out of that fierce competition and the struggle for existence in those predatory simian ancestors that slowly acquired an upright posture, thereby freeing the arms from locomotion so that the hands could be the first tools of labor (after the teeth). He also stressed the social importance of both eating meat and using fire. Over an extended period of time, a bipedal gait as well as toolmaking and mutual group behavior resulted in a superior brain, articulate speech, and social activity that is both creative and purposeful.

Engels emphasized that his social theory of human labor did not exclude but, in fact, complemented Darwin's explanatory mechanism of natural selection in organic evolution. Labor transforms both the human animal and material nature itself. With tools and fire, consciousness exerts both a constructive and destructive influence on the further evolution of matter in motion and society in development. More and more through cooperative activity (collective labor), global man is achieving mastery over nature and establishing dialectical harmony between human society and the material world. In brief, this dynamic universe, including organic history and human progress, is grounded in dialectical materialism (not idealism or spiritualism). Even today, one may read Engel's unfinished manuscript with benefit. Particularly relevant is his evolutionary perspective and concern for the environment.

In step with the ideology of Marx and Engels, the Russian revolutionary V. I. Lenin (1870–1924) also grounded his worldview in pervasive materialism. Of special importance is his book *Materialism and Empirio-Criticism: Critical Comments on a Reactionary Philosophy* (1920).[33] It may be argued that dialectical materialism and scientific evolutionism are mutually inclusive explanations for dynamic reality. This synthesis is still a rich area for critical inquiry.

Early in this century, Russian biology suffered stagnation under a Lamarckian theory of evolutionary genetics dogmatically upheld by Trofim Denisovich Lysenko (1898-1976) for vested sociopolitical interests.[34] Today, the success of neo-Darwinism is a victory for science and reason over vitalism and spiritualism.

In sharp contrast to collectivism, atheist philosopher Friedrich Nietzsche (1844-1900) favored the creative individual and envisioned the coming of the superman beyond good and evil as that superior intellect free from the false values of a decadent society.[35] The future superman will be as far beyond both man and ape of today as they are above the lowly worm! Nietzsche's awesome idea of the eternal recurrence of the same argued for an endless cosmic series of finite but absolutely identical evolutionary cycles. Because each of the infinite number of finite epochs is identical, there is no evolution as such from one cosmic cycle to the next, but only that monotonous repetition of the same events (although there is the identical evolution within each finite epoch).

In England, Herbert Spencer (1820-1903) was the first philosopher of cosmic evolution. He presented an early version of his ideas in *Essay on Evolution* (1852). Spencer viewed this universe and the history of biology, society, psychology, and ethics within an evolutionary framework. His mature ideas are presented in a ten-volume work, *Synthetic Philosophy* (1860-1898).[36]

Spencer's universal law of progressive evolution maintains that all things in nature develop from simplicity to ever-greater complexity until equilibrium is reached and then devolution occurs. As such, dynamic reality is an endless series of finite cosmic cycles. Each cycle is the same in principle but different in content; contrast Spencer's evolutionary worldview with Nietzsche's cosmic doctrine of the eternal recurrence of the same events.

Viewing human society as a developing superorganism, Spencer extended the Darwin/Wallace explanatory mechanism of natural selection or the "survival of the fittest" from biology into sociology. Going even beyond Lamarck, Spencer held to the inheritance of acquired knowledge, experience, and morality as well as physical characteristics. Consequently he wrote that, given enough time and without human intervention, progressive evolution itself will perfect our own species. The resultant Social Darwinism argues simplistically for the success of the individual without any help from the government. For the continued improvement and perfection of our own species, the Spencerian viewpoint taught that no measures should be taken to encourage the preservation of mentally and physically imperfect beings, who would only pollute humankind; Darwin himself never supported this ruthless social theory that unfortunately came to bear his name. This obvious overextension of a descriptive concept from biology into sociology is unwarranted and has been correctly labeled "pseudoevolutionism" by the philosopher Marvin Farber (1901–1980).

In the United States, Spencer's cosmic philosophy was championed by John Fiske (1842–1901), while Social Darwinism was advocated by William Graham Sumner (1840–1910). In *Outlines of Cosmic Philosophy* (1874), Fiske popularized Spencer's worldview.[37] Later, however, his own thoughts embraced a theistic interpretation of the universe: the goal of evolution is the perfection of a spiritualized humankind. However, at Yale University, Sumner, as an ordained Episcopal minister and staunch conservative, vigorously advocated Social Darwinism. He taught that competition is the law of nature and that the survival of the fittest is the law of civilization. His conception of sociocultural evolution is grounded in an unmitigated individualism. Fortunately, others in anthropology, sociology, and psychology have offered critical insights and theoretical frameworks for understanding and appreciating sociocultural evolution (including the development of religious belief systems and behavior patterns) entirely free from Social Darwinism.

Having spent his entire life in Yugoslavia, Bozidar Knezević (1862–1905) speculated on the nature of the universe and wondered about the meaning, purpose, and ultimate destiny of humankind within a cosmic scheme of things. As a philosopher of evolution, he was both a rationalist and an empiricist. His two major books are *Succession in History* (1898) and *Proportion in History* (1901); these two works are jointly combined in a single volume titled *The Principles of History*.[38] A third important book consists of his 876 aphorisms published as *Thoughts* (1902, 1931).

Knezević was primarily a historian who regarded astronomy as that quintessential model of exactness and rational precision. He was firmly committed to science, mathematics, and the brotherhood of humanity. This remarkably creative visionary achieved an original worldview that synthesized both facts and concepts within a cosmic perspective of objects and events. He offers a unique interpretation of the place our own species occupies within the determining laws of first an evolving and then a devoluting universe. Briefly, his worldview is a dynamic scheme outlining cosmic ascent and subsequent universal descent (including the evolution and devolution of humankind).

Knezević rejected all views supporting geocentrism, zoocentrism, and anthropocentrism. He also refuted astrology, alchemy, and phrenology in favor of scientific inquiry and rational reflection. In general, he speculated about the immensity of our dynamic universe. In particular, he recognized three major stages of earth history: inorganic, organic, and psychic levels of development. Despite its antiquity, humankind is seen as a recent product of (and totally within) organic evolution; its terrestrial destiny is bound to the general laws of material reality.

Knezević held that cosmic reality is an infinite series of finite evolutionary cycles: all broadly analogous to, but each uniquely different from, this particular universe containing the ephemeral history of the planet and our own inevitably doomed species. He recognized forward evolution and subsequent backward devolution as the two essential aspects of both celestial and terrestrial history (material reality is full of semicircles).

This semicircular scheme of earth history consists of an evolutionary ascent from primordial chaos, through life, to the fulcrum of a temporary intellectual and moral convergence and future unification of a free humankind on a global scale in a strictly secularized sense. This ascent will be followed by a reverse process of descent from orderly but ephemeral equilibrium back to an ultimate material chaos.

According to Knezević, earth history has been an evolution from the simplicity of material chaos through a sequential emergence of more and more complex life forms (fishes, amphibians, reptiles, and mammals) to the appearance of our own species. The future devolution of the planet will see the human animal vanish first, followed by a series of extinctions: all other mammals, then the reptiles and amphibians and fishes, and finally those lowest organic forms that had emerged first on earth will be the last remnants of life to disappear.

For Knezević, the material universe is utterly indifferent to the fleeting incident of human existence; even if all human beings suddenly disappeared from the face of the earth, the stars and planets of this galaxy and others would undoubtedly still endure. The assessment of the proper place of mental activity and our species within the totality of nature is a necessary precondition for any defensible understanding of, and acceptable appreciation for, the overall cosmic scheme of things. Since all things must die, then the human animal with its civilization is merely a luxury of nature that will inevitably disappear within the endless flux of material reality (as will next the organic and finally the inorganic realms of the planet).

Knezević did envision other universal processes within the infinity of superspace and the eternity of supertime, each analogous to that semicircular history of the physical cosmos and our earth (characterized by the inexorable movement from inorganic chaos through increasing organic complexity and then back to material chaos). He conceived of those histories of other material planets, stars, cosmic systems, and even entire universes as general phases or particular episodes of the immeasurable and incomprehensible process of God as nature itself.

Consequently, Knezević did not rule out the rare occurrence of alternative forms of life and intelligence existing elsewhere on other celestial bodies. Yet, he did not foresee the human animal ever escaping the finite bonds of this earth. In fact, he held that the eventual end of our own species is inseparably linked to that fatal destiny of our perishable planet, which is an integral part of an ephemeral solar system.

Knezević maintained that time is the force that reduces epochs to moments, galaxies to atoms, and all life (including humankind) to a handful of ashes. From this awesome cosmic perspective, the seemingly eternal and infinite universe is not the ultimate whole but only a small part of the superspatial and supertemporal process of material reality. He refers to this ongoing process as the unfinished biography of God. His vision sees our own species within the natural history of life on this planet without any religious preferences or presuppositions; no personal God or mystical apotheosis will save humankind from devolution and entropy within this particular finite cosmic epoch of material reality.

The free-thinking Knezević, as pantheist and humanist, held God to be nature itself. Concerning the future of our own species, he claimed that (at the peak of maturity) a united humankind will be able to reconcile within itself all those successive errors of its earlier phases of evolution. Likewise, he foresaw a cult of the sun as that final religion of nature, marking the end of conscious science on earth (just as it had marked that beginning of unconscious religion in the remote past).

Unfortunately, Knezević has forced a rigid semicircular structure upon dynamic reality. Furthermore, humankind may in fact leave the confines of this planet before our sun burns out. No dogmatic closure should be placed on human progress or material nature.

Not all evolutionists emphasized the struggle for existence and the survival of the fittest. The Russian naturalist and humanist Peter A. Kropotkin (1842–1921) devoted his extensive scientific studies to a serious inquiry into the geology, geography, botany, zoology, anthropology, and sociology of eastern Siberia and northern Manchuria. Although greatly influenced by the writings of Charles Darwin, Kropotkin investigated and extolled those roles as important adaptive mechanisms throughout both nonhuman and human history that mutual aid and mutual support or group cooperation play in the living world. His significant findings were presented in the volume *Mutual Aid: A Factor of Evolution* (1902, 1904);[39] this book presents a social interpretation of animal behavior, which does not contradict Darwinian evolution with its major explanatory mechanism of natural selection but merely supplements it by pointing out the value of group cooperation throughout animal evolution.

Differing from Darwin, Kropotkin stressed the survival value of mutual aid in social behavior throughout the animal kingdom (including human evolution) rather than the explanatory mechanism of natural selection, which emphasizes the reproductive results of only individual struggle. He saw group cooperation as a progressive factor in organic history. With its ideas of mutual aid and mutual support as crucial aspects in the animal kingdom, his biology represents an important advance in ethology for modern evolutionary thought.

In his *Ethics: Origin and Development* (1922), Kropotkin saw humankind in all its aspects as totally within the evolving matter of concrete reality.[40] Yet, unlike Huxley and others, he rejected a sharp distinction between animal behavior "red in tooth and claw" and the realm of human conduct grounded in ethics, morals, values, and vested interests. For this utopian social philosopher as anarchistic Communist, the ethical and social progress of our own species is the direct outcome of both instincts and feelings transformed into morality and love: justice, freedom, and solidarity emerge from and depend upon the factor of mutual aid in this evolving animal world.

Early in this century, the French philosopher Henri Bergson (1859–1941) reacted strongly against the mechanistic materialism of Charles Darwin and his followers.* In his major work *Creative Evolution* (1907), Bergson presented a vitalistic view of humankind within organic history.[41] Claiming that a strictly scientific explanation of process reality is insufficient to account for the emerging novelty and pervasive creativity continuously manifested within the diverging process of biological evolution, he introduced into his systematic interpretation of planetary development the metaphysical concept of a living force running throughout the otherwise static matter of this physical universe. In the human animal this cosmos of matter and spirit is linked through perception.

This *élan vital* causes matter to exhibit organic properties and, consequently, to evolve toward ever greater complexity and consciousness. Diverging evolution displays ongoing emergent novelty as a direct result of this vital force. Organic creativity in earth history has taken three major directions: plants with torpor, insects with instinct, and animals with consciousness.

For Bergson, intuition and mysticism are superior to science and reason. According to him, the vital force may not explain much, but it is at least a metaphysical principle affixed to human ignorance as an occasional reminder that the use of both science and reason has its limits, while mechanism and materialism invite one to ignore these limits.

In *The Two Sources of Morality and Religion* (1932), Bergson adopted a more Christian view of man's place in this universe.[42] He now held that creation will appear to a mystic as a God undertaking to create humankind so that he may have, besides himself, beings worthy of his love. Clearly,

---

*For a more detailed treatment of Bergson's philosophy of evolution, refer to chapter 4.

his final vision of evolution goes far beyond mechanism and materialism and even diverging vitalism to embrace theistic mysticism.

Sigmund Freud (1856–1939) extended the evolutionary framework to account also for the origin and history of the human psyche, but not without persistent and severe criticism from both a conservative public and those myopic scholars who preferred to save at least the alleged immortal soul or private mental world of our own species from evolutionary speculation and scientific inquiry. Nevertheless, an undaunted Freud pursued his study of the human mind. He linked the libido or unconscious energy of the inherited id back to our emergence from some earlier animal, while focusing on the sexual as well as religious etiology of both neurotic and psychotic behavior in modern civilization.

Pragmatists (Charles S. Peirce, William James, and John Dewey) and emergent evolutionists (C. Lloyd Morgan, Samuel Alexander, Jan Christiaan Smuts, Roy Wood Sellars, and Alfred North Whitehead) offered various interpretations of evolution ranging from idealism to materialism.

Pragmatism has been the most influential philosophy in America, representing both an evolutionary framework as well as a science-oriented attitude. It was founded in 1878 by Charles Saunders Peirce (1839–1914), who had established the Metaphysical Club at Old Cambridge in 1871. At Harvard, Peirce's own philosophy (which he dubbed pragmatism) grew out of his study of the phenomenology of human thought and the use of language. He hoped to work out a general theory of signs, a method for clarifying those meanings of intellectual signs (words, ideas, or concepts) used in communication for the purpose of helping to solve scientific or philosophical problems. For Peirce, a sign must have some conceivable practical consequence; he presented a method of inquiry grounded in a pragmatic theory of truth.

Peirce was interested in the physical sciences (especially astronomy), mathematics, and logic. His philosophical development passed through four distinct systems. The first system (1859–1861) was a form of extreme post-Kantian idealism grounded in a threefold ontological classification of all reality into matter, mind, and God: the three categories of it, thou, and I respectively. Following this early idealism, Peirce began the serious study of logic. The second system (1866–1870) argued that his three ontological categories of matter, mind, and God are necessary for signhood within converging inquiry. As such, he gave a doctrine that was both phenomenalistic and realistic. Then, the third system (1870–1884) presented a modern logic of relations supporting utilitarian pragmatism and his doubt-belief theory of converging inquiry. Pragmatism was now presented as a theory of meaning particularly attributable to scientific definitions concerning the real as a permanent possibility of sensation, and the doubt-belief theory of converging inquiry was set in the context of biological evolution. Finally, the fourth system (1885–1914) involved a complete revision of his earlier three categories, which are now rendered as simply three classes of relations: monadic, dyadic, and triadic. He held that all thought is in the form of signs and therefore irreducibly triadic.

Peirce taught that a sound philosophy of nature requires rigorously acknowledging the theory of evolution.[43] Writing at the close of the nineteenth century, he recognized and incorporated three basic theories of biological evolution: Lamarckian, cataclysmic, and Darwinian. They represent evolution by creative love or mechanical necessity or fortuitous variations, respectively (i.e., agapastic or anacastic or tychastic evolution). Peirce himself favored the agapastic interpretation of organic evolution. He presented a worldview supporting both an evolutionary cosmology and an evolutionary espistemology.

In 1905, Peirce referred to his own doctrine of pragmatism as pragmaticism to clearly distinguish it from William James's religious form of pragmatism.

Peirce presented an intriguing, if unconvincing, evolutionary cosmology ultimately rooted in pervasive idealism. In fact, his interpretation of cosmic evolution was based upon the Lamarckian doctrine of the inheritance of acquired characteristics. To begin with, Peirce held to the historical continuity of reality and taught that the universe as a whole is a living and evolving organism of pervasive feelings and habits; dynamic nature itself is a creature of habit. The laws of nature describe the habits of the universe. He argued that a human mind contains innate adaptive knowledge accumulated through centuries of experience as a result of the inheritance of acquired characteristics. Therefore, judgments of common sense are more likely than not to be a source of true hypotheses.

To account for the origin of this universe, Peirce held that there had been an initial condition in which the whole cosmos was nonexistent, i.e., a state of absolute nothingness. From this first state of pure zero or unbounded potentiality, he taught that there then originated by evolutionary logic (the logic of freedom or potentiality) a state of potential consciousness or an undifferentiated continuum of pure feelings completely without order. From this primordial chaos of unpersonalized feelings and pure chance, the dynamic universe had arbitrarily evolved brute force or action followed by living intelligence. Cosmic evolution has manifested three successive but different modes of metaphysical being: feeling, antirational action, and rational thought. This sequence represents the evolutionary perfecting of habits toward greater order or harmony. This cosmic view supported both the temporal continuity and absolute chance within the universe. Even now, there is chance within this universe and, as such, natural laws are not yet completely exact.

For Peirce, cosmic evolution is a general continuous tendency of increasing habit and decreasing chance developing toward an inevitable, rational, absolutely perfect, and symmetrical goal. Consequently, this universe continues to evolve from its past state of homogeneity (chaos and absolute chance) to a future state of heterogeneity as the complete order of concrete reasonableness and maximum beauty. Briefly, Peirce clearly held that this universe is mind rather than matter and that both human life and scientific inquiry must adapt to and remain in harmony with the direction of cosmic evolution.

William James (1842–1910) remained, for the most part, a science-oriented thinker. As a young man, he traveled widely and developed a profound sensitivity for the breadth and depth of human experience (especially aesthetic and religious experiences). But abandoning his desire to become an artist, James devoted himself to the study of several natural sciences, e.g., chemistry, comparative anatomy and physiology, and medicine. During his scientific training, he adopted an evolutionary and naturalist orientation. In June 1869, he received his medical degree from Harvard and then began teaching anatomy and physiology (1873), psychology (1875), and philosophy (1879).

Turning from the natural sciences, James now devoted his attention to that relationship between psychology and philosophy. He was particularly concerned with questions about the human condition and offered a philosophical defense of free will, morality, and religious beliefs. His form of pragmatism is primarily a theory of meanings (rather than truths) grounded in the world of pure experience. Although his pragmatism rejected the possibility of a pure phenomenological description of inner experience, it nevertheless supported naturalism and the possibility of a metaphysics relevant to both scientific and theoretical inquiry.

James's pragmatism is an openended method for settling metaphysical disputes within a practical context. It is grounded in an empiricist attitude recognizing the relationship between science and philosophy. Pragmatism is concerned with consequences (theories thus become instruments): "the truth is the name of whatever proves itself to be good in the way of belief, and good, too, for definite, assignable reasons" (*Pragmatism*, 1907, p. 59).[44]

As an evolutionary naturalist, James rejected the traditional mind-body dualism found in philosophical literature. However, he did distinguish between mental states and those underlying physical states of the brain. He saw psychology as providing an adequate introspective description of mental states, and pragmatic philosophy as a critical analysis of those meanings in human experience. Above all, James focused upon spontaneous religious experience. For him, evolution suggests that there must be a meaningful adaptive function to religious feelings. He then developed his doctrine of the will to believe, a vague philosophical defense of religious beliefs within his otherwise naturalistic orientation.

James rejected traditional supernaturalism, merely holding that religious experience suggests that there is a higher part of this universe. His pragmatism was an attempt to formulate a metaphysics grounded in natural processes, the special sciences, and humanism. Since he recognized the richness of human experience, his vision mediates between tough-mindedness and tender-mindedness.

Adopting a psychological orientation, James proposed that the meaning or truth of ideas and beliefs and values rests in their "cash value" or usefulness. Therefore, he defended the will to believe within his own interpretation of pragmatism.

As the most recent and major figure in the school of American pragmatism, John Dewey (1859–1952) had been influenced by Kantian and Hegelian idealism.[45] However, he became greatly inspired by the works of Charles Darwin and William James. As a result, Dewey abandoned his early idealism for a dynamic biological-anthropological orientation within a naturalistic evolutionary framework. Although adopting an organismic perspective, he developed no philosophical system in the traditional sense. Unlike Kant and Hegel, Dewey left speculative philosophy in order to concentrate his work on a practical consideration of psychology and education within a democratic community of creative intelligence. He was concerned with warranted assertions derived from scientific inquiry that is grounded in a naturalist logic; his instrumentalism focused on both science and values (education and morality).

Dewey emerged as America's most significant philosopher. His form of pragmatism advocated naturalism and humanism within a practical context uniting thought and action. Influenced by Darwinism (though not Social Darwinism), he taught that ideas are instruments or tools to solve problematic or indeterminate situations. Ideas allow for human adaptation and survival within a social context. For Dewey, ideas are guides to action; true learning takes place only when the individual is free to explore and test this world. In brief, his philosophy advocated individual self-realization within a harmonious democratic society.

As a rigorous naturalist, Dewey held the natural world to be prior to the emergence of human existence. Nature is in continuous change, process, and interaction. Human experience is a product of and totally within dynamic nature (it is both reflective and nonreflective) and growth is through experience and education. Education results from an interaction (dialectical relationship) between a human knower and those knowable events of nature. Action is central to Dewey's philosophy. It results in the creative self-realization of an individual within a democratic community.

Dewey emphasized both the scientific method and rigorous reflection. He called the empirical method the denotative method. His naturalistic philosophy rejected traditional dualisms: the knowable and the unknowable (becoming and being), philosophy and science (ideas and facts), rationalism and empiricism (intellectualism and sensationalism), mind and body, as well as the theoretical and practical (thought and action). Also, he rejected both supernaturalism and innate ideas. For Dewey, there are only methodological truths. Ideas are meaningful, valuable, and true if they result in fruitful inferences and/or consequences. The results of thought and action determine pragmatic significance. Unfortunately, his metaphysics (ontology and cosmology) was never clearly developed.

As an evolutionist, Dewey recognized three major natural levels of development within the continuity of planetary history: the physicochemical, psychophysical, and human experience levels. Six major points emerge from his instrumentalism or experimentalism: (1) the intellect or reason is on a higher plane than the other human qualities; (2) ideas emerge out of one's need to establish an equilibrium between oneself and one's environment—this point illustrates Darwin's influence on Dewey's form of pragmatism; (3) ideas are meaningful if and only if they are applicable to problematic or indeterminate situations; (4) ideas are true if and only if they solve problematic or indeterminate situations; (5) the scientific method is applicable to ethics and morals, human principles and social behavior, as well as to nature; and (6) there is no need to have recourse to any form of supernaturalism.

In summary, Dewey's pragmatism held that metaphysics should be descriptive and not merely speculative (he called for a naturalistic philosophy); subjectivism results in an unwarranted dualism (he rightly rejected the traditional mind-body dichotomy); philosophy must recognize the facts and values of the special sciences, e.g., physics, chemistry, biology, anthropology, and psychology (he taught that the study of anthropology is an ideal preparation for philosophy); knowledge and the awareness of problems are obtained through experience; there are no absolute truths; ideas are meaningful if they are applicable to human problems and true if they eliminate such problems; education does not necessarily liberate a person but it does develop his creative intellect; there is no distinction between knowing and doing (to know is to act); and it is the aim of future philosophy to clarify those meanings and truths of human ideas and human actions. In the final analysis, Dewey's pragmatism is essentially a theory of inquiry.

Emergent evolutionists presented a new interpretation of this universe in general and the history of living things on earth in particular. Some presented a creative cosmology, while others focused only on planetary history. As evolutionists, they were especially concerned with three major concepts: emergence, levels, and novelties. They rejected outright preformationism, materialism, reductionism, vitalism, and finalism. The biological theory of preformationism maintained that the development of an organism or the history of a species is simply the enlargement of all parts or all individuals already existing in the germ cell or cells, respectively. Finalism is the claim that the advance of evolution is progressing toward a preestablished end or goal.

The differing schemes of the emergent evolutionists emphasized the unpredictable appearance of variety, diversity, and complexity within the discontinuous history of creative advancement; each new emergent quality engendered by evolution is held to represent a new level of existence, and there may even be sublevels within levels. Emergent qualities are novel and therefore unpredictable. These emergent evolutionists presented speculative descriptions of cosmic and organic evolution rather than explanatory demonstrations. However, perhaps somewhere else in this vast universe an evolutionary process identical with (or very similar to) our own planetary history may have taken place, is taking place, or will take place.

Biologist and philosopher C. Lloyd Morgan (1852–1936) presented a unique metaphysical system in which he described the historical emergence of new kinds of intrinsic relatedness among evolving spatiotemporal events. As a result of his comparative psychological studies, he rejected both radical behaviorism and teleology. His conceptual framework is neither mechanistic nor finalistic. For the most part, his metaphysical system supports a naturalistic vision of evolution.

In his major work *Emergent Evolution* (1927), Morgan distinguished among three kinds of relatedness or natural systems:[46] matter, life, and mind, i.e., physicochemical, vital, and conscious events. Matter, life, and mind represent distinct emergent levels within which there are qualitative differences. For Morgan, evolution is a progressive advance with novelty or the actualization of enfolded possibilities. There is an ascending hierarchy of kinds or orders of relatedness ranging from those in the atom to those in reflective consciousness (this hierarchy of emergent relations presents itself as a pyramid with an atomic base and reflective consciousness near its apex).

Morgan acknowledges an independent evolving physical world as a system of events and God as the ultimate source and directing activity on which the continuity and progress of emergent evolution are ultimately dependent: by acknowledgment, he meant a judgment whose verification lies beyond the range of such positive proof as naturalist criticism rightly demands. Furthermore, Morgan taught that such an acknowledgment of God was necessary to supplement a strictly scientific interpretation of evolution (an ultimate synthesis of interpretation and explanation would result).

God is presented as directive activity, i.e., the creative source of emergent evolution. God is at once dependence, causality, and efficiency. God explains not only the directive activity in evolution but also that which from above draws all things and all persons upwards. That is, God is immanent as the all-embracing activity and transcendent as the goal of emergent evolution.

Morgan claimed that only through intuition may one enjoy the creative activity in evolution as well as feel at one with existence. He held that emergent evolution manifests a threefold relatedness of involution, dependence, and correlation. Involution refers to the basic fact that higher events involve the existence of lower events from which they have emerged, i.e., conscious events (mind) presuppose physiological events (life), and organic events (life) presuppose physicochemical events (matter). In short, there is no mind without life and no life without matter. Dependence refers to the position that the existence of events also depends upon supervenient events: involution and dependence supplement each other within the space-time continuum of emerging events, i.e., as one descends the pyramid of evolution, one is concerned with dependence, but as one ascends the pyramid of evolution, one is aware of involution. As such, the coherency of evolution is guaranteed from above and below, respectively. Correlation refers to the inseparable union of the physical and the psychical. That these three categories are a priori is a judgment of natural piety or an acknowledgment. Emergent evolution recognizes creativity as the novelty of new kinds of events or relations in space-time, but the immanent activity and the transcendent source are not susceptible of scientific proof.

In *The Emergence of Novelty* (1933), Morgan held that the historical advance of all cosmic events may be plotted on an ascending curve with God as its final cause.[47] However, he never held to a future mystical union of the activity and the source of emergent cosmic evolution. Instead, he stressed the significance of organization. Hence, evolution is the pervasive advancement of creative organization rather then merely natural aggregation. He is the only emergent evolutionist to resort to the influence of a transcendent God to account for that creative advance throughout the process of evolution.

Samuel Alexander (1859–1938) was a realist interested in biology, physiology, and psychology. In his major book *Space, Time, and Deity* (1920), he gave a comprehensive metaphysical analysis of the emergence of qualities within a space-time continuum.[48] His work distinguished between space-time or motion as an objective, continuous, infinite whole and the finite homogeneous units of space-time or motion that fills this infinity (these units are the very stuff or substance of this infinite universe as space-time or motion of which finite space-times or motions are merely fragments).

It referred to these cosmic units as point-instants, finite extensions, durations, or pure events; they are not homogeneous units of psychic energy. In brief, this metaphysical scheme holds that reality is a four-dimensional continuum of interrelated complexes of instances of spatiotemporal motions.

Alexander taught that within this plenum there is a nisus or creative trend. As a result of this innate tendency, space-time or motion generates a chronological hierarchy of new, finite, empirical qualities. The successive order of emergent qualities is said to be as follows: motions, materiality with primary or constant qualities, materiality with secondary or variable qualities, life (organic world), mind (consciousness), and deity.

As an emergent evolutionist, Alexander saw this cosmos as a hierarchy of qualities differing in kind. He distinguished between emergent qualities in general and those various levels of reality differing in degree, i.e., there are levels of degree within each of the major levels that differ in kind. He emphasized that mental functions succeeded but are dependent upon biological structures (just as living functions succeeded but are dependent upon physical structures).

Alexander also attempted to reconcile theism and pantheism within an evolutionary framework. He gave priority to time, holding that it had generated space-time. In turn, space-time generated an order of successive, finite, empirical qualities. The latest emergent quality is consciousness. He then made a distinction between God and the deity within this irreversible creative process or nisus of space-time. God is the infinite world as space-time or motion (this is the pantheistic aspect of God). God is also dynamic, for this infinite universe is evolving toward deity; i.e., God is "pregnant" with deity. What is deity? Deity is the next higher, finite, empirical quality that has yet to emerge from the productive movement or nisus of time. Within the infinite series of levels of existence, deity is always that angelic quality next to emerge in the scheme of things. God is the infinite universe itself, while the deity is only a future portion of God's nature. The next deity to emerge will differ in kind (not merely in degree) from mind or spirit or personality; deity is God's divine quality still to emerge ahead of this present state of emergent evolution. In short, God is infinitely immanent here and now while the next deity is finitely transcendent in that time to come.

This attempt to reconcile theism with pantheism (transcendence with immanence) is grounded in religious sentiment or natural piety. Since the quality of mind or consciousness emerged from the quality of life, the divine quality of deity will emerge out of consciousness. Religious experiences are those feelings for the vague future quality of deity. To use an analogy, this universe is the infinite body of God while the future deity is his finite mind yet to emerge (this view is both organismic and teleological). God is this whole universe engaged in process toward the emergence of this new quality, while religion is that personal sentiment that draws one toward it. A person is caught in the movement of our world toward this higher level of existence, that next emergent quality as deity.

Jan Christiaan Smuts (1870-1950) proposed a philosophy of holism as an attempt to reconcile those conflicting views and implications of mechanism and vitalism. In *Holism and Evolution* (1925), he presented a creative and comprehensive interpretation of planetary evolution within a non-Euclidean universe, i.e., within a curved or warped space-time continuum.[49] Within the creative synthesis of evolution, he taught that matter, life, mind, and personality represent those successive major advances of cosmic energy. He stated that this universe is ultimately intensely active energy. Matter is concentrated structural energy that is not only active but also elastic and transmutable. It is capable of manifesting creative forms or arrangements or patterns and values.

Smuts taught that the fundamental factor of the universe is its wholeness. Wholeness is that basic character and tendency of events within the geometric progression of cosmic development from necessity to greater degrees of freedom; holism is the ultimate category from which are derived those physical, chemical, organic, psychic, and personal categories of nature. Therefore, Smuts spoke of the evolution of wholes (creative reality is caused by the continued production of new wholes from preexisting ones). Thus arise those physical, chemical, organic, psychic, and personal categories that are all expressive of holistic activity at its various levels and reducible to terms of holism. Briefly, mind structures presuppose life structures while the latter presuppose energy structures, which are themselves graded according to the various forms of physical and chemical grouping.

Smuts held that the quality of mind represents a new product of emergent evolution. The superstructure of mind is immeasurably greater than the brain or neural structure on which it rests; mind is something of a quite different order that marks a revolutionary departure from the organic order out of which it originated. Smuts viewed mind as a wholistic structure in keeping with his philosophy of holism. His view of evolution is naturalistic with no resort to teleology, spiritualism, or a transcendent absolute. Although his pervasive metaphysical factor of holism remains inscrutable, he had nevertheless provided a naturalistic framework.

Roy Wood Sellars (1880-1973) adopted a position of critical realism or evolutionary naturalism. His doctrine of emergent evolution was a reaction against Platonism and Kantianism. In terms of epistemology, it advocated a position of critical and physical realism. In terms of metaphysics, it held to emergent materialism. Sellars took time, change, evolution, creative synthesis, novelty, and accumulative growth seriously. He asserted that there is a discoverable massiveness, orderliness, and immanent executiveness in nature. Nature itself represents a four-dimensional manifold of space-time and manifests four primary categories: space, time, thinghood, and causality. Evolution implies both novelty and genetic continuity; it has resulted in distinct levels of development but different levels of reality. Matter is active and capable of high levels of organization and accumulation resulting in the emergence of new properties by degree.

Sellars's major work is *Evolutionary Naturalism* (1922).[50] In it, he wrote that the logical structure of nature or emergent evolution reveals a hierarchy of a pyramid or tierlike construction representing steps or levels of complexity of organization and degrees of freedom (as well as new laws and categories). Historically, these general levels are represented by matter, life, mind, and society. However, Sellars does not hold to a rigid teleology, i.e., nature is neither evolving according to any predetermined design or order nor moving toward an end or goal. Yet, it is unfortunate that he never developed his principle of organization within a systematic explanation for emergent evolution.

As a process philosopher, Sellars held that new properties emerge as a result of the evolution of matter into new organizations; humanity has emerged out of organic evolution, and mind has emerged by degree as a new property or category of highly complex organization. As a consequence of his critical realism, he maintained a naturalist or psychobiological view of the mind-body problem. The whole of humankind must be included in nature, and nature itself is so conceived that this inclusion is possible; thought is retrospective and supervenes upon material reality. Sellars argued that consciousness is an activity totally dependent upon the proper functioning of a brain. Therefore, human immortality is clearly impossible since there cannot be a personal consciousness without a brain; all emergent properties are grounded in matter. Sellars's evolutionary cosmology and nonreductive, materialistic ontology are in line with those advancements of the special sciences, and his democratic socialism and scientific humanism are in the best interests of our human condition.

Roy Wood Sellars upheld a humanist philosophy advocating a materialist ontology and evolutionary cosmology. He grounded knowing in biology and saw our own species existing totally within the flux of emergent nature. His other books include *Critical Realism* (1916), *The Philosophy of Physical Realism* (1932), and *Philosophy for the Future: The Quest of Modern Materialism* (1949).

Alfred North Whitehead (1861–1947) was interested in the new facts and concepts within relativity physics, quantum mechanics, and early evolutionary biology. No longer were space, time, and motion considered independent absolutes, simple causality and determinism applicable to alleged material atoms, or biological species fixed entities within a static nature. Whitehead's intellectual life passed through three successive but distinct stages: (1) mathematics and logic, (2) history and philosophy of science, and (3) process metaphysics. Although he was both empiricist and rationalist in his philosophical quest, Whitehead emerged in this century as a major speculative and systematic thinker devoted to an aesthetic appreciation for the creative advance of this cosmos. He always held that philosophy is an attempt to express the infinity of this universe in terms of those limitations of language. Yet, for a major philosopher emphasizing experience and activity, Whitehead's personal life seems to have been remarkably uneventful. His own satisfaction came from that inner world of creative thought and rigorous reflection.

Whitehead's *Science and the Modern World* (1925) is a series of lectures devoted to the development of Western science and philosophy during the last three centuries (a consideration of epistemology is admittedly entirely excluded).[51] Extending from Copernicus and Giordano Bruno to modern ideas in physics and biology, his book emphasizes that the growth of these special sciences directly influences conceptual frameworks. He held that the creative development or evolution of this cosmos represents a fundamental unity, and that cosmology must do justice to both objective and subjective realms of inquiry.

Whitehead's major work is *Process and Reality: An Essay in Cosmology* (1929).[52] It is concerned with speculative philosophy as an endeavor to frame a comprehensive, coherent, logical, and applicable scheme of things. It relies upon knowledge gained through reason and experience, stressing the elucidation of immediate experience and those resultant feelings of satisfaction. His philosophy of organism gives a unique metaphysical description of process and reality. It treats the fluidity and plurality of this world in a most general way, as well as those eternal structures assumed necessary to give unity and rational order to our particular cosmic system.

Whitehead presented the fallacy of dogmatic finality: no permanently fixed, sufficient explanation for this entire universe is ever possible. Similarly, he presented the fallacy of misplaced concreteness: a crucial distinction is required between the ontological status of ideas and the objective reality that they purport to reflect. He also upheld the ontological principle: there is a rational explanation for the existence of all objects or events; something comes from something; all existing things emerge from preexisting things. The ontological principle can be summarized as such: all existence has an ontological source or if there is no actual entity, then there is no reason.

Whitehead presented a categorical scheme covering the ultimate, existence, explanation, and obligation. For him, this world is composed of both actual entities and actual occasions or events within the becoming of things. Each entity manifests a dipolarity, a physical pole as well as a mental pole. Likewise, there are eternal objects, i.e., potentialities or possibilities (forms, qualities, and values). God's primordial nature is the realm of eternal objects, while God's consequent nature is the realm of events. God experiences this world in a reciprocal relationship, deriving feelings and understanding as his own superjective nature (mental satisfaction).

For Whitehead, God is an actual entity necessary for a comprehensive explanation of nature. God relates to this world and thereby gives it existence, stability, order, creativity, novelty, purpose, and value. God is held to be the chief exemplification of all metaphysical principles: He is eternal, immanent, self-consistent, but transcends any finite cosmic epoch. As an actual entity, God is also dipolar. The primordial nature or conceptual experience of God is that "object of desire" of the eternal urge of this universe. This aspect of God is his unlimited potentiality, i.e., it is the realm of those external objects, while the consequent nature or physical experience of God is that finite, fluent multiplicity of the world. As such, God and the world are ontologically interrelated through an eternal reciprocal relationship of everlasting creativity seeking a perfect unity. Clearly, to a rigorous scientific naturalist, this is a form of panentheism that remains less than completely convincing despite its impressive intellectual loftiness and unquestionable personal integrity.

For Whitehead, this world is an extensive continuum of internal and external interconnectedness. To reject this position is to commit the fallacy of simple location. This world is a hierarchy of societies or enduring objects (nexus). The eternal objects are ingressed in or participate in these enduring objects, i.e., those eternal forms in the mind of God are mirrored by the objects of this world. Societies evolve into new societies, giving objective immortality to actual entities (however, this does not imply personal immortality). This preservation of new aggregations of events in novel societies throughout cosmic emergence is referred to as the theory of objectification.

The ultimate category of this world is creativity, that process by which the one and the many teleologically develop toward novelty. According to Whitehead's doctrine of universal relativity, no sharp distinction is possible in describing universals and particulars. This present cosmic epoch manifests its own peculiar laws and creative order. Feelings (conformative, conceptual, simple comparative, or complex comparative) or prehensions (physical or conceptual, positive or negative) are pervasive throughout this actual world. They are experienced by a human knower through the mixed mode of symbolic reference, i.e., through both causal efficacy and presentational immediacy.

Each entity is a subject-superject within the process of becoming. Likewise, each entity manifests an innate subjective aim or appetition (subjectivist principle). This world is a preestablished harmony of societies moving toward new creative unities or concrescences (principle of satisfaction). In short, each object or event of this world is at the same time an external subject and a subjective form for every other experiencing actual entity. Transmutation refers to the act of deriving macrocosmic perceptions from microcosmic prehensions.

In *Process and Reality*, Whitehead exhibited an analytic and synthetic mind. His original views called for new perspectives and a unique terminology. He gave priority to experience and feelings, resulting in an idealist cosmology. Nevertheless, he attempted to systematically reconcile the temporal realm of becoming with the eternal realm of being. He gave methodological preference to mathematics as well as ontological preference to experience and its resulting feelings. Whitehead desired to construct a modern process cosmology. His dynamic system has the advantage of being open-ended to infinite possibilities (potentialities). For him, this universe is an infinite series of unique cosmic epochs, each particular cosmic epoch having its own rational order; nature is a creative continuum. God and the world are never complete. There is no final achievement or goal or end; no final cosmic satisfaction resulting from the creative urge of the world toward a novel synthesis. Furthermore, each future cosmic epoch will have its own laws since they, too, evolve through time.

In general, Whitehead's process cosmology is remarkably open but idealistic as a result of its residually anthropocentric and anthropomorphic orientation. By going beyond the basic facts of empirical experience, his speculations are open to criticism. However, he will be remembered for his contributions to logic, mathematics, and the history and philosophy of science, as well as his emphasis on the creative development of nature, the principle of causal efficacy, and the method of extensive abstraction.

These emergent evolutionists rejected mechanistic, crude materialistic, and reductionist interpretations of the process of nature. Instead, they emphasized that dynamic creativity and unpredictable novelty are pervasive throughout cosmic and/or earth history. However, their views are aligned more closely with metaphysics than with the special sciences and reason (Sellars's naturalist interpretation of evolution is a welcome exception to this predominantly speculative orientation in recent process thought).

Of lasting value are those rewarding insights of recent evolutionists in sociology, anthropology, and natural philosophy.

The interdisciplinary sociologist Lester Frank Ward (1841–1913) viewed human civilization from a sweeping cosmic perspective. His two major works are *Dynamic Sociology* (1883, 1897), and *Glimpses of the Cosmos* (1913–1918).[53] Ward's ideas are strikingly modern: earth is merely a fragment of cosmic history, life itself originated from creative chemical syntheses, and our species is an emergent product of organic evolution.

Ward made a sharp distinction between the purposeless genetic evolution of earth with its plants and animals on the one hand, and the teleological evolution of human activities on the other. Although there is no cosmic purpose in this physical universe, there is at least human purpose in the social realm. Progress is achieved through the triumphs of science and reason over the errors of theology and superstition.

Furthermore, Ward advocated the prudent engineering of human society with its creative energies through the educated intellect. He foresaw the future formation of a "sociocracy" as the direct result of collective telesis (the purposeful democratic control of society by society as a whole).

Edward Burnett Tylor (1832–1917), the father of cultural anthropology, wrote *Researches into the Early History of Mankind and the Development of Civilization* (1865), *Primitive Culture* (1871), and *Anthropology* (1881).[54] He was primarily concerned with problems about the origin and history of religion and the fact that early humans had misinterpreted those cause-and-effect relationships of nature in their beliefs and rituals. For Tylor, religion is essentially a belief in spiritual beings. The central theme of all primitive religions is the concept of animism, that belief in the existence of spiritual beings (souls are also held to exist in plants and nonhuman animals). Mana is a universal impersonal force, while animatism is the attribution of life to those inanimate objects of nature.

Tylor taught that the origin of animism resided in a misunderstanding of life, death, hallucinations, and especially dreams (this misunderstanding resulted in a duality between matter and spirit that gave rise to the belief in an immortal human soul).

In reply to those rigid religionists who criticized the evolutionary framework of science, Tylor was quick to point out that, in fact, complex sophisticated religious beliefs and practices involving an assumed supernatural realm have themselves actually developed through the sociocultural stages of human history from their elementary or simplistic beginnings in animism and animatism. He advocated the following evolutionary sequence: from totemism, through polytheism, to finally monotheism. Furthermore, he saw that cultural evolution manifests a decreasing reliance on magic and religion as a direct result of the increasing success of science and reason (perhaps resulting in the eventual demise of both magic and religion).

Tylor held that the human mind differs in kind (not merely in degree) from that of all other animals; a mental gulf divides the lowest savage from the highest ape. He further claimed that our early ancestors thought and acted more or less like those members of hunting/gathering societies now existing in the remote areas of the world. In general, sociocultural evolution has passed through three major changes of development: from savagery through barbarism to civilization, and there have also been numerous minor levels of evolution within each of these three major stages.

Tylor also spoke of cultural "survivals" as those bygone but once useful traits of technologically primitive and nonliterate societies or "living fossils" that, analogous to vestigial organs in biology, remain in modern civilization although they no longer serve any necessary function or purpose, e.g., the making of pottery.

In summary, Tylor held to the following positions: (1) the unilinear, predetermined, and progressive evolution of society and culture; (2) the sociocultural evolution of explanations from magicoreligious beliefs to scientific and rational understanding; (3) the parallel invention of cultural traits resulting in similar specific evolutionary histories; (4) the simple diffusion of cultural traits through social contacts; (5) the interpretation of contemporary so-called primitive societies as "living fossils"; (6) the existence of "survivals" in present-day cultures; and (7) the psychic unity of humankind. His major contribution to the history of anthropology is a general evolutionary framework within which sociocultural phenomena may be understood and appreciated in terms of change and history.

Lewis Henry Morgan (1818–1889) also envisioned human history as consisting of three major predetermined evolutionary periods: first savagery, then barbarism, and finally civilization. This simplistic view is not supported by empirical evidence and it is therefore not considered a meaningful framework. Morgan's major work remains *Ancient Society* (1877).[55] Its view of social history greatly influenced both Marx and Engels. Yet, it must be stressed that the evolutionary theories of both Tylor and Morgan are unable to explain satisfactorily those wide variations or conflicting details throughout cultural evolution that are now known to modern anthropologists.

James George Frazer (1854–1941) saw both magic and religion as a means by which irrational primitive humans might control natural forces. His evolutionary theory emphasized psychology and was presented in *The Golden Bough: A Study of Magic and Religion* (1890–1915).[56] Concerning magic, Frazer taught the law of sympathy; it distinguished between demopathic magic (law of similarity) and contagious magic (law of contact). He claimed that magic is a pseudoart and a pseudoscience. Magic is grounded in nature itself, but religion requires supernatural beliefs and subordinate behavior. Whereas magic and science take an impersonal view of this universe, religion takes a personal view of the cosmos. Frazer clearly distinguished among magical, religious, and scientific behavior. He presented the following generalization: human views of nature pass from magic through religion to science.

Leslie A. White is the most important exponent of cultural evolution in this century. His major work is *The Evolution of Culture: The Development of Civilization to the Fall of Rome* (1959).[57] This book is very significant because it reinstates the evolutionary perspective into modern ethnology. Appropriately, it appeared in the year that celebrated the centennial of the publication of Darwin's *On the Origin of Species* (1859).

For White, the theory of evolution is as relevant to culturology as it is to biology. The human animal is unique because it symbolizes and has a culture. Culture may be divided into four categories: its philosophical or ideological, sociological, sentimental or attitudinal, and technological aspects. White holds that culture is extrasomatic or suprabiological, i.e., the superorganic is a single system nonbiological in character. He also distinguishes between human culture in general and specific cultures. Technology is held to be the basis and determinate of cultural systems; as technological control increases, the belief in supernaturalism decreases.

White's view represents an organismic theory of culture. As such, cultural evolution is held to be nonrepetitive and irreversible. Like the evolution of life, the development of human culture is presented as an ongoing energy-harnessing process, antithetical to the second law of thermodynamics (entropy).

Each culture is a functioning whole of interrelated elements in a moving equilibrium of sequential forms. Originally, early humans were the principal source of energy. After millions of years of Stone Age technology, the cultivation of plants and domestication of animals had dramatically increased the energy potential for human use. This origin of agriculture was followed by a rapid advancement in metal technology. Clearly, White gives preference to technological development in his depiction of cultural evolution within a materialist framework.

Concentrating on human social systems, White distinguishes among the following three major stages of cultural development: from hunting/gathering/fishing societies, through pastoral/horticultural societies, to agricultural societies. Each stage represents an increase in scope, complexity, solidarity, and technological dependence. Human social organizations and relationships are determined by nutrition, various forms of protection, and reproduction.

According to White, prehistoric humans were primarily weak and defenseless against their harsh environment. One may safely assume that they lived in a few nomadic, male-dominated, polygynous groups. Four cultural factors had elevated these early hominids above their hominoid contemporaries: the use of symbols, toolmaking, social cooperation, and language as articulate speech. These four factors allowed for adaptation and survival against incredible obstacles. For White, a human being is essentially the symboling animal. It is this unique ability that has allowed our own species to develop culture and subsequently to dominate this planet (at least for the present time).

Although wild chimpanzees do make simple tools, only the symboling human animal is able to use simple tools to make far more complex implements. This mental gulf between ape and human, whether considered to be a difference in kind or merely in degree, is due to the fact that the latter has a superior central nervous system and splendid brain.

Continued social evolution resulted in the formation of mutual-aid groups of monogamous family units with incest, endogamous (inbreeding), exogamous (breeding across clans), and residence rules for securing social solidarity and effectiveness. A kinship system is based upon biological relationships as well as social relationships. It is a means of organizing individuals upon ties of consanguinity and affinity. Kinship groups may be divided into clans or moieties. Primitive cultures may even extend kinship relationships beyond the boundaries of human society by embracing plants, animals, and even inanimate objects into their lineages. Such totemic systems relate our own species directly to nature and have a magicoreligious significance (this phenomenon is both pervasive and precious in primitive societies).

White points out that the evolution of culture results in the progressive differentiation of integrated structures and functions. Since the agricultural revolution, social complexification and technological specialization have increased at an accelerated rate. Where subhuman societies are primarily determined by biology, human societies are now essentially determined by culture, with technology as the major determinant.

There has also been an evolution of concepts. Primitive human philosophy was primarily supernaturalistic, while our modern worldview is becoming ever more naturalistic. The previous reliance on myth and ritual has been replaced by the present reliance upon science and technology. This crucial shift in outlook is grounded in the human ability to conceptualize and experiment; the symbol is the universe of humankind.

In the evolution of culture, the revolutionary change from primitive society to modern society has been clearly grounded in scientific and technological advances. Following the Stone Age, our Space Age represents a striking if not alarming continuation of tool use, toolmaking, and tool ownership. However, it must be remembered that a primitive technology does not necessarily imply that all of the other elements of such a society are also simplistic. In fact, an alleged primitive society may exhibit a complex language, kinship system, and magicoreligious worldview. Likewise, modern science and technology do not guarantee that a society is ethical or moral; the distinction among knowledge, intelligence, and wisdom is crucial.

Leslie A. White represents an interpretation of human sociocultural development that is diachronic and synchronic, structural and functional, analytic and synthetic. He upholds an organismic view of society within an evolutionary framework. Culture is held to be an extrasomatic and suprabiological process of events superimposed upon the physical and biological levels of phenomena. For White, the transition from early hominoids to true hominids was basically due to the emergence of symboling, not bipedality. Although he overemphasizes the technological element of culture, his own evolutionary orientation and anthropological synthesis do assure him a permanent place in the history of ethnology.

According to anthropologists Marshal Sahlins and Elman Service, one may distinguish between universal or general evolution (Leslie A. White) and multilinear or specific evolution (Julian H. Steward).[58] At present, an evolutionary view of cultural materialism is expounded in the writings of Marvin Harris.[59]

As an anthropologist, Raoul Naroll (1921–1985) focused on several social problems plaguing the entire human situation and contributed to the perspective of sociocultural evolution. He emphasized the enormous value of both the nuclear family and strong relationships; they provide a moral network so that people are less likely to stray from accepted social norms. His major work is *The Moral Order: An Introduction to the Human Situation* (1983).[60]

However, Naroll's grand vision goes far beyond the microcosm to formulate a general-systems-theory model of cultural evolution grounded in the natural selection of behavioral patterns; there are significant analogies with neo-Darwinism in modern biology. Although it does not explain everything, he stresses that cultural selection is nevertheless a powerful process that should not be ignored by social scientists. His dynamic model of cultural evolution is based upon three factors: blind variation (servos), selective retention (darwins), and self-reinforcement (snowballs). This model embraces naturalism, humanism, and extends the modern punctuated equilibria hypothesis from organic evolution to sociocultural development. Raoul Naroll even speculated on the cosmic future of our own species as it ventures throughout deep space long before the sun burns out.

Marvin Farber (1901–1980) never underestimated the predictive, explanatory, and exploratory powers of the theory of evolution. As a philosophical materialist, he grounded his own naturalist viewpoint in the advances of the special sciences within a cosmic framework that also recognized the need for rigorous reflection. Consequently, he saw all religions as being essentially grounded in the psychosocial wants, needs, and desires of our very vulnerable species that totally exists within this evolving material universe.

Farber clearly presented a scientific, naturalistic, and rational humanist viewpoint that is ontologically grounded in an uncompromising and unapologetic materialism. To a significant degree, this pervasive materialism was the direct result of his steadfast commitment to the far-reaching theoretical implications and, at times, sobering if not devastating physical consequences of both the theory of evolution and sociocultural development (as well as his serious adoption of a truly cosmic perspective). There can be no doubt that he championed the process philosophers and scientific evolutionists, unequivocally maintaining a materialist explanation of this dynamic universe. His cosmic materialism is in sharp contrast to those various idealist interpertations of reality itself also present in the serious philosophical literature.

Farber spoke of the philosophical quest. He especially respected anthropology as an academic discipline and thought it to be a highly valuable preparation for philosophy. Likewise, his writings stressed those perennial "themes of inquiry" in the history of philosophy. Briefly, he maintained that the fundamental questions in the great tradition of reflective thinking are always open to critical analysis and rigorous reevaluation in light of the advancing special sciences and a plurality of cooperative, logical, and methodological procedures.

Farber's *Naturalism and Subjectivism* (1959) is a major statement clarifying the crucial distinction between ontology and epistemology.[61] He wrote that the process of human experience occupies only an infinitesimal part of this material universe and it occurs only when there are sentient beings in action. Furthermore, his pervasive materialism taught the all-important "principle of independence": this natural world is prior to and independent of the human knower and its sociocultural environs. As such, the whole range of human experience is seen to be a relatively recent emergence in the vast organic history of this planet (not to mention sidereal reality). He also wisely and astutely appealed to the awesome cosmic perspective for both a sound interpretation and proper evaluation of humankind's fleeting place within the flux of this physical universe.

For Farber, the existence of the material world is not contingent upon human or divine experience. He rightly pointed out that philosophy is a human activity that requires presuppositions, and he drew attention to the basic fact that the human animal as knower and doer can never really get outside of its natural, sociocultural, and mental environs. Those who do not suffer from what he refers to as the error of "illicit ignorance" realize that methodological and logical pluralism are always required for adequately handling that spectrum and diversity of human problems.

Surely, there is a need to clarify the true place of our own species within this material universe and to assess that proper status of mental activity in the context of biological evolution. As a pervasive materialist, Farber would sometimes zestfully refer to the human animal as a "bag of bones" to draw attention to the relatively insignificant place that humankind occupies within the sidereal depths of the physical cosmos, which for all practical purposes is eternal in time and infinite in space as well as endlessly changing.

Farber warned that the emergence of the theory of evolution in the special sciences and natural philosophy in the last century as well as process theology in this century did not bring to an end the old type of idealist and fideist worldviews (but such residually subjectivist and theistic perspectives did cease to be as important as they once were). In his wise and learned opinion, materialism, supported by an impressive array of empirical evidence from evolutionary biology to scientific cosmology, is a major blow to all forms of metaphysical idealism and lingering supernaturalism.

Farber stressed that no closure should ever be placed on either the reach or direction of human inquiry as long as it is ethically defensible. According to his viewpoint, philosophical activity has four major functions: a clarification of perennial problems (questions) and basic ideas concerning material reality, a recurring attempt at synthesizing the findings of the special sciences, a rigorous analysis of all methods and human experiences, and the ongoing critical examination of all values within the context of sociocultural development as well as natural history (evolution) and the cosmic perspective.

Finally, Marvin Farber never condoned nonsense, cruelty, fanaticism, or totalitarianism of any sort. He held human concepts and ideas to be symbolic instruments of adaptive value (in a moment of wit, Farber called a belief the principle of sufficient wishing). Yet, in establishing values within a cosmic evolution for which time is brutally real, he stressed that one must never underestimate human fallibility and finitude.

The year 1959 not only marked the centennial celebration of the first appearance of Charles Darwin's On the Origin of Species, but it also represented a crucial turning point in paleoanthropology. In July of that year, Dr. Mary D. Leakey had discovered a hominid skull and teeth in Bed I (the lowest layers of Olduvai Gorge), thus ending a thirty-year search for early humans in central East Africa.[62] This find—at that time the oldest known fossil hominid remains—was analyzed, described, and interpreted by Dr. Louis S. B. Leakey. He classified this eighteen-year-old skull as Zinjanthropus boisei but then mistakenly held it to be the maker of those crude Oldowan pebble tools/weapons found nearby in the same rock strata. The use of the potassium/argon radiometric dating technique determined this remarkable fossil hominid evidence to be about 1.75 million years old! At last, the Leakeys had made an enormously significant breakthrough in comprehending the emergence of humankind in this area of the world.

Indeed, the worldwide publicity and scientific attention given to this robust australopithecine specimen resulted in an increase in funds now made available for the continued research by these and other professional anthropologists in quest of more empirical evidence needed factually to document the beginning and early evolution of these very early ancestors of our own species. Since this first find, both earlier and later fossil hominid material has been discovered within the Gregory Rift Valley areas. For geological perspective, paleontological and archeological sequence, and scientific significance in terms of human evolution, there are very few fossil hominid sites that can match Olduvai Gorge in Tanzania. However, interpretations of that wealth of hominid paleontological evidence from these sites in this part of the world vary greatly. At the present, this growing fossil record clearly points to central East Africa as the birthplace of our hominid family.

Only two short years later, in 1961, Louis Leakey discovered an even more hominid fossil skull from Olduvai Gorge that he classified as Homo habilis;[63] it had a larger brain and smaller dentition than the "Zinj" skull. In fact, this specimen is remarkably hominid (far more humanlike than all those previously discovered australopithecine specimens). Homo habilis is nearly two million years old, exhibits a cranial capacity of about 750 cc, was the actual maker of those Paleolithic artifacts associated with the Oldowan culture, and has both hand and foot bones confirming bipedality. The significance of the Zinjanthropus skull, now adjudged to be only an example of the Paranthropus (Australopithecus) robustus form, has faded into the background; this find remains a reminder to prehistorians not to make hasty and premature scientific decisions about the taxonomic position, biosocial importance, and ultimate evolutionary implications of a particular fossil hominid discovery before sufficient evidence is available and critical reflection has been given.

Following in his parents' footsteps and rewarded with early success, Richard E. F. Leakey took that ongoing quest for the origin of humankind in central East Africa to the Koobi Fora area on the eastern shore of Lake Turkana in northern Kenya. There, in 1972, he, too, discovered fossil evidence of Homo habilis (including the now famous Skull No. KNM-ER 1470 with a brain size of 775 cc and recently dated to be about 1.9 million years old, although its age was first thought by Leakey to be considerably earlier).[64] Despite some honest mistakes in specific interpretations of this find that have been capitalized upon and reputably exploited by certain academic competitors in a rather puerile manner, Richard Leakey's own overall contributions to human paleontology as well as wildlife ecology remain very impressive and highly respectable.

Also, a diseased fossil hominid was found in Area 104 at Koobi Fora; this unique example of paleopathology was caused by an excess of vitamin K in the individual due to his eating those poisonous livers of older carnivores (which had fed on herbivores).

In light of all this cumulative fossil evidence, those early Pleistocene hominids of the Villafranchian time represent at least three different genera: the robust *Panathropus* form that was large and rugged in appearance, the gracile *Australopithecus* type that was small and delicate in appearance, and *Homo habilis* (the most hominid of this group). Of course, there may be other hominid genera of this same time span yet to be discovered, studied, and evaluated. Nevertheless, there still remained a need to unearth the skeletal fragments and concomitant traces of the ancestors of even these hominids, the earliest known to science at that time.

In 1974, paleoanthropologist Dr. Donald C. Johanson discovered "Lucy" (A.L. 288–1) or *Australopithecus afarensis* at Hadar site 288 in the Afar Triangle of Ethiopia.[65] This 40 percent of a fossil skeleton belonged to a short (about 3½ feet tall) and slight (55 pounds) adult female from the Plio-Pleistocene junction. She is about 3.5 million years old, and both her jaws and teeth suggest an omnivorous diet.

According to its discoverer, Lucy is representative of the small-brained but erect-walking ancestors of humankind. This specimen is hominid in every general anatomical feature of its postcranial skeleton. However, her primitive skull and small cranial cavity as well as dental characteristics are more hominoidlike than hominidlike.

Johanson's multidisciplinary and systematic team approach has also found a "First Family" as old as Lucy at Hadar site 333; this Plio-Pleistocene hominid evidence (including jaws, teeth, and a knee joint) represents at least thirteen individuals (adults and juveniles of both sexes) close to four million years old. All this fossil material probably represents one single species at Hadar. Be that as it may, this hominid evidence from Ethiopia is helping to shed more light on the provenance of our whole human zoological group.

In December 1974, Dr. Mary D. Leakey discovered fossil hominid jaws and teeth in the volcanic ash bed at Laetoli; this remote site is twenty-five miles south of Olduvai Gorge, in Tanzania. Since then, similar fossil hominid material has been found not only at sites near Lake Turkana in Kenya, but also from areas in the Awash River Valley of Ethiopia. Isolated fossil hominid evidence, including robust jaws and large teeth, from the Omo Valley in southwest Ethiopia may date back over three million years. All these new findings add empirical support to the Darwinian hypothesis that our human species originated in Africa (not Asia). In fact, some fossils suggest that early hominids stalked the African woodlands and savannahs over four million years ago: even these individuals may have hunted, used stone implements, and controlled fire!

As if Johanson's early and awesome luck were in itself not enough to increase excitement in the science of physical anthropology, in 1976 Mary Leakey had the outstanding good fortune to unearth three tracks of hominid footprints about 3.75 million years old from Site G at Laetoli in Tanzania, near Lake Eyasi, south of the Olduvai Gorge.[66] These late Pliocene Laetoli Beds have also yielded numerous hominid fossils, chiefly lower jaws and teeth, as well as a wide variety of nonhuman animal remains. As yet, no stone tools/weapons have been found at Laetoli. Johanson suggests that both these footprints and his own late Pliocene findings from those Hadar sites in Ethiopia belong to the same distinctive hominid form *Australopithecus afarensis* (although this model and interpretation of early hominid phylogeny in Africa is not shared by the Leakey family). Nevertheless, one significant generalization can safely be made: this early australopithecine form stood erect and walked fully upright with a free-striding bipedal gait long before its descendants used and later made Paleolithic tools/weapons. Only with the recent emergence of *Homo sapiens* did the human animal acquire a very large brain concurrent with articulate speech and complex culture.

In 1984, one of the earliest *Homo erectus* specimens was discovered by Kamoya Kimeu at site Nariokotome III on the western shore of Lake Turkana in Kenya.[67] During the subsequent years, ongoing work at this fossil site by Richard E. F. Leakey and his team has yielded a nearly complete hominid skeleton (KNM-WT 15000) of this individual, now dated at about 1.6 million years. In fact, this male specimen is far more complete than the Lucy skeleton from Ethiopia. This male specimen is essentially human, except for the chipanzeelike vertebrae, and has a cranial capacity of about 900 cc. Not only is the discovery of this *Homo erectus* specimen remarkable in its own right but also, had this twelve-year-old boy lived to adulthood, his anatomy suggests that he would have been about six feet tall! This important fossil skeleton shatters the long-held opinion that all early hominids were short in stature.

In summary, all this evidence clearly substantiates the claim that our oldest ancestors were walking erect on two legs with a small cranial capacity of about 400–750 cc for several million years; the modern cranial capacity of nearly 1500 cc was first obtained by *Homo sapiens neanderthalensis* about 150,000 years ago. Of course, more fossil evidence will have to be found before those remaining questions concerning the origin and emergence of humankind are finally answered, and conflicting scientific interpretations are settled once and for all.

One may safely assume that the paleoanthropologists of the future will discover additional fossil hominid evidence that could very likely clarify those basic issues still surrounding the origin and early history of our own species. Such finds would greatly help to fill in those admitted gaps now inherent in the current scientific overview of human development within a naturalist framework of organic evolution. Likewise, it is absurd to maintain that the present (or even lasting) incompleteness of the geopaleontological record in regard to any empirical knowledge of man's place within nature warrants rejecting the fact of human evolution. The symboling animal as tool user, toolmaker, and tool owner is merely one of the millions of species that have inhabited this dynamic planet, whose biological history is one of speciation and extinction. What is clearly needed is more scientific evidence and rational interpretation. Perhaps in the future, scientists will find an early hominid preserved in ice or discover a cave mural that does depict in detail our prehistoric ancestors.

In light of the growing fossil evidence, Darwin was prophetically right to select Africa as the cradle of humankind; Haeckel had thought Asia to be the birthplace of our own species.

Comparative studies of the Miocene hominoid crania and teeth from the Dryopithecinae complex in Africa point to a major separation that occurred more than twelve million years ago within this large and diversified group of higher primates: one line represents those pongidlike dryopithecines that are ancestral to the present living great apes, while another line consists of those hominidlike kenyapithecines that are directly antecedent to the later true hominids and eventually conducive to *Homo sapiens sapiens* of today.

Some physical anthropologists now claim a much later date for this pongid-hominid divergence based on a comparative study of both the biochemistry and postcranial anatomy of the human animal with the two African great apes (especially the pygmy chimpanzee). Of course, in this complex history of the hominoids, there were probably numerous unsuccessful lines of evolutionary descent, as represented by the extinction of the pongids *Oreopithecus* and *Gigantopithecus* as well as the disappearance of at least two early hominid forms in Africa, *Paranthropus* and *Australopithecus*.

Homo erectus appeared about 1.7 million years ago and may have been responsible for the extinction of certain ground-dwelling pongids and possibly even other hominid forms.

Some molecular biologists now claim that all living human beings are descended from a single African woman dubbed "Eve" who lived about 142,000 years ago (those lineages from all the other women became extinct). This controversial hypothesis is based upon only recent research in molecular biology. Evidence is from 37 genes that mutate at a constant rate and they are found only in female mitochondria. It suggests that our own species migrated north from southern Africa about 75,000 years ago, finally disposing the assumed genetically incompatible Neanderthal people of Europe about 35,000 years ago. Surely, more research and evidence are needed. Even so, this scientific viewpoint does not support those religious beliefs of biblical fundamentalist creationism concerning the origin of the human animal.

As in the case of other great thinkers, Darwin's speculations on human evolution were not always either theoretically or factually correct in every detail. In light of subsequent discoveries, he turned out to be in error for holding that the modern cranial capacity of about 1500 cc was reached in our prehistoric ancestors before they acquired an erect posture and subsequently made stone implements for defense. As one can see today, the fossil evidence clearly indicates that just the opposite was the case; the human cranial capacity did not start to expand greatly until about 150,000 years ago. Although the neurophysiological complexity of the early hominid brain must be taken into consideration, as well as its sheer size, a sufficient explanation for the relatively rapid expansion of the human cranium roughly between 150,000 and 45,000 years ago is still forthcoming. Obviously, a considerably smaller hominid cranial capacity had been sufficient for merely adapting to and surviving on those open habitats of our Villafranchian ancestors in central East Africa.

Despite the fact that there is still no consensus among physical anthropologists as to the specific classification of all this fossil hominid material (not to mention that lingering so-called Neanderthal problem), a general outline of human forms is possible. It is neither *Australopithecus africanus* nor *Paranthropus robustus* (there may even have been other genera), but the more hominid form *Homo habilis* that is ancestral to the later *Homo erectus* stage (e.g., Java Man and Peking Man). The following Neanderthal phase or *Homo sapiens neanderthalensis* gave rise to the Cro-Magnon man who, in turn, preceded the appearance of modern man (both later stages are referred to as *Homo sapiens sapiens*). With modern man, a hunting/gathering lifestyle was soon replaced by the coming of agriculture (the cultivation of plants and the domestication of animals), then metallurgy, and soon afterward civilization.

Today, however, the anthropologist may safely make this generalization about the emergence of our own species: our remote hominid ancestors acquired an erect position and walked upright with a bipedal gait long before making stone implements, while the modern cranial capacity was reached only recently. However, science may never know when the use of fire, articulate speech, and monogamy with an incest taboo first appeared in human evolution.

During the past thirty years, primatologists have made a concerted effort to study the wild apes in their natural habitats. The pongids are those animals closest to the human being from an evolutionary viewpoint but, unfortunately, each of the three great apes is now an endangered species (a vanishing animal threatened with extinction) because of the encroachment of civilization.

Dr. Birute Galdikas is studying wild orangutans in Borneo, where the "red ape" of Asia lives a solitary and peaceful life in a tropical rain-forest environment on this island in Indonesia; this pongid is also found on the island of Sumatra.[68]

In Africa, Dr. Jane Goodall has been living among the wild chimpanzees of Tanzania. Her long-range and intensive observations have shed enormous light on the remarkable similarities between this great ape and a human being (especially in terms of toolmaking, intelligence, and group behavior).[69]

In the footsteps of zoologist Dr. George B. Schaller,[70] the late Dr. Dian Fossey had studied the mountain gorillas on the slopes of the Virunga Volcanoes in central East Africa. Her dedicated work and sympathetic approach brought the plight of this largest of the three pongids to the attention of the entire world.[71]

Such primate research studies clearly demonstrate that our own species is closer to these three great apes than Huxley, Haeckel, or even Darwin himself thought in the last century (particularly in terms of biochemistry and genetics).

However, a discrepancy does exist concerning a scientific interpretation of the empirical evidence. The hominid fossil record suggests to some paleontologists that the crucial split between man and the apes occurred about fourteen million years ago, whereas the results of comparative studies in molecular biology indicate that this separation happened less than seven million years ago (the facts show that the human animal is closest to the pygmy chimpanzee of central Africa). No doubt, further research will answer the questions now surrounding the evolutionary relationships among the pongids and our own species.

Since the middle of this century, neo-Darwinism or the synthetic theory of organic evolution has been built upon genetic variation and natural selection (including the discovery of both the DNA and RNA molecules as well as those principles of population studies). Even so, perhaps neo-Darwinism is necessary but not sufficient to account for organic evolution; there may be other natural mechanisms that influence the history of life on this planet that are still unknown to modern science.

Of invaluable help in explaining and defending this modern synthesis in organic evolution are the popular writings of George Gaylord Simpson in paleontology,[72] Julian Huxley in biology,[73] Ashley Montagu in anthropology,[74] and Theodosius Dobzhansky in genetics (among many others).[75] Of course, all future interpretations of evolution must always remain open to new facts, concepts, hypotheses, and perspectives. Ongoing areas of research include systematics, coevolution, exobiology, paleoecology, sociobiology, genetic engineering, and the punctuated equilibria hypothesis as well as speculations on the celestial and terrestrial influences on both organic evolution and mass extinction.

The geopaleontologist and religious mystic Pierre Teilhard de Chardin

(1881–1955) presented a tentative and personal interpretation of cosmic evolution.* He boldly sought to synthesize the special sciences, dynamic philosophy, and process theology. Although acknowledging this universe as a cosmogenesis, his worldview actually focuses on earth history and the human zoological group. Teilhard attempted to both understand and appreciate the essential meaning and ultimate purpose of evolution. Specifically, he dealt with the direction, survival, and fulfillment of our own species on this finite planet. His holistic vision with its pervasive sympathies is impressively reflected in numerous letters, essays, articles, and three books published posthumously: *The Divine Milieu* (1927), *The Phenomenon of Man* (1940), and *Man's Place in Nature* (1949).

Like Darwin's, Teilhard's scientific interests encompassed geology, paleontology, biology, and anthropology. He researched the Cenozoic prehistory of China and is especially associated with the discovery of Peking Man (*Sinanthropus pekinensis*) in the Western Hills at Zhoukoudian, a most fortunate circumstance that established his reputation worldwide. In 1947, Teilhard was elected to the French Academy of Sciences for his outstanding work in geopaleontology throughout his long, distinguished career as a natural scientist; he held a position as corresponding member of mineralogy in this prestigious society.

Teilhard's comprehensive worldview rests upon four basic assumptions: a pervasive spiritual monism, the accelerating and involuting law of complexity-consciousness, a progressive series of critical thresholds, and the forthcoming omega point. Matter, life, and thought are the three major stages of ongoing planetary evolution (representing geogenesis, biogenesis, and noogenesis respectively). Because of the spherical geometry of this finite earth, converging evolution has formed three distinct layers around the planet: the geosphere, the biosphere, and the noosphere. The geosphere includes those prebiotic layers of earth; the biosphere is that envelope of plants and animals covering the globe; and the emerging noosphere is that human realm of mind, society, culture, science, technology, and especially spirit now encircling the planet.

Teilhard argued that the phenomenon of humanity is due to two facts: the human animal has stayed a single species while surrounding this planet (physical reflexion) and at the same time is the only self-conscious being to emerge in nature (psychic reflection). As a futurist, this Jesuit priest envisioned ongoing mental evolution resulting in a collective spiritual unity of our own species at the end of this world. He believed the future terrestrial goal of human evolution to be the formation of an ultimate creative synthesis at the omega point. The final outcome is a mystical union of

---

*For a more detailed treatment of Teilhard's life and thought, refer to chapter 5.

this point omega with a personal God beyond space and time, forming the theosphere.

Teilhard has presented a provocative and profound synthesis of facts, concepts, and beliefs. The lasting contribution of his exquisite but controversial vision is its bringing a serious treatment of the scientific truth of organic evolution into modern religious thought as process theology.

In essence, science and theology are incompatible (unless one holds to a pantheistic view of the cosmos). There is only one objective reality, and our species is totally immersed in it. Materialists as naturalists, evolutionists, and humanists are aware of this brute fact and its sobering implications. Concerning discovery and understanding, even a cursory overview of the history of human inquiry clearly shows that science and reason are always victorious over religion and metaphysics. The successes of empirical and formal inquiry are both pervasive and profound, giving humankind the best reliable knowledge about the natural world that is now available to it. No doubt, our exploration of outer space will continue to accelerate the growth of the special sciences.

In principle, the naturalistic attitude or scientific viewpoint rejects all supernatural assumptions and idealistic claims about the ontological status of this universe. Instead, it stresses empirical facts and the use of reason while religious beliefs and preternatural causes are avoided. Of course, all scientific information is open to revision if new evidence about objective reality warrants changes. However, perhaps some questions concerning the origin and end of this universe will forever remain unanswered (although what is held to be unknowable at this time may actually be merely unknown).

The difficulty of clinging to biblical fundamentalism in our present age is being greatly exacerbated by ongoing discoveries in the special sciences. No modern scientist or philosopher or enlightened theologian takes as objective facts the religious myths of Creation, a Garden of Eden, the Noachian Deluge, and those miracles and mysteries in the Old Testament (among many other such theological beliefs, particularly those espoused by the anachronistic biblical fundamentalist movement of today). For an evolutionist, the spurious stories in Genesis are simply absurd. Yet, they do represent a conceptual framework from the undisciplined imagination of a prescientific age.

An anthropologist places myths, legends, opinions, and beliefs in their historical and sociocultural contexts. It is the modern scientific standpoint that provides, more and more, an adequate explanation for the true nature of this material cosmos and the recent emergence of our species. Overwhelming evidence and rational thought places humankind completely within this dynamic universe.

As a philosopher/scientist, I am optimistic that our species will eventually embrace both naturalism and humanism as it evolves throughout this expanding cosmos.

In *Wonderful Life* (1989), Stephen Jay Gould offers a fascinating story about science and evolution.[76] As a professor of paleontology and historian of science at Harvard University, he is richly equipped to represent the scientific community as well as tackle head-on the critics who spend much time and money in their fruitless attempts to discredit the theory of biological evolution and its far-reaching consequences.

Evolution is a fact. Yet, interpretations of dynamic reality do vary greatly among scientists as well as philosophers and theologians. Explanations to account for the origin and history of life on earth range from mechanistic materialism through creative vitalism and finalistic spiritualism to cosmic mysticism.

The specific topic of Gould's present volume is the Burgess Shale of British Columbia, certainly one of the most significant areas for paleontological study on the earth. What makes this particular locality so unique is not only those strange fossils found there but also, and more importantly, the implications all this detailed evidence holds for comprehending organic evolution in general. The persons and events surrounding this shale's exciting drama shed more light on the complexities of scientific research and conceptual interpretation; these two human activities are, of course, influenced by vested interests and entrenched values.

Now 8,000 feet above sea level, this small limestone quarry in Yoho National Park of the Canadian Rockies was formed 530 million years ago. Its bed of fossilized mudslides holds an enormous variety of marine life that appeared after the so-called Cambrian explosion of multicellular species.

In 1909, Dr. Charles Doolittle Walcott (1850–1927), the world-famous but ideologically conservative geopaleontologist and expert on fossil trilobites, discovered the Burgess Shale with its exquisitely preserved soft-bodied invertebrate animals. Unfortunately, his own investigation of these early life forms was inadequate, resulting in false conclusions. A victim of strong preconceptions and administrative burdens, Walcott classified these peculiar ancient creatures by forcing them into those standard taxonomic categories of extant arthropods and worms. Consequently, he did not recognize the unconventional model of life history suggested by these incredible fossil forms.

Forty years later, however, three devoted paleontologists (among others) began to painstakingly reexamine this Burgess Shale evidence. The team effort of Harry B. Whittington, Derel Briggs, and Simon Conway Morris especially contributed to a reinterpretation of these fossils and their message. A conceptual revolution resulted from the free inquiry of these dedicated scientists who used only a camera lucida, binocular microscope, and a very fine vibro-drill. Surely a creative imagination also played a crucial role, particularly concerning anatomical description and taxonomic placement.

This spectacular fauna of the Burgess Shale contains an abundance of both odd and ordinary anatomies (about twenty-five basic morphological plans). The strange soft-bodied organisms of this site include such enigmatic coelomates as the fearsome two-feet-long predator *Anomalocaris* with its circular jaw; the bizarre two-inch *Opabinia* with five eyes, a frontal nozzle, and gills above lateral flaps; and the weird inch-long *Hallucigenia* with a bulbous head, seven tentacles, and long cylindrical trunk supported on the floor of an ancient sea by its seven pairs of struts. Perhaps even more surprises will be discovered at this same site in the future.

Besides addressing himself to the strictly empirical evidence from the Burgess Shale rocks, Gould does not hesitate to explore the deeper conceptual aspects that this fossil site implies for understanding and appreciating organic evolution. He focuses on these major issues: progress, contingency, and predictability. His own examination of earth history in terms of life argues for a new model of organic evolution. In fact, this unique revision of evolutionary thought presents a remarkable worldview radically different from the traditional interpretation of earth history in paleontology and biology.

Early naturalists pictured organic evolution as a ladder, branching tree, growing coral, cone, pyramid, or even a series of circles (each with the same or different content). These views were usually earth-bound and human-centered, giving a privileged place to man in nature, i.e., the Mosaic cosmogony.

Gould claims that all these unusual fossil forms of the Burgess Shale disprove the conventional paradigm of organic evolution as a cone of increasing diversity representing anatomical types. In sharp contrast, he argues that biohistory should be represented by a revised picture: an inverted cone that stresses the early decimation of prolific designs of life followed by the later diversification of fewer and fewer different groups.

Overcoming the "tyranny of the cone" as he puts it, Gould offers a new pattern of evolution that stresses the global decimation by lottery of species throughout biohistory: after the early Cambrian explosion of life on earth involving experimental diversification in anatomical designs, most of these forms became extinct. As a direct result, the continuation of evolution is based on merely a few surviving lineages. Briefly, as represented by the Burgess Shale fossil evidence, great anatomical disparity was followed by an organic diversity of only a small number of distinct body plans: from rich potentiality through unpredictability to limited actuality.

Of course, this pattern of biohistory is based primarily on the Burgess Shale material. The discovery of other such sites would put this reinterpretation of organic evolution to the scientific test.

Nevertheless, conspicuously absent in this story of life is any application of the Gould-Eldredge hypothesis of punctuated equilibria to shed more light on this new model of organic evolution (although I for one cannot see how the Cambrian explosion of life, with its awesome number of biological types, could be accounted for only in terms of population isolation and rapid speciation).

Evolution and extinction go hand in hand. The ongoing elimination of specific species leaves their now-vacant habitats available for future occupancy by those emerging forms of new life. Actually, extinction is the rule rather than the exception througout organic evolution; most species have vanished like the trilobites. In fact, there have been several mass extinctions on a global scale throughout the sweep of geological time. For example, the demise of all dinosaurs paved the way for an adaptive radiation of the mammals. No doubt, both celestial influences and continental drift have altered earth history, contributing to evolution and to extinction on this planet.

Some evolutionists have been quick to see life on earth as a progressive march forward and upward, from simplicity to ever-increasing complexity. They argue that there is a design in nature, and therefore the final end to this assumed directive movement of organic history is held to be consciousness or personality. Such interpretations of evolution center on the human animal, and are grounded inevitably in a religious belief that upholds the cosmic uniqueness of our own species. In short, alleged teleology in nature as well as orthogenesis in paleontology and embryology are used to support these dogmatic tenets of theistic evolution.

However, Gould makes it very clear that one of his major objectives in science is to debunk such a progressive picture of organic evolution. For him, evolution is both unpredictable and irreversible.

Charles Darwin himself knew very well that the primary objection to his own scientific theory of organic evolution would be its devastating implications for our own species. For this reason and others, he deliberately left out any serious discussion of the human animal in his major work, On the Origin of Species (1859). Darwin had merely written: "Light will be thrown on the origin of man and his history." This is certainly the scientific understatement of the last century. Predictably, a bitter controversy over Darwinism did ensue because of its obvious threat to all human-centered worldviews in general and to Christianity in particular.

In his Newsweek (November 20, 1989) review of Gould's Wonderful Life, Jerry Adler misrepresents the secular humanist position. For the record, secular humanism does not maintain that organic evolution on this planet has been a progressive process unfolding in a straight line direction from primeval bacteria up to the recent appearance of our own species. In truth, it is just this view that is claimed by theistic evolutionists. The spiritualists (not the materialists) believe that humankind is the necessary end-product of four billion years of teleological evolution.

From a cosmic perspective, naturalistic humanists maintain that the human animal is a product of and totally within organic history (just like all the other plant and animal species). Humanists support science and reason rather than blind faith and metaphysical flights of wishful thinking. For them, the uniqueness of *Homo sapiens sapiens* as the symbolizing animal is its complex central nervous system. A superior brain allows for self-consciousness and human language as articulate speech as well as rational thought and moral conduct. Furthermore, within this materialist framework, all aspects of our own species are open to the wise purview of scientific research.

All present life, including the human being, is a result of millions upon millions of years of contingent events throughout earth history. As Gould emphasizes, change any of these events and the outcome of evolution would be different. Jolted from our presumed cosmic comfort, we are indeed lucky to be here.

Even today, the theory of evolution remains controversial. In part, this is due to the disturbing fact that most individuals are not well informed in those special sciences that provide the empirical evidence to support the evolutionary framework, e.g., geology, paleontology, comparative morphology, biochemistry, taxonomy, and population genetics. Likewise, the empirical truth of organic evolution still challenges those entrenched beliefs of traditional theology concerning a god, free will, immortality, and a divine destiny.

Concerning the creation of life, there is the ongoing anachronistic conflict between close-minded, biblical fundamentalists and open-minded, scientific evolutionists. Clearly, all those who hold to a strict and literal interpretation of Genesis will be hard-put to fit the Burgess Shale evidence into their own dogmatic and narrow religionist beliefs about both the origin and history of life on this planet.

The facts of material nature itself embody neither ethical principles nor theological assumptions. If there is no personal God, then humankind is not the inevitable outcome of a divine design in this dynamic universe. Rather, the emergence of our own species is an improbable event in a seemingly endless cosmos.

In principle, no evolutionary worldview is ever complete. There are always more facts and new ideas. Future scientific evidence and critical thought will probably modify neo-Darwinism, the present evolutionary synthesis founded on genetic variation and natural selection. Mechanisms now unknown to modern science may influence organic evolution, and additional sites may be found that will provide rocks and fossils to help answer those questions still raised by geologists and paleontologists.

Finally, in this vast universe, earth may not be the only planet that harbors organic beings. Exobiology may hold a cascade of consequences for science, philosophy, and theology. I have coined the new term "exoevolution" to distinguish this word from the emerging science of exobiology, which is the search for and study of merely the existence of life elsewhere in the universe. One can envision the cosmic science of comparative exoevolution. Within this plausible context, we may indeed live in a world of wonderful life.

The philosophy of secular humanism upholds both a cosmic perspective and the fact of evolution. It is clearly and convincingly expressed in the writings of Corliss Lamont[77] and Paul Kurtz.[78] Lamont stresses that our species must affirm life within nature, which is enough for all enlightened thinkers. Likewise, Kurtz argues that a human being is able to live a full life without religion. In step with the advances of science and emphasizing the use of reason, secular humanism remains open to new facts and the inevitable challenges of our future.

One may speak of evolution from those rational speculations in the naturalist cosmologies of antiquity, through the pivotal writings of Charles Darwin, to its present status as an established fact of modern science (with its often devastating implications for philosophy and theology).

The fact of evolution had to overcome the metaphysics of essentialism found in the thought of Plato and Aristotle as well as the worldviews of Aquinas, Kant, and Hegel (among many others). Instead of maintaining the eternal fixity of forms, which ignores historical relationships in terms of organic evolution, one must now accept the reality of both pervasive change and transitory species throughout geological time.

Today, the edifice of neo-Darwinism is a comprehensive theory of scientific evidence and rational concepts rich in exploratory, descriptive, explanatory, and predictive powers. Evolution is an inescapable challenge to all the beliefs and values that uphold a static view of the material universe. Indeed, evolution remains a very sobering fact for natural philosophy and process theology.

The continued advancement of evolutionary thought clearly requires that we distinguish among scientific facts, philosophical concepts, and religious beliefs. Likewise, no dogmatic closure should be placed on human inquiry or critical reflection. The theory of evolution is always subject to both rigorous reexamination and reevaluation in light of new empirical evidence, the expanding horizon of experience, and ongoing logical procedure. The naturalistic humanist knows that science and reason are the only means of reaching truths about matter and energy at this time. For the materialist, mind is grounded in life, while life itself is grounded in matter and energy. In fact, to deny the scientific theory of material evolution in terms of its cosmic and planetary aspects, it would be necessary to both undermine the entire structure of the physical universe and uproot the whole "coral of life" from earth history. Lastly, the theory of evolution is indispensable for both believers and nonbelievers if they are to achieve a sound understanding of and proper appreciation for the true place of humankind within this dynamic universe.

Today, modern evolutionary thought must overcome both geocentrism and cosmocentrism. Furthermore, exobiology implies exoevolution and other universes may have existed before, are existing, or will exist after this particular cosmic evolution. The present absence of empirical evidence for life elsewhere in this dynamic universe is not to be trusted. In the future, there may even be the cosmic science of comparative exoevolution. Likewise, some problems concerning the origin and destiny of this universe and life within it may tease the human mind forever (these questions having no possible solutions).

Reminiscent of Darwin but ages from now, one may imagine a critical observer traveling throughout this material universe: not as a naturalist aboard a survey ship sailing among those prehistoriclike islands of a volcanic archipelago isolated in a sea of space, but rather as a scientist on a sidereal exploration among the countless galaxies scattered everywhere in an endless ocean of time. What new facts, values, ideas, and perspectives in terms of organic history will this brave questor add to the sound understanding of and proper appreciation for both life and thought within this dynamic world?

As a consequence, one may even anticipate the cosmic science of comparative exoevolution, an empirical comprehension of biology far more sweeping in scope and richer in theory than any previous scientific philosopher or natural theologian could ever have conceived or believed, respectively.

Furthermore, one may also imagine a future theologian similar to Teilhard grappling with the awesome implications of a material universe filled with organic creativity. No doubt, this religionist will critically reflect on the geology and paleontology as well as botany and zoology of other planets among the galaxies (not to overlook the probability that intelligent beings exist elsewhere in this dynamic world); one may be certain that this universe is full of surprises and wonders now unknown to our own species. Overcoming all earth-bound and human-centered conceptual frameworks, this bold individual of ages to come will surely envision a truly cosmic god as that collectivity of all evolutions and all universes.

In the final analysis, Charles Darwin brought about a revolution in science while Teilhard sought to update theology. To a significant degree, our modern age belongs to their bold visions.

ENDNOTES AND SELECTED REFERENCES

1. Cf. Shree Purhit Swami and W. B. Yeats, eds., *The Ten Principal Upanishads* (London: Faber and Faber, 1970); and Clive Johnson, ed., *Vedanta* (New York: Bantam Books, 1974).

2. Cf. Betty Radice, ed., *Lao tzu* or *Tao te ching* (New York: Penguin Classics, 1981).

3. Cf. John Blofeld, ed., *I Ching: The Book of Change* (New York: E. P. Dutton, 1968).

4. Cf. G. S. Kirk and J. E. Raven, *The Presocratic Philosophers: A Critical History with a Selection of Texts* (Cambridge, England: Cambridge University Press, 1962).

5. Cf. Benjamin Farrington, *Aristotle: Founder of Scientific Philosophy* (New York: Prager, 1970); Marjorie Grene, *A Portrait of Aristotle* (London: Faber and Faber, 1963); W. K. C. Guthrie, *The Greek Philosophers: From Thales to Aristotle* (New York: Harper Torchbooks, 1960); G. E. R. Lloyd, *Aristotle: The Growth and Structure of His Thought* (Cambridge, England: Cambridge University Press, 1968); Richard McKeon, ed., *The Basic Works of Aristotle* (New York: Random House, 1941); John Herman Randall, Jr., *Aristotle* (New York: Columbia University Press, 1960); and A. E. Taylor, *Aristotle* (New York: Dover, 1955).

6. Cf. Lucretius, *De Rerum Natura*, 2d ed. (Cambridge, Mass.: Harvard University Press, 1982).

7. Cf. A. H. Armstrong, *Plotinus* (New York: Collier Bros, 1953).

8. Cf. Lloyd L. Morain,"Avicenna: Asian Humanist Forerunner," *The Humanist* 41(2):27–34.

9. Cf. Jean Paul Richter, ed., *The Notebooks of Leonardo da Vinci.* 2 vols. (New York: Dover, 1970, esp. vol. 2, chaps. 14, 15, and 16); and V. P. Zubov, *Leonardo da Vinci* (Cambridge: Harvard University Press, 1968).

10. Cf. M. A. Dynnik, "Man, Sun, and Cosmos in the Philosophy of Giordano Bruno," *Soviet Studies in Philosophy* 6(2); Sidney Greenberg, *The Infinite in Giordano Bruno* (New York: King's Crown Press, 1950); Irving Louis Horowitz, *The Renaissance Philosophy of Giordano Bruno* (New York: Coleman-Ross, 1952); and Dorothea Singer, *Giordano Bruno: His Life and Thought* (New York: Henry Schuman, 1950).

11. Cf. Baruch Spinoza, *The Collected Works of Spinoza* (Princeton, N.J.: Princeton University Press, 1985).

12. Cf. Gottfried Wilhelm Leibniz, *The Monadology and Other Philosophical Writings* (Oxford, England: Oxford University Press, 1965).

13. Cf. Wilfred Blunt, *The Compleat Naturalist: A Life of Linnaeus* (New York: Viking Press, 1971); and Heinz Goerke, *Linnaeus* (New York: Charles Scribner's Sons, 1973).

14. Cf. Ernst Cassirer, *The Philosophy of the Enlightenment* (Boston: Beacon Press, 1965).

15. Cf. Julien Offroy de La Mettrie, *Man the Machine* (LaSalle, Ill.: Open Court, 1961).

16. Cf. Otis E. Fellows and Stephen F. Milliken, *Buffon* (New York: Twayne, 1972).

17. Cf. Marquis de Condorcet, *Sketch for a Historical Picture of the Progress of the Human Mind* (New York: Noonday Press, 1955).

18. Cf. George Simpson, *Auguste Comte: Sire of Sociology* (New York: Thomas Y. Crowell, 1969).

19. Cf. Immanuel Kant, *Universal Natural History and Theory of the Heavens* (Ann Arbor: University of Michigan Press, 1969).

20. Cf. Donald M. Hassler, *Erasmus Darwin* (New York: Twayne, 1973); Desmond King-Hele, *Erasmus Darwin* (London: Macmillan, 1963); and Hesketh Pearson, *Doctor Darwin* (New York: Walker, 1930).

21. Bentley Glass, Owesei Temkin, and William L. Straus, Jr., eds., *Forerunners of Darwin: 1745–1859* (Baltimore, Md.: Johns Hopkins Press, 1968).

22. Refer to chapter 7, "Evolution: Science or Theology?"

23. Cf. John Farley, *The Spontaneous Generation Controversy from Descartes to Oparin* (Baltimore, Md.: Johns Hopkins Press, 1977).

24. Cf. J. B. Lamarck, *Zoological Philosophy: An Exposition with Regard to the Natural History of Animals* (New York: Hafner, 1963).

25. Cf. Arthur Schopenhauer, *The World as Will and Representation*, 2 vols. (New York: Dover, 1958).

26. Cf. Marx W. Wartofsky, *Feuerbach* (Cambridge, England: Cambridge University Press, 1977), esp. pp. 397–401.

27. Cf. Robert Chambers, *Vestiges of the Natural History of Creation*, 4th ed. (New York: Wiley and Putnam, 1846).

28. Barbara G. Beddall, ed., *Wallace and Bates in the Tropics: An Introduction to the Theory of Natural Selection* (Toronto: Macmillan, 1969); and John Langdon Brooks, *Just Before the Origin: Alfred Russel Wallace's Theory of Evolution* (New York: Columbia University Press, 1984).

29. Cf. Thomas H. Huxley, *Man's Place in Nature* (Ann Arbor: University of Michigan Press, 1959). Also refer to Cyril Bibby, *The Essence of T. H. Huxley* (New York: St. Martin's Press, 1967).

30. Cf. Ernst Haeckel, *The Riddle of the Universe at the Close of the Nineteenth Century* (New York: Harper & Brothers, 1905).

31. Cf. Sidney Hook, *From Hegel to Marx: Studies in the Intellectual Development of Karl Marx* (Ann Arbor: University of Michigan Press, 1962).

32. Cf. I. L. Andreyev, *Engel's "The Part Played by Labour in the Transition from Ape to Man"* (Moscow: Progress Publishers, 1985).

33. Cf. V. I. Lenin, *Materialism and Empirio-Criticism: Critical Comments on a Reactionary Philosophy* (New York: International Publishers, 1927).

34. Cf. Loren R. Graham, *Science, Philosophy, and Human Behavior in the Soviet Union* (New York: Columbia University Press, 1987).

35. Cf. Friedrich Nietzsche, *The Portable Nietzsche* (New York: Viking Press, 1968).

36. Cf. Herbert Spencer, *Synthetic Philosophy*, 10 vols. (New York: D. Appleton, 1862–1893). In particular, refer to Herbert Spencer, *First Principles* (1862) (New York: De Witt Revolving Fund, 1958).

37. Cf. John Fiske, *Outlines of Cosmic Philosophy Based on the Doctrine of Evolution, with Criticism on the Positive Philosophy*, 4 vols. (Boston: Houghton Mifflin, 1874).

38. Cf. Bozidar Knezević, *History, the Anatomy of Time: The Final Phase of Sunlight* (New York: Philosophical Library, 1980). Also refer to H. James Birx, "Knezević and Teilhard: Two Visions of Cosmic Evolution," *Serbian Studies*, 1(4):53–63.

39. Cf. Peter A. Kropotkin, *Mutual Aid: A Factor of Evolution* (Boston: Extending Horizons Books, 1914).

40. Cf. Peter A. Kropotkin, *Ethics: Origin and Development* (New York: Tudor, 1924).

41. Cf. Henri Bergson, *Creative Evolution* (New York: Modern Library, 1944).

42. Cf. Henri Bergson, *The Two Sources of Morality and Religion* (Garden City: Anchor Books, 1935).

43. Cf. Justus Buchler, *Philosophical Writings of Peirce* (New York: Dover, 1955).

44. Cf. William James, *Pragmatism and Four Essays from 'The Meaning of Truth'* (New York: Meridian Books, 1955).

45. Cf. John Dewey, *The Influence of Darwin on Philosophy, and Other Essays in Contemporary Thought* (Bloomington: Indiana University Press, 1965).

46. Cf. C. Lloyd Morgan, *Emergent Evolution* (New York: Henry Holt, 1927).

47. Cf. C. Lloyd Morgan, *The Emergence of Novelty* (London: Williams & Norgate, 1933).

48. Cf. Samuel Alexander, *Space, Time, and Deity*, 2 vols. (New York: Dover, 1966).

49. Cf. Jan Christiaan Smuts, *Holism and Evolution* (New York: Viking Press, 1961).

50. Cf. Roy Wood Sellars, *Evolutionary Naturalism* (Chicago: Open Court, 1969).

51. Cf. Alfred North Whitehead, *Science and the Modern World* (New York: Macmillan, 1967).

52. Cf. Alfred North Whitehead, *Process and Reality: An Essay in Cosmology* (New York: Free Press, 1979).

53. Cf. Lester Frank Ward, *Glimpses of the Cosmos*, 6 vols. (New York: G. P. Putnam's Sons, 1913–1918).

54. Cf. Edward B. Tylor, *Anthropology* (Ann Arbor: University of Michigan Press, 1960); *Primitive Culture: Researches into the Development of Mythology, Philosophy, Religion, Language, Art, and Custom*, 2 vols. (New York: Harper Torchbooks, 1958); and *Researches into the Early History of Mankind and the Development of Civilization* (Chicago: Phoenix Books, 1964).

55. Cf. Lewis Henry Morgan, *Ancient Society, or Researches in the Lines of Human Progress from Savagery through Barbarism to Civilization* (New York: Meridian Books, 1963).

56. Cf. James George Frazer, *The Golden Bough*, 3d ed. (London: Macmillan, 1911–1915).

57. Cf. Leslie A. White, *The Evolution of Culture: The Development of Civilization to the Fall of Rome* (New York: McGraw-Hill, 1959).

58. Cf. Marshall D. Sahlins and Elman R. Service, eds., *Evolution and Culture* (Ann Arbor: University of Michigan Press, 1961).

59. Cf. Marvin Harris, *Cultural Materialism: The Struggle for a Science of Culture* (New York: Vintage Books, 1980), and *The Rise of Anthropological Theory: A History of Theories of Culture* (New York: Thomas Y. Crowell, 1970).

60. Cf. Raoul Naroll, *The Moral Order: An Introduction to the Human Situation* (Beverly Hills, Calif.: Sage, 1983).

61. Cf. Marvin Farber, *Naturalism and Subjectivism* (Albany, N.Y.: State University of New York Press, 1968).

62. Cf. Mary D. Leakey, *Disclosing the Past* (New York: Doubleday, 1984).

63. Cf. Sonia Cole, *Leakey's Luck: The Life of Louis Seymour Bazett Leakey (1930–1972)* (New York: Harcourt Brace Jovanovich, 1975).

64. Cf. Richard E. F. Leakey, *One Life* (Salem, Mass.: Salem House, 1984).

65. Cf. Donald C. Johanson and Maitland A. Edey, *Lucy: The Beginnings of Humankind* (New York: Simon and Schuster, 1981). Also refer to Donald C. Johanson and James Shreeve, *Lucy's Child: The Search for Our Beginnings* (New York: William Morrow, 1989).

66. Cf. Mary D. Leakey, *Disclosing the Past* (New York: Doubleday, 1984).

67. Cf. Kenneth F. Weaver, "The Search for Our Ancestors," *National Geographic* 168(5):560–623, and especially Richard Leakey and Alan Walker, "Homo Erectus Unearthed," *National Geographic* 168(5):624–29.

68. Cf. Birute Galdikas, "Orangutans: Indonesia's People of the Forest," *National Geographic* 148(4):444–73.

69. Cf. Jane Goodall, *In the Shadow of Man* (Boston: Houghton Mifflin, 1983) and *The Chimpanzees of Gombe: Patterns of Behavior* (Cambridge, Mass.: Belknap Press of Harvard University Press, 1986).

70. Cf. George B. Schaller, *The Mountain Gorilla: Ecology and Behavior* (Chicago: University of Chicago Press, 1963), and *The Year of the Gorilla* (New York: Ballantine Books, 1965).

71. Cf. Dian Fossey, *Gorillas in the Mist* (Boston: Houghton Mifflin, 1983). Also refer to Farley Mowat, *Woman in the Mists: The Story of Dian Fossey and the Mountain Gorillas of Africa* (New York: Warner Books, 1987); and Harold T. P. Hayes, *The Dark Romance of Dian Fossey* (New York: Simon and Schuster, 1990).

72. Cf. George Gaylord Simpson, *This View of Life: The World of an Evolutionist* (New York: Harcourt, Brace & World, 1964).

73. Cf. Julian S. Huxley, *Evolution: The Modern Synthesis* (New York: John Wiley & Sons, 1964).

74. Cf. Ashley Montagu, *Darwin: Competition and Cooperation* (New York: Henry Schuman, 1952).

75. Cf. Theodosius Dobzhansky, *Mankind Evolving: The Evolution of the Human Species* (New York: Bantam, 1970).

76. Cf. Stephen Jay Gould, *Wonderful Life: The Burgess Shale and the Nature of History* (New York: W. W. Norton, 1989).

77. Cf. Corliss Lamont, *The Philosophy of Humanism*, 6th ed. (New York: Ungar, 1982), esp. p. 145.

78. Cf. Paul Kurtz, *Eupraxophy: Living Without Religion* (Buffalo, N.Y.: Prometheus Books, 1989).

FURTHER READINGS

Allman, William F. "The First Humans." *U.S. News & World Report* (February 27, 1989), 106(8):52–59.

Boorstin, Daniel J. *The Discoverers.* New York: Random House, 1983.

Bowler, Peter J. *Evolution: The History of an Idea*, rev. ed. Berkeley: University of California Press, 1989.

Clements, Tad S. *Science vs. Religion.* Buffalo, N.Y.: Prometheus Books, 1990.

Cohen, I. Bernard. *Revolution in Science*. Cambridge, Mass.: Belknap Press of Harvard University Press, 1985.

Farrington, Benjamin. *Greek Science: Its Meaning for Us*. Baltimore, Md.: Penguin Books, 1953.

Fothergill, Philip G. *Historical Aspects of Organic Evolution*. London: Hollis and Carter, 1952.

Gillespie, Charles Coulston. *Genesis and Geology: The Impact of Scientific Discoveries upon Religious Beliefs in the Decades Before Darwin*. New York: Harper Torchbooks, 1969.

Glass, Bentley, Owesei Temkin, and William L. Straus, Jr., eds. *Forerunners of Darwin: 1745–1859*. Baltimore, Md.: Johns Hopkins Press, 1968.

Huggett, Richard. *Catastrophism: Systems of Earth History*. New York: Edward Arnold/Hodder & Stoughton, 1990.

Knight, David M. *The Age of Science: The Scientific World-View in the 19th Century*. Oxford, England: Blackwell, 1986.

Lovejoy, Arthur O. *The Great Chain of Being: A Study of the History of an Idea*. New York: Harper Torchbooks, 1960.

Madigan, Tim. "John Dewey: Educator and Innovator." *Free Inquiry* (Winter 1987/1988):54-55.

Mayr, Ernst. *The Growth of Biological Thought: Diversity, Evolution, and Inheritance*. Cambridge, Mass.: Belknap Press of Harvard University Press, 1982.

McMenamin, Mark, and Dianna McMenamin. *The Emergence of Animals: The Cambrian Breakthrough*. New York: Columbia University Press, 1990.

Osborn, Henry Fairfield. *From the Greeks to Darwin: The Development of the Evolution Idea Through Twenty-four Centuries*. New York: Charles Scribner's Sons, 1929.

Santillana, Giorgio de. *The Origins of Scientific Thought: From Anaximander to Proclus*. New York: Mentor Books, 1961.

Schoenwald, Richard L., ed. *Nineteenth-Century Thought: The Discovery of Change*. Englewood Cliffs, N.J.: Prentice-Hall, 1965.

Charles Darwin

CHARLES ROBERT DARWIN (1809–1882). As a result of extensive travel and rigorous research, Darwin gave both empirical facts and rational arguments to support the scientific theory of organic evolution. A mechanist and materialist, he relied primarily upon that explanatory principle of natural selection to account for the mutability of species throughout biological history. His penetrating insights are presented in numerous publications, especially *On the Origin of Species* (1859) and *The Descent of Man* (1871). Yet, these prolific writings avoided the philosophical questions and theological issues raised by the truth of evolution. *Courtesy of American Museum of Natural History.*

3

Charles Darwin:
The Origin of Species

Charles Robert Darwin (1809–1882) is the scientific father of organic evolution. Yet, he had neither the first word nor the last word on this theory of life and truth of nature. His pivotal writings contain facts, concepts, perspectives, and hypotheses that still demand research and inspire reflection. Without exaggeration, all of the special sciences from astronomy and geopaleontology to anthropology and social psychology owe their present status to the Darwinian framework: our modern worldview is grounded in an evolutionary interpretation of things.

In the serious development of this idea of evolution from the philosophical speculations in antiquity to the present scientific synthesis, Darwin is undoubtedly the key figure among the intellectual giants of the special sciences.[1] He had boldly asked a crucial question: How do new plant and animal species appear within organic history on this planet? His own answer, impressively supported by massive empirical evidence and unique personal experience as well as intuitive insights and rational reflections, brought about a scientific revolution that gravely challenged traditional Judeo-Christian beliefs and influenced forever all conceptualizations of earth history and the dynamic universe.

Darwin's remarkable glimpse of process reality, which focused on rocks and fossils as well as floral and faunal populations, now encompasses ev-

113

erything from the origin of sidereal galaxies to the foundation of human ethics. It is to his lasting credit that modern biology rests upon a comprehensive and intelligible view of life within an evolutionary framework. Likewise, a static cosmology is no longer permissible in modern science, philosophy, or theology: nothing is fixed in nature.

Charles Robert Darwin was born at The Mount in Shrewsbury, England, on 12 February 1809; he was the fifth of six children. His mother died in July 1817. Like the natural philosopher Aristotle, whom he greatly admired, Charles was raised in a wealthy family that contained a line of physicians. Dr. Robert Waring Darwin, his atheist father, had a domineering character that molded the psychological development of this inquisitive youth. Interestingly enough, Charles seems never to have been greatly influenced by his siblings, although he shared a teenage interest in chemistry with his older brother Erasmus (who later accompanied Charles at the University of Edinburgh to study medicine).

Fortunately, Charles was born into both family prosperity and social prestige, in sharp contrast to the biologists Huxley, Mendel, and Wallace. Both his paternal grandfather and father were successful medical doctors with lucrative practices, which allowed for educational advantages and experiential opportunities that would have otherwise been closed to the young enthusiast of unspoiled nature.

Charles enjoyed the wonderous complexity of the material world. Despite the pervasive conservativism of Victorian England, his upbringing was not religious; his tolerant father kept his own atheism silently to himself. Charles was a very ordinary boy, presumably somewhat below the common standard of intellect. There was no indication of his latent creative genius. As he developed, Charles immersed himself in those natural beauties of the quaint Shropshire countryside. Not dedicated to book learning, he was a mediocre student. Instead, he was to acquire a lifelong dedication to earth science and natural history (especially geology and biology).

Charles had an insatiable curiosity about the objects of nature. At the age of ten, he exhibited an interest in insects (especially beetles, as well as moths and butterflies). His early concern for entomology would soon extend itself to include historical geology and marine zoology. He had a strong innate passion for collecting things, which proved to be a significant asset in later years when the descriptive analysis of empirical evidence played such a crucial role in supporting his own scientific theory of biological evolution ("descent with modification").

Erasmus Darwin (1731–1802) was a versatile genius who, besides his medical career, had found time to write poetry and speculate on nature.[2] In his book Zoönomia (1796), this naturalist anticipated the scientific theory of biological evolution; in fact, this volume was put on the Roman Catholic

Index of forbidden works. He wrote that sensibility or irritability pervades the entire organic world and that all life had its origin from a single primordial filament that existed millions of years ago in the ancient sea of our early earth. Charles's growing curiosity and intellectual propensity clearly owed more to his paternal grandfather than to his living father; the latter was not interested in science as such.

Jean Baptiste de Lamarck (1744–1829), a natural philosopher and invertebrate specialist, was the first thinker to write a volume solely for the purpose of presenting a speculative theory of biological evolution as it applies to the organic history of the animal kingdom on earth.[3] Titled *Zoological Philosophy* (1809), it appeared in the year of Darwin's birth and exactly fifty years before his publication of *On the Origin of Species* (1859). Lamarck's controversial arguments for organic evolution were metaphysical rather than empirical and, as a result, went unaccepted and unappreciated by most naturalists of his time. Instead, many early biologists supported divine intervention, special creations, forms of vitalism, or even spontaneous generation. Briefly, Lamarck based his interpretation of animal evolution on speculations not supported by empirical evidence or logic while Darwin would do as best he could to ground his theory of descent with modification in the science and reason of the middle of the nineteenth century.

Charles progressed in his passion for geology and entomology. More and more, he turned his interest to natural history. He collected rocks and minerals, fossils and shells, plants and insects (particularly beetles), and spent time catching rats and shooting birds (e.g., pheasants and partridges) for sport, taking the time to study their habits and eggs. Charles enjoyed fishing, hunting for foxes with dogs, and taking long, solitary walks. He learned ornithology and taxidermy—even thought of becoming an ornithologist—studied the variability of plants, and took note of glacial striations and erratic boulders. In all this, he was developing an intense devotion to the accuracy of naturalist study and scientific thought grounded in empirical facts, rigorous logic, and hypothetico-deductive inferences.

In October 1825, having finished school and with no definite plan for a career, Charles Darwin was sent by his father to the University of Edinburgh, Scotland, to study medicine. Not interested in committing himself to this prestigious profession, Charles often cut classes to attend lectures in science or to participate in field trips. He also attended meetings on natural science and natural history.

Several factors were decisive in destroying what slight medical ambition Charles had: he could not stand the sight of blood or bear the agony of having to watch two very bad operations, one on a child patient undergoing the amputation of a limb without an anesthetic. To avoid this ghastly circumstance, he fled from the operating room, convinced he could never practice medicine. After less than two years at the university, he returned home. Darwin simply had no stomach for human suffering. Without a heart for surgery, his mind now gravitated even more to natural history (especially geology).

Disappointed, Dr. Robert Waring Darwin next recommended that his son attend Christ's College, Cambridge, to prepare for the ministry. For three years (1828–1831), Charles studied theology as well as mathematics and the classics. For the most part, what today are the special sciences were, at that time, considered to be areas in natural philosophy relegated to private pursuits. Perhaps fortunately, his academic training excluded geology, paleontology, taxonomy, botany, zoology, and comparative ecology, those subjects in which he would later excel and distinguish himself. Consequently, he was free to doubt the biological ideas of the time: spontaneous generation and the eternal fixity of species. Furthermore, young Charles never had laboratory training in either invertebrate or vertebrate morphology, although on his own he achieved an uncanny skill for analyzing and describing living things from apes to worms. He now acquired a taste for art, music, and poetry but always remained poor in languages, mathematics, and metaphysics.

At this time, Charles enjoyed reading Shakespeare and several other poets (e.g., Scott, Byron, Thomson, and later Milton, Coleridge, and Wordsworth). He would always find pleasure in essays and novels.

While at Cambridge, Charles made important contacts with two prominent professors of natural science, the geologist Adam Sedgwick and the botanist John Henslow. Both were religionists, and Henslow became Darwin's lifelong intimate friend. Numerous field trips with these two naturalists reinforced Darwin's love of nature. These excursions provided an excellent preparation for both his forthcoming five-year voyage around the world on the HMS *Beagle* and subsequent scientific research.

During these years of formal study, Charles was also influenced by five books: *The Elements* (c. 300 B.C.E.) by the mathematician Euclid, *Paradise Lost* (1667) by the poet John Milton, *Natural Theology: or, Evidences of the Existence and Attributes of the Deity, Collected from the Appearances of Nature* (1802) by the theologian William Paley, *Preliminary Discourse on the Study of Natural Philosophy* (1830) by the astronomer Sir John Herschel, and especially *Personal Narrative of Travels to the Equinoctial Regions of the New Continent* (1805–1834) by the naturalist Alexander von Humboldt.

Archdeacon Paley argued for the existence of God from the alleged design in nature. Humboldt was the father of scientific exploration as well as one of the pioneers in ecology, geography, and meteorology; he visited Darwin later in the century.

Apparently, Charles Darwin did not take the writings of George Louis Buffon, Baron Georges Cuvier, Denis Diderot, Jean Baptiste de Lamarck, and Étienne G. Saint-Hilaire seriously; he often ignored the important contributions of the French naturalists. He seems to have also dismissed the ideas of his grandfather Erasmus Darwin and, later, the speculations of his contemporary Robert Chambers.

Finally, Charles received a bachelor of arts degree from Cambridge; he had been an undistinguished student at school, in the university, and at this college. At this time, he believed in a strict and literal interpretation of the Holy Bible. As a rural parson, when not reading the biblical story of creation as found in Genesis or preaching the Christian beliefs of the Anglican Church to his devout parishioners, he imagined there would be leisure time to spend in the English countryside as an enthusiastic naturalist. The would-be minister could study beetles, collect rocks and fossils, and absorb natural history; he was already having religious doubts about points of traditional theology. However, this self-made scientist was never to become a country clergyman (in retrospect, an ironic idea). Yet, one wonders why he never pursued a legal career, which was the chosen occupation of some important nineteenth-century naturalists (e.g., Charles Lyell and Lewis Henry Morgan).

During the summer of 1831, a trip to Wales awakened in Charles an exquisite delight for the scenery of nature. Nevertheless, this enjoyment, among others, waned in the coming years.

Charles Darwin was shy, gentle, inquisitive, and tolerant. This admirable young man of patience and zeal would soon display emerging genius and sensitive wisdom. He had an insatiable curiosity and penetrating intellect. Even so, at the age of twenty-two, this graduate of Cambridge still had not decided on a career. He had no formal degree in geology or biology and, in fact, would never teach these subjects at a college or university. As such, however, he was not indoctrinated into any of the scientific hypotheses of the time that dogmatically precluded the idea of evolution. For Darwin, experience was the best teacher and nature itself was the classroom. Also, from our perspective and in the light of our benefit of hindsight, many of the very instruments he used in his scientific investigations were faulty. To his immortal credit, Charles always remained open and hospitable to both the theoretical implications and physical consequences of his own experiments and experiences as well as the relevant discoveries of other naturalists.

During his life, Darwin was influenced by a wide variety of accidental incidents and ironic twists of fate. In retrospect, it was as if an unlikely series of improbable events was to guide him inevitably to a position aboard the HMS *Beagle* and, subsequently, to develop his theory of evolution; his own ideas are expressed in sixteen books and 152 other publications.

Darwin, the disillusioned medical student and half-hearted theologian, was not to remain for long back at The Mount. On August 29, in a letter from Henslow dated 24 August 1831, he received an invitation to become both the unpaid naturalist and a traveling companion aboard the HMS *Beagle* (actually, the young gentleman had been nominated for this dual position by George Peacock, a mathematician and astronomer at Cambridge; four other individuals had already turned down this unique appointment). It would not be the only letter to suddenly change both the course of Darwin's life and the history of science. Since he still had no definite plans for a career, Darwin was elated at this prospect of taking an extensive scientific trip around the world. Such a voyage would provide him with enough opportunities to satisfy the most ardent naturalist. Unfortunately, however, Dr. Robert Waring Darwin was reluctant to support this astonishing undertaking. Yet, should the now-disappointed amateur scientist find someone whom his father respected who would hold this daring venture to be worthwhile, then he would be allowed to go.

When visiting Josiah Wedgwood II ("Uncle Jos") of the famous pottery family at Maer Hall in Shropshire, only 20 miles from Shrewsbury, Darwin found such a sympathetic friend who thought this global trip would be beneficial to the emerging scientist. In truth, perhaps, Uncle Wedgwood may have had the fate of his two young daughters more at heart than Darwin's naturalist ambitions. Be that as it may, both returned to The Mount and persuaded its successful physician that this unusual invitation should be accepted and monies provided. In finally granting permission, the father was hoping that this voyage would mature his indecisive son; after all, Dr. Darwin would be financing the trip. In retrospect, this unique journey proved to be a very worthwhile endeavor, indeed!

Having overcome his first hurdle, Charles now encountered a second one: Captain Robert FitzRoy of the Royal Navy's HMS *Beagle* (a military ship renovated for scientific surveys). FitzRoy was an illegitimate aristocrat, a conservative member of the Tory political party, an outspoken strict biblical literalist, and a young eccentric with both manic-depressive and suicidal tendencies. He was also a pioneer meteorologist and hydrographer,* who wanted this particular global voyage to be the most successful scientific survey of the nineteenth century.

At first, FitzRoy was reluctant to allow Darwin to fill the unique position of companion/naturalist. FitzRoy claimed that one could determine the mental qualities and character of a man by his outward appearance (especially facial features), and he disliked Darwin's nose. Nevertheless, after several days, this devotee of the then-fashionable alleged science of physiognomy consented to have an amateur geologist aboard as his shipmate. Therefore, Charles overcame the second hurdle.

---

*One who scientifically studies and maps bodies of water.

Actually, Captain FitzRoy's personal desire was to have a naturalist as traveling companion aboard his ship to amass empirical evidence to support scientifically a strict and literal interpretation of the story of creation, as presented in Genesis. Then, ironically enough, this task fell to the would-be minister but now-confirmed naturalist Charles Darwin, who viewed the upcoming worldwide voyage as an incredibly fortunate opportunity to pursue further his love of geology and various related sciences. The approaching journey of discovery through space and time would not only determine his own thoughts and writings but also represent a major turning point in the scientific ideas of Western thought. It would alter forever our view of the age of the planet, the origin and development of life on earth, and the antiquity of humankind within natural history: rocks, fossils, and artifacts were becoming the subject matter of serious inquiry. Of course, the young geologist from Shrewsbury did have a degree in theology. Soon, a dogmatic adherence to the eternal fixity of species in nature would be superseded by a growing awareness of the mutability of plant and animal forms throughout organic history on planet earth.

Before this voyage, Charles Darwin was primarily an amateur geologist who accepted the then-taught doctrine of the immutability of species and believed in a biblical account of the origin of things (although he never took religion seriously). After sailing around the world, he became a specialist in biology who developed an evolution theory of organic history and espoused agnosticism (if not maintaining silent atheism). This astonishing shift in his intellectual outlook from amorphous opinions to reasoned convictions was due primarily to three major influences: his careful reading of Sir Charles Lyell's three-volume tome *Principles of Geology* (1830–1833), which argued for the steady-state theory of uniformitarianism (the gradual but continuous alteration of the stratigraphic structures of the crust of our planet as a result of those same pervasive natural forces operating throughout earth history); the scientific circumnavigation of the HMS *Beagle* (especially its five-week visit to the Galapagos Islands during the fall of 1835); and a chance but beneficial reading of Thomas Robert Malthus's classic work *An Essay on the Principle of Population* (1798, 1803).

After several unsuccessful attempts to leave Plymouth harbor due to severe storms, as if material nature did not wish to easily surrender its evolutionary evidence to this young scientist, the *Beagle* on 27 December 1831 finally set sail from England to the Canary Islands.[4] At this time, the energetic Darwin had no idea just how remarkably influential this unique adventure would be on his own view of life and the history of science.

Charles had looks, brains, and a pleasant personality as well as social status, financial security, and considerable good fortune. His powerful intellect and enlarged curiosity, unencumbered by theology or metaphysics, were free to observe keenly and to record meticulously the details of nature. With an open mind during this entire expedition, Charles would analyze evidence and ponder experience. More and more, he would realize that the implications of this evidence and experience were challenging science, philosophy, and theology; he would begin to question traditional facts, values, and beliefs. Unknown to him at the time, he was already synthesizing a vast range of facts and concepts conducive to bringing about a major revolution in science and natural philosophy in terms of an evolutionary perspective.

In the four decades immediately preceding this voyage of the HMS *Beagle*, the impact of scientific hypotheses and empirical discoveries in both geology and paleontology as well as archeology and taxonomy had seriously challenged the religious beliefs and social values of the Judeo-Christian tradition. As was previously mentioned, an essential contribution to the shift in Charles Darwin's own conceptual view of earth history was his reading of Sir Charles Lyell's *Principles of Geology*. Volume one had been given to the young naturalist by his friend Henslow, while the remaining two volumes were sent to him from England. Actually, this materialist theory of geological evolution had first been conceived by James Hutton (1726–1797) and published in his book *Theory of the Earth* (1795); it had earlier appeared as an article titled "Investigation of the Laws Observable in the Composition, Dissolution, and Restoration of Land Upon the Globe," published in the transactions of the Royal Society of Edinburgh and, seven years later, as this two-volume work that marked the beginning of dynamic concepts in historical geology.

Hutton had appealed directly to nature for an understanding of geological changes in terms of rational thoughts and natural forces.[5] With a sweeping imagination, this Scottish theorist and physician offered the earliest comprehensive treatise providing a geological synthesis: this Huttonian steady-state worldview taught the endless, gradual, and imperceptible cycles of rising and falling continents as well as the antiquity of life on earth. This vulcanist* espoused the ongoing uniformity of geological dynamics, although they vary in intensity from age to age. His lofty vision maintained the uninterrupted continuity of past and present natural causes and effects throughout everlasting time and cyclic change, especially the force of internal heat from a central subsurface source. In fact, not until 1912 would the German meteorologist Alfred Wegener father the scientific theory of continental drift based on plate tectonics.

Hutton's thoughts represented a process view in natural philosophy

---

*One who holds that fire (not water) has modified the surface of the planet throughout history.

supporting a world-machine interpretation of planetary changes: our earth is a terrestrial system subject to both physical laws and pervasive alterations, but it ultimately exists for the benefit of humankind. Those enigmatic fossils in rock strata are the remains of life forms in previous ages that were deposited in sediments over vast periods of planetary time. Summarizing, subterranean heat has slowly but repeatedly transformed the surface of the planet (the principle of decay), resulting in a continuous and unending series of new worlds (the principle of renovation) during the eternality of our earth (the principle of endurance). In contemplating planet earth, this great geological theorist found no vestige of a beginning and anticipated no prospect of an end.

Hutton did not openly deny the teachings of the Sacred Scriptures to advocate atheism or, on the other hand, accept the beliefs of biblical literalism. Clearly, he did not support a strict and literal belief in the biblical account of divine creation. For him, there was simply no empirical evidence that demonstrated a supernatural origin of earth. His materialistic interpretation of world changes and earth cycles in the emerging science of geology dealt a fatal blow to biblical literalism with its belief in a rigid acceptance of the Mosaic cosmogony as described in Genesis. As such, Hutton instilled an attitude conducive to the further advancement of both descriptive and theoretical geology. Two other significant books in this new area of study were William Smith's *Stratigraphical System of Organized Fossils* (1817) and Leopold von Buch's *Description of the Canary Islands* (1825).

In his two works, *Illustrations of the Huttonian Theory of the Earth* (1802) and *Outlines of Natural Philosophy* (1812), John Playfair, a vulcanist and mathematician at the University of Edinburgh, rigorously popularized this Huttonian steady-state worldview. Scientific inquiry boldly opposed religious dogmatism: the traditional six days of theistic creation as believed by the majority were challenged by a naturalist theory supporting millions of years of earth history as documented by rock strata and fossil evidence, which so far convinced only a minority. Hutton's own geological uniformitarianism with its endless time and mechanistic operations left no room for spiritual intervention but it did encourage scientific inquiry into those workings of material nature.

In Scotland, with ironic appropriateness, Sir Charles Lyell (1797–1875) was born in the year Hutton died.[6] Although he also had an interest in zoology, Lyell became the distinguished founder of modern historical geology. In his *Principles of Geology* (which rejected outright all mystery, global cataclysms, supernatural violence, and theological fantasies), he provided empirical evidence as best he could at that time to substantiate a theory of uniformitarianism, the slow and continuous alteration of geological structures on the surface of our earth over vast periods of time due to the action of constant and existing physical causes of change within and upon this planet itself. Eventually, this Hutton-Lyell viewpoint with its sweeping geological framework of endless time and pervasive change suggested to Darwin the mutability of species throughout organic history. Likewise, a growing awareness of changes in the starry heavens above resulted in some naturalists considering the probability of a flux of life on earth (fossils in rock strata were now recognized to be traces of prehistoric life). The deeper scientists dug into the earth's crust and the farther they searched into the depths of outer space, the more apparent it became to them that cosmic history spans across millions if not even billions of years.[7]

Darwin became a thoroughgoing uniformitarian or actualist who overcame the orthodox geological creed of global castastrophism and, later, even those biological beliefs in vitalism, creationism, essentialism, and spontaneous generation. Agassiz, Blyth, Buckland, Cuvier, Henslow, Owen, and Sedgwick were among those who resisted the emerging evolutionary framework. Ironically, although he had a profound influence on Darwin, Lyell resisted the evolutionary movement until near the end of his life; he had not taken seriously the insights of Lamarck, Chambers, or even Darwin. Lyell's years of reluctance were grounded in a mental residue of the human-centered geotheological drama popular among some naturalists of the last century, who saw no evidence for a materialist explanation for the origin of new plant and animal species (particularly the human animal).

The ensuing debate in France between Georges Cuvier, the theistic catastrophist as defender of special creations, and Étienne G. Saint-Hilaire, the deistic protoevolutionist, foreshadowed the bitter controversy that would later emerge between the rigid biblical fundamentalists and the open scientific evolutionists. The Lyellian worldview now provided Darwin with a vast temporal framework of ongoing geological change that suggested, if only unconsciously at first, the mutability of species (even Immanuel Kant, whom Darwin read without benefit, suggested that our earth could be millions of years old but this German philosopher rejected organic evolution).

Unlike FitzRoy, Darwin acknowledged the awesome age of this planet. He realized that the geology of the earth is of immense duration: it is not a solid orb of static elements but a world of ceaseless change both violent and gradual. This radically new conception of time and events implied that floral and faunal types could also be gradually and continuously transformed throughout countless epochs due to natural causes discoverable by the human intellect. In short, a dynamic explanation for biological history was clearly forthcoming in terms of science and reason.

After a two-month delay, the Beagle finally sailed from Plymouth Sound with about seventy crew members along with Darwin and FitzRoy. The numerous practical purposes of this unexpected five-year global voyage

(1831–1836) included: charting the extensive coastlines of South America, including Patagonia, Tierra del Fuego, and more especially the Straits of Magellan; taking ocean soundings and getting more accurate longitudinal measurements via worldwide chronological reckonings; and both surveying and mapping sea currents as well as the rivers of numerous volcanic islands and continents. These undertakings were sponsored by the British Admiralty in order to improve navigational maps for obvious economic and mercantile reasons.

Robert FitzRoy (1805–1865), age twenty-five and only three years older than his companion/naturalist, looked forward to achieving a splendid survey of the Southern Hemisphere. (In 1828, Philip King, the previous captain of the *Beagle*, had committed suicide in the Straits of Magellan; in retrospect, an ironic coincidence.) Only twenty-two and largely self-taught, Charles Darwin could not envision how this lengthy trip around the world with its visit to a prehistoric-like microcosm would radically alter both his view of life and the development of natural science as well as undermine forever both Thomism and Aristotelianism.

Through various circumstances, young Darwin had managed to attain this unique opportunity to expand his inquisitive intellect while aboard the *Beagle*. In a brief time, he was transformed from an English gentleman in proper attire (suit with double-breasted waistcoat and tails, high-collared shirt and cravat, top hat, and an elegant topcoat) walking the deck of this ship with a marine collecting net or telescope in his hand to an explorer riding horesback across the extensive pampas of South America and sharing an evening campfire with rugged gauchos or, later, an excited naturalist astride the back of a giant tortoise on the Galapagos Islands in the Pacific Ocean. Deep under a projecting forehead with its bushy eyebrows, Darwin's clear and alert gray-blue eyes steadily surveyed all of nature with incredibly disciplined observational powers. He gave patient, steadfast, and practical attention to details in observing, analyzing, and describing the objects of a dynamic world. This inexperienced naturalist had an uncanny ability to determine which geological features and biological characteristics are of scientific value in terms of crucial inferences for understanding and appreciating earth history. Despite his constant seasickness, he exuded determination and energy. One marvels at his stamina, humanism, and scientific perceptiveness. The crew referred to him as "the dear old philosopher" (Darwin himself greatly venerated Aristotle,[8] although the latter was not an evolutionist).

This amateur scientist came prepared with the following: a compass, magnifying glass, geological hammer, dissecting instruments, binoculars, telescope, notebooks, jars for preserving specimens, and even a case of pistols. He quickly became aware of the stupendous dimensions of our planet in terms of space, time, and change. His unending discoveries were made against a panorama of endless ocean, awesome complexity, staggering diversity, and essential unity.

Darwin seems to have been interested in everything that moved in water, crawled on land, and flew in the air. His interests in geology encompassed icebergs and glaciers; atolls, coral reefs, fringing reefs, and barrier reefs; rocks and minerals; fossils and even volcanic ash. He collected and examined invertebrates, marine life (especially luminous fish), insects, amphibians, reptiles, birds, mammals (including vampire bats), and various plants as well as exotic seeds. With great zeal this industrious and curious young naturalist found time to explore the jungles of Brazil, cross the windswept grasslands of Patagonia, climb the impressive Andes of Chile, and traverse volcanic islands with their unusual flora and fauna. He was not one to pass up an opportunity to widen his grasp of nature, and these great efforts were certainly not doomed to oblivion.

During this voyage, Darwin became aware of the possibility that differentiated but similar varieties inhabiting ocean islands may share a common ancestral origin from a form on the nearby continent, as demonstrated in the isolated populations of the Cape Verde Islands off the coast of Africa as well as, later, the independently varying plants and animals among the Galapagos Islands off the coast of South America. This geographical distribution of representative types (biogeography) would, in retrospect, argue for the adaptive radiation of life forms from common sources throughout organic history on earth.

Darwin loved the biological exuberance of the South American tropics; in the jungle, his inquisitive mind was a chaos of excitement and delight. He was enthralled with the overwhelming novelty in that teeming life of a Brazilian rain forest (particularly the hummingbirds and beetles), the richest area of organic species on earth, and became increasingly aware of the important roles of camouflage and mimicry in the living world. Likewise, Darwin developed a deep and lasting respect for life itself. He abhorred slavery, vivisection, and cruelty. He also opposed violence, barbarity, ideological controversy, and the infliction of physical suffering of any sort (as such, he found naval discipline distasteful). In fact, he even lost his earlier passion for hunting as a sport.

Free from the intellectual confines of Victorian England, Darwin was now able to reflect on the empirical evidence he was amassing and its scientific implications; he was continually sending unique data back to Henslow at Cambridge. Again, in sharp contrast, as a strict biblical literalist, FitzRoy had his own personal religious motive for bringing along a gentleman naturalist: Darwin was to collect empirical evidence to substantiate scientifically as true a strict and literal interpretation of the story of creation as offered in the book of Genesis. Unable to predict the ironic consequences of his desire, the captain had given this task to the would-be minister from Cambridge. It is unfortunate indeed that Darwin did not record for posterity his personal conversations with FitzRoy during the meals they had together throughout this five-year voyage in the southern hemisphere. No doubt, the religious captain often expressed his own fundamentalist beliefs about creation to his doubting but silent naturalist companion.

While in South America, Darwin became aware of the startling resemblances between the fossil remains of certain extinct animals preserved in rock strata and those living species now surviving on the earth. This evidence suggested both the historical continuity and biological unity of all life on our planet. With enlarged curiosity, Darwin unearthed the giant fossil remains of prehistoric mammals from the cliffs of Punta Alta in Argentina. Two visits to the paleontological site resulted in his discovering numerous specimens: *Mylodon, Toxodon, Megalonyx, Macrauchenia, Megatherium,* and *Scelidotherium.* At such times, he must have pondered the sobering fact of extinction.

In the same deposits, he found fossils of both giant extinct quadrupeds and existing species (e.g., the *Chlamys* scallop). This fact at least discredited the idea that global catastrophies periodically wipe out all life on earth.

For Darwin, these huge fossils of an earlier epoch represented an impressive and undeniable record of previous existence; they shed light on the appearance and disappearance of life on earth. There was a "secret" held in these remains of organic creatures from the remote past, and the ship's clever naturalist rejected their being interpreted as representing antediluvian quadrupeds that had been suddenly drowned as a result of the Noachian Deluge less than six thousand years ago (a religious position dogmatically defended by the intellectually myopic FitzRoy).

Darwin must have wondered about the origin, adaptation, survival, reproduction, distribution, and history as well as the extinction of these enormous animals of distant eons. Gradually, he realized that our planet is a grand museum containing an incomplete record of the history of life on earth. Even the sea-worn pebbles strewn across the open pampas of Argentina suggested the geological activity of prehistoric times. Yet, the arid pampas are now a land of rodents, ostriches, guanacos, and condors where once mighty dinosaurs had freely roamed.

FitzRoy was displeased and dismayed at Darwin's interpretation of rocks and fossils, which challenged a theological account of planetary history. As months passed, the captain's religious fundamentalism became even more pronounced. Giving preference to science, however, the naturalist's immersion into earth history seriously questioned the divine creation and worldwide flood of biblical beliefs. Inevitably, the conflicting viewpoints of Darwin and FitzRoy became diametrically opposed. A serious personal conflict ensued. One may safely argue that the intensifying war between open science and blind faith was personified by Darwin the naturalist and FitzRoy the religionist. Even today, this conflict of ideologies continues with no sign of resolution.

FitzRoy believed the divine creation of our material planet to have occurred in the year 4004 B.C.E. (a figure that had been determined by Archbishop James Ussher, an Anglican prelate in seventeenth-century Ireland). In sharp contrast, however, Darwin now thought that the world could be millions of years old. As a result of his new perspective, he asked: Is there throughout organic history a biological series of changes from those past forms of life to plants and animals now inhabiting our earth? The growing evidence was clearly suggesting an ongoing succession of floral and faunal types from epoch to epoch without any supernatural interventions or sudden discontinuities.

After a brief visit to the Falkland Islands in January 1833, the *Beagle* set sail for the bleak world of Tierra del Fuego at Cape Horn on the southern tip of South America. Darwin considered the naked and perhaps cannibalistic natives who barely eked out an existence in this desolate environment of ice, snow, harsh winds, and freezing waters to be the most savage and primitive people on earth. He claimed that these paleolithic and nonliterate Fuegians represented human degradation. He argued that they demonstrate how wide the distance is between savage natives and civilized human beings and wondered how humans could live under such inhospitable conditions. On at least this point, Darwin was less an anthropologist than Alfred Russel Wallace, who held to the psychic unity of humankind and took a deeper interest in the sociocultural differences among distinct human populations around the planet.

An overzealous evangelical fundamentalist, FitzRoy had hoped to convert these Fuegian bands to Christianity and thereby start a religious community of civilized natives among the so-called backward peoples of Tierra del Fuego. In fact, the captain was now returning to this rugged world three Fuegians whom he had already taken to England in 1829 to be civilized in the beliefs, practices, and values of modern Christianity. The three passengers were York Minster, Jemmy Button, and Fuegia Basket; a fourth Fuegian, Boat Memory, had died of smallpox years earlier. Unfortunately, in a short period of time after they were returned to their homeland, these three hostages reverted to their primitive ways. As a result, FitzRoy's religious plan had failed miserably; he was now understandably depressed. Unlike FitzRoy, Darwin had a far more humane tolerance for the wide range of sociocultural thoughts and behavior patterns he experienced while traveling around the world.

When at Cape Horn, he gave special attention to surveying the Straits of Magellan. During this stay, Darwin was awed by the impressive icebergs and massive glaciers of this foreboding area of the world. After overcoming a vicious storm while rounding the tip of South America, the scientific vessel next headed to Chile (Darwin was still a victim of unending seasickness).

Charles Darwin was easily persuaded to ride muleback into the Andes above Valparaiso, crossing this magnificent mountain range 13,000 feet above sea level. Reminiscent of Leonardo da Vinci in the Alps, Darwin developed a strange assemblage of ideas: geological gradualism and the web of life, along with a growing awareness of the enormous periods of time in earth history and the pervasive damages throughout the living world, clearly suggested to the young scientist a dynamic worldview that radically departed from the biblical story of a recent divine creation. In brief, to the immensity of space revealed by astronomy was now being added the immensity of time revealed by geopaleontology. For Darwin, this grasp of rock and life was intensified when he discovered petrified conifer trees in situ and fossil seashells in the peaks of the Andes. Without doubt, natural forces had slowly and continuously elevated this massive range of mountains from beneath a prehistoric ocean. Such an astonishing geological change on the surface of our malleable planet gave Darwin additional empirical evidence to support Lyell's scientific theory of global uniformitarianism; perhaps it even implied to this acute naturalist (if only unconsciously) the transformation of living things over vast periods of time and change.

Returning from the Andes, Darwin became seriously ill due to the bite of a Benchuca bug, a large black insect of the pampas that carries Chagas's disease. It may have been this particular sickness that undermined his health throughout the remainder of his life.

Next, at Valdivia and later at the towns of Concepción and Talcuhano as well as the surrounding areas, Darwin saw the devastating effects of a series of violent earthquakes. He also became more aware of the destructive forces of volcanic eruptions, hurricanes, and tidal waves. It was now very obvious to this geobiologist that our world is not solid and secure but always subject to physical changes (whether slight or major) and certainly insecure.

After a brief stay in Peru, the Beagle then set sail for the Galapagos Islands. One wonders what ideas and visions must have passed through Darwin's imagination during the many hours he paced the deck of the small survey ship as it crossed the expansive ocean beneath a night sky of bright stars in the seemingly fixed heavens. With a touch of irony, the Southern Cross dominated the horizon.

The remote and primeval Galapagos Islands, also known as the Encantadas or the Enchanted Isles, do seem to be removed from both the ravages of time and the destructive encroachment of human civilization.[9] These islands take their name from the Spanish word galapagos, i.e., giant tortoises, those huge reptiles that once thrived throughout this most extraordinary archipelago and had reminded Darwin of antediluvian monsters or creatures from another planet. This unique archipelago, isolated form the South American mainland, consists of fifteen major islands along with numerous islets, uplifts, and reefs located on the equator about 600 miles west of Ecuador in the Pacific Ocean. Sharks, swift currents, and mild winds isolate most organisms among these islands. Darwin's eerie "cradle of life" appeared as an inhospitable world of lava flows and desolate vistas: a cluster of igneous islands looking like a group of enormous chunks of lunar landscape haphazardly floating upon the swift Humboldt Current. When surrounded by mists and low clouds, these deceptively foreboding islands present a vivid scene reminiscent of our reconstruction of prehistoric times. In fact, this unusual archipelago actually represents a delightful anachronism: present-day life forms eking out an existence in a pristine environment. Fortunately, the inhabitants of these islands enjoy a subtropical climate that is cool, due to fresh tradewinds and several capricious ocean currents (especially the Humboldt Current).

The static first appearance of the Galapagos Islands is an illusion. An extremely strange but very special place on earth, this slowly changing archipelago testifies to the pervasive influences of evolutionary forces. This area of our planet is a geological hot spot of intense and frequent volcanic eruptions reflecting the convulsive and restless entrails of our earth: one sees collapsed cliffs, craters, and calderas everywhere. About fifteen years from now, wind erosion should finally cause the famous and official geological sentinel of these islands (Pinnacle Rock of Bartolome Bay, which rises 200 feet above sea level) to topple back into the ocean. In fact, this entire archipelago is almost imperceptibly drifting eastward on the Nazca Plate due to tectonic movement. The eastern uplift islands are the oldest and may sink back into the sea, while the western volcanic islands may suddenly explode. The ages of these islands range from six hundred thousand to at least three million years; the discovery of fossil invertebrates and fishes in uplifted marine limestone layers between lava flows allows geologists to date the ancient strata of these rocks. Yet, further geological building is likely to result in the eventual formation of new islands. Although creatures do differ distinctly from island to island throughout this geologically recent archipelago, they nevertheless bear a marked resemblance to those species found on the vastly older South American mainland.

At the end of summer in 1835, the HMS *Beagle* reached the Galapagos Islands; it was exactly three hundred years after their first discovery by man! From September 15 to October 20, as the ship sailed from island to island, Darwin explored and familiarized himself with this puzzling world of unusual geological structures as well as unique plants, peculiar birds, and fascinating animals. He especially marveled at the giant tortoises, land and marine iguanas, and those intriguing varieties of ground finches and mockingbirds that held the key to the understanding of and appreciation for that "mystery of mysteries"—the origin of new species on our earth. No doubt, Darwin's philosophical curiosity and scientific exactitude were put to the test. Within only five weeks, he discovered over 200 new species. Actually, in retrospect, his visit to this unique archipelago proved to be the most significant event of the entire trip: his conceptions of space, time, and change were abruptly altered, while his doubts, puzzlings, and misgivings about divine creation were once again reawakened.

From the summit of a volcano on Bartolome Island, one may experience the breathtaking panoramic view of a moonlike landscape of craters and spatter cones against the blue ocean. Populations of surface colonizer plants and lava-cacti need only soil, moisture, and oxygen to survive in this rugged and demanding volcanic environment. To the north, Tower Island is a veritable ornithologist's paradise. This isolated island supports a rich bird population that includes the mockingbird, great and magnificent frigate birds, the yellow warbler, red-footed ground dove, black lava heron, yellow crowned night heron, and the world's only nocturnal swallow gull (which is now found throughout this archipelago). On the other islands, Darwin found the penguin, pelican, waved albatross, and vermilion flycatcher as well as the flightless cormorant, red-billed tropic bird, and even the flamingo. The Galapagos microcosm also has its own hawk and two species of owl. Yet the most beloved birds are the masked, red-footed, and blue-footed boobies. Their comical appearance along with the humorous sounds and behavior patterns of the blue-footed booby have endeared this particular bird to every visitor of these islands. Today, like some other animals, the boobies often seem to be deliberately posing for a camera.

Throughout this archipelago, the traditionally recognized fourteen species of Galapagos finches of Darwin's time actually represent a far more complex situation of about fifty-two distinct overlapping types with physical and behavioral similarities or differences attributed to adaptive radiation (genetic variation and natural selection) or microevolution and extinction. These enigmatic "Darwin's finches" (as they are now called) remain a puzzle to modern evolutionists, particularly taxonomists and geneticists, because certain populations prematurely defined as species are in fact capable of interbreeding. Likewise, other groups of finches are reproductively indifferent and still others even display altruistic behavior. However, the iguanas and tortoises do exhibit clear biological variations to the point of speciation. Darwin had haphazardly collected specimens of finches among these islands, not being careful to label each according to its own particular niche. Ironically, it was FitzRoy who had made detailed notes of both these birds and their environments. Only later did Darwin himself comprehend the evolutionary significance of population variation among the species of finches throughout this archipelago.

Until recently, all the wild birds and animals had few if any predators and as a result were at first unafraid even of man. As a sad demonstration of their capacity to learn and thereby acquire a new instinct, one may now observe that some finches and iguanas already do exhibit a fear of man (as do other animals among the once-fearless wildlife of these islands).

This struggle of life on the Galapagos Islands demonstrates both the creative and destructive forces always at work in nature. The fragile ecosystems and precarious niches of this changing archipelago have been drastically altered by recently introduced plants and animals that are harmful to those endemic botanical and zoological specimens. Such harmful feral animals include Norway rats, mice, pigs, goats, dogs, cats, horses, cattle, and burros. There are already an estimated 200,000 wild goats on James Island. Of course the most dangerous species is the human animal, and over 8,000 people are now inhabiting the Galapagos Islands.

About 600,000 years ago violent eruptions formed the impressive Alcedo Volcano of Isabela Island, and Darwin was the first to describe the resultant double crater (strangely enough, FitzRoy's stone engraving marking the historic visit of the *Beagle* to this island has never been found). A population of giant tortoises now lives among the steaming fumaroles in this volcano's wet and grassy caldera valley one mile above sea level. In fact, it is possible that a tortoise still survives that was living when the young Darwin explored this captivating area in the last century! With driving curiosity, the "unfinished naturalist" probably climbed to the top of a high, rugged outcrop to obtain an awesome view of this island. Here, on the lava flows at the base of this shield volcano, Darwin came close to experiencing the beginning of our earth.

The world's only seagoing iguana is found at the Galapagos Archipelago and may even inhabit the same island with the land iguana. The black-red marine iguanas are found in seething colonies camouflaged against lava rocks while basking in the warm equatorial sun. These "imps of darkness" represent an evolutionary shift from land to water adaptation, and they feed on marine algae and seaweed to a depth of thirty-six feet. They may remain underwater for up to one hour. However, retaining a land reptile metabolism, the marine iguanas eject salt water from their systems. The larger brown-yellow land iguanas feed primarily upon the buds and spiny pancake pads that fall from the giant prickly-pear *Opuntia* cactus. No doubt, Darwin was baffled and disappointed by the scarcity of insects in this archipelago (probably due to strong winds).

The clear Galapagos waters are inhabited by fur seals, sea lions, and green turtles. There are even rays, sharks, dolphins, and whales farther out at sea. Coastlines are usually spotted with red Sally Lightfoot crabs. At Tagus Cove (Isabela Island) one finds soft-colored sponges, beautiful cup corals, and brown sea anemones.

To paraphrase the words of Darwin, at the Galapagos Islands, both in space and time, one seems to have been brought somewhat nearer to those first appearances of new beings on earth. Under the Southern Cross life has been evolving on these islands for millions of years. One may even speak of the evolution of the Galapagos Islands themselves. Yet, if incisive measures are not immediately taken, this unique archipelago, as the great naturalist once knew it, will soon no longer exist. This loss would be irreplaceable.

In retrospect, the Galapagos Islands were the catalyst that finally convinced Darwin that plant and animal species are in fact mutable. After the worldwide voyage and as a consequence of his serious investigation of this archipelago and subsequent critical reflection upon its biohistorical significance, he was transformed from an amateur naturalist into a rigorous scientist. Darwin rejected both mutually contradicting and incompatible biblical myths of divine creation as recorded in Genesis as well as the religiously motivated belief in a series of ascending and improving acts of special creation (along with the then-popular biological explanations of spontaneous generation and vitalism). He became convinced that descent with modification does take place within the biosphere and that, throughout the sweep of organic evolution, survival is the exception while extinction is the rule.

The Charles Darwin Research Station was established in 1964 at Academy Bay on Santa Cruz Island. Despite its status as a legally protected national park and wildlife preserve since 1959, the ongoing conservation efforts of this scientific station have not prevented the archipelago from being exploited for personal gain and profit. Actually, international scientists are more destructive than tourists in upsetting the normal natural conditions of the delicately balanced ecosystems and fragile niches throughout these renowned islands. In touching and banding rare birds as well as netting and branding land iguanas, scientists have altered the total behavior of these life forms and as such have even contributed to many needless deaths.

Fortunately, the Charles Darwin Research Station is successful in taking steps to restore the populations of giant tortoises. One may weigh up to 1,000 pounds and live to be 200 years old. In the past, they were killed by the thousands for their meat and the fresh water from their stomachs. Today, there may be fewer than 15,000 wild tortoises living on these islands. At the research station, tortoise eggs of this endangered species are incubated and hatched. The young tortoises are raised until they are from five to ten years old and then released only when they are capable of protecting themselves from predators in their appropriate natural habitats. This scientific station also helps eradicate the introduced pests, conserve other unique species and the differing natural environments of these islands, and maintain *Beagle III* for essential patrolling and communication throughout the Galapagos Islands.

Leaving this fascinating archipelago, the *Beagle* was now en route to Tahiti, New Zealand, Australia, Tasmania, and several islands in the Indian Ocean (including the Cocos-Keeling Island). Darwin observed the vividly tatooed Maoris of New Zealand and later, in Australia, was fascinated by the duck-billed platypus as well as the kangaroo, other marsupials, and the aborigines of this geographically isolated part of the world. During this time, he also developed a correct theory to explain the sequential origin of atolls or lagoon islands of coral reefs (a view that differed from Lyell's account). At Cape Town, Darwin met the astronomers Sir John Herschel and Sir Thomas Maclear. However, Darwin never explored the continent of Africa; this area would later yield that fossil evidence so crucial to substantiating human evolution in terms of science and reason. Then, after a brief return to Brazil, the scientific survey headed back to England; nearly five years had elapsed since the beginning of its great and exhilarating voyage of discovery.

By the time Charles Darwin arrived at Plymouth Sound on 2 October 1836, ending a global journey that covered about 40,000 miles, his own invaluable collections and scientific letters had already earned him a reputation as one of the greatest naturalists of the day. Clearly, the voyage of the HMS *Beagle* proved to be the most important event in Darwin's life; it determined his entire career in the natural sciences. He had glimpsed the gradual emergence, creative force, staggering diversity, historical continuity, and essential unity of all life on earth: nature is an ongoing theater of birth, change, struggle, and death.

Only in retrospect, after returning to England, did Darwin appreciate the significance of his five weeks of discovery at the Galapagos Islands. The marvelous display of creative adaptations throughout this archipelago now represented to the naturalist a living laboratory as an isolated microcosm in nature that demonstrates the organic results of evolutionary forces. The complexity and diversity of life could be easily seen in those populations of giant tortoises whose shells varied in size and shape from island to island, those groups of land and sea iguanas that varied in coloration from environment to environment, and especially among the fourteen species of finches that varied in the size and shape of their beaks. This curious but successful adaptive graduation among the beaks of these finches is directly related to their different diets and feeding habits within a range of specific ecological niches throughout the Galapagos Archipelago.

Darwin was the first naturalist to study seriously rocks, fossils, plants, insects, birds, and animals from a worldwide perspective (not even Alexander von Humboldt or Alfred Russel Wallace came close to his extensive voyage and intensive research). Actually, Darwin had seen more of the world in terms of science and natural philosophy than any other human being up to his time. As a result, he could no longer ignore the evolution and adaptive radiation of life on earth. Interestingly enough, this naturalist never made another major trip during his lifetime. Today, geological structures and biological species throughout the world bear Darwin's name and thereby attest to his global trip of scientific discovery and its lasting significance, e.g., Mount Darwin at Tierra del Fuego, the rare ostrich *Rhea darwini* of the pampas in South America, and, especially appropriate, a giant tortoise subspecies *Geochelona elephantopus darwini* of the Galapagos Islands.

Darwin probably did not become an evolutionist until March 1837, about six months after the historic voyage of the *Beagle* had ended. At last, this naturalist had come to the inescapable conclusion that species are mutable. Interestingly enough, however, for many years he avoided using the word evolution because of its general use to denote process as progress (teleology).

Without doubt, Darwin had assembled the most unique collection of scientific specimens in the world. Yet, only after reflecting on all this data and the experiences of a five-year global journey, did he finally recognize and accept the evolutionary perspective that had slowly emerged in his creative genius. Darwin had finally come to reject totally both the Mosaic story of divine creation and the Aristotelian account of earth history as a monotonous continuance of the same without creation or evolution or even extinction. Likewise, vitalism and especially those fantastic claims of spontaneous generation must have now seemed truly ludicrous to him: maggots from rotting apples, tadpoles from fallen leaves, rats from decaying wheat, mice from old clothes, and crocodiles from river mud!

Fortunately, Darwin as scientist had had the time to ponder the larger consequences of a dynamic overview of earth history and organic development. Consequently, he accepted evolution as a mechanist and materialist.

Free from all preconceptions of an eternal nature as a static hierarchy of fixed types in which both pervasive changes and frequent extinctions were inconceivable, Darwin was able to recognize the evidence for and patterns in biological evolution. He became preoccupied with the overwhelming diversity of life on earth as a creative unity of mutable plant and animal species. As a true scientist, Darwin was rapidly developing a comprehensive and intelligible vision of the organic history of our entire planet (in religion, however, disbelief crept over him at a slow rate but was eventually to become complete as silent atheism).

In May 1837, Darwin started to write his first of several early notebooks on his hypothesis about the transmutation of species (1837–1839). He filled pages with facts, observations, and critical opinions: the theory of evolution was clear in his mind, although the working out of its mechanisms and implications still awaited completion. Later, in July, Darwin first envisioned organic history as an irregularly branched tree of life and, as a result, had developed the principle of divergence before Alfred Russel Wallace.

In October 1838, Darwin happened by chance to read casually but seriously Thomas Robert Malthus's monograph *An Essay on the Principle of Population* (1798).[10] The Malthusian treatise on population growth claimed as follows: animal numbers (including humans) have a potential to increase at a geometrical rate, whereas the food resources from plants have an optimum potential to increase at only an arithmetical rate. Therefore, since geometric progression outstrips arithmetic growth (and despite the checks on endless human population growth such as disease, famine, and war), this English economist and social philosopher argued that there will inevitably be a pervasive struggle for existence in the living world. Malthus had also emphasized the sober facts of the human condition. Suddenly, upon reading this first of two studies, Darwin intuited the explanatory mechanism of natural selection as the primary driving force of biological evolution, with random chance variation and severe blind competition playing their roles. This discrepancy between the rate of increase in means of subsistence and the number of animals guaranteed an unending ruthless struggle for existence in which only a favored few organisms are destined to survive and thereby perpetuate their kind within the limited economy of changing nature. Likewise, the social theory of the political economist and philosopher Adam Smith (1723–1790) also influenced Darwin's interpretation of life on this planet.

To Darwin, any beneficial-because-it-is-useful slight variation or major modification enhancing the adaptation of an organism to its own niche will provide the organism with a survival advantage that will (more likely than not) allow for its producing offspring to insure future generations. Therefore, Malthus's schema was the theoretical catalyst that gave to Darwin the explanatory mechanism of natural selection; his contemporary Herbert Spencer had already referred to this selective principle as the "survival of the fittest" to adapt and reproduce (among others, both Erasmus Darwin and Comte de Buffon had also glimpsed this struggle for existence).

In the human realm, the most important argument against Malthus is that he had completely underestimated the gain in food productivity made possible by the Industrial Revolution (not to mention the coming of birth control procedures, to which he objected, having favored voluntary moral restraint). Be that as it may, Malthus did give to Darwin his major causal principle or explanatory mechanism of natural selection to account in part for the origin of species. For Darwin as brilliant biologist, the task was to synthesize facts and concepts and perspectives in such a way as to demonstrate clearly the truth of organic evolution on earth.

Darwin's scientific theory of biological evolution is essentially a synthesis of Lyell and Malthus: Lyellian historical geology provided the vast temporal framework of slow changes and natural causes, while the Malthusian principle of population growth offered the major causal explanation of natural selection that crystalized Darwin's thoughts on descent with modification, i.e., the transformation of species understood in terms of science and reason. Briefly, geological change implied biological evolution. The pervasive competition in life and ruthless struggle for existence ensure that only the strongest and most vigorous organisms in populations will adapt, survive, and reproduce. Throughout time, there is the survival of the fittest within changing environments as a result of some individuals having by chance favorable or beneficial-because-useful mutations in physical characteristics and/or behavioral patterns. Since habitats are always subject to change, most organisms must also change to adapt and thereby survive to reproduce or face extinction. Furthermore, these useful variations are inherited by the offspring. With its explanatory mechanism of natural selection, the Darwinian evolutionary framework was at least a plausible working theory to be taken seriously by scientists, natural philosophers, and enlightened theologians.

In 1838, Darwin was elected secretary of the Geological Society of London (a position he held until 1841) and became a ten-year member of the Philosophical Club. During his later years, however, he was never tied to any professional or social duties.

On 29 January 1839, Charles Darwin married his first cousin Emma Wedgwood; the Darwin-Wedgwood ancestry represented the Utilitarian and Whig traditions. Charles was a loving father and devoted husband, but Emma did not share his scientific preoccupations and religious doubts. In August of that same year, Darwin's first book, *The Voyage of the Beagle*, was published.[11] Written in a lucid literary style, it is the most readable of his naturalist volumes (the second edition was dedicated to his friend Sir Charles Lyell). This journal of research was published independently of FitzRoy's own manuscript, causing more ill feelings between the two.

Eventually, in 1842, the Darwins left London to settle at Down House in the village of Downe in the Kent countryside about twenty miles southeast of the city.[12] This move proved to be more suitable to Darwin's health and work. Here, the young couple found refuge from the endless demands of a sometimes thoughtless world. They lived a remote and secluded life of peace and security, raising a large family that provided them with comfort and happiness.

A timid and gentle genius, as well as a semi-invalid and quasi-recluse, Charles Darwin remained for the rest of his life more or less isolated from the scientific community. He limited his personal associations to only a few intimate friends (e.g., Lyell, Hooker, and Huxley) and, at least on one occasion, a visit from FitzRoy. There was now ample time to consider the evolutionary implications of his panoramic glimpses into natural history during the five-year global voyage of scientific discovery: the teeming life of the primeval jungles of Brazil, the gigantic fossil mammals from a cliff in Argentina, the vast grassy plains of Patagonia, the primitive natives of Tierra del Fuego, the awesome geology of the Andes, that strange zoology of the Galapagos, the unique species of Australia, and that impressive coral reef of the Cocos-Keeling Island.

Darwin the naturalist was always content to work on his own estate, carrying out experiments in his gardens and greenhouses and the surrounding fields. He worked slowly but steadily, devoting only four hours each day to writing manuscripts or completing research on a wide range of various subjects in the natural sciences; his three books on geology were published during this time.[13]

Darwin settled into a comfortable routine. Each morning he took a solitary stroll down his private path, which became known as the sandwalk, and afterwards devoted part of the day to science and family life. Free from menial tasks and mundane responsibilities, and having the advantage of being independently wealthy, he accomplished an enormous amount of work, even though chronic illness limited the time he devoted to serious writing or scientific research. For the most part, Darwin seems to have enjoyed his intellectual pursuits in quiet isolation. At this time, the world at large was still unaware of the evolutionary theory emerging in this creative naturalist. Later, however, the impact of evolution would devastate the beliefs and values of the Christian worldview.

Although he was not a professional scientist, Darwin did search for the truth. This now "finished naturalist" had time methodically to examine in minute detail his painstaking research, which yielded an impressive body of empirical evidence. He worked cautiously and slowly, paying scrupulous attention to his investigations and recording everything in copious notes. There was also time to become involved in worldwide correspondence with the naturalists who shared his controversial ideas on evolution and accepted his dynamic vision of natural history.

Darwin devoted years to empirical research that covered a broad array of subjects. His geological writings contributed to understanding the origin of coral formations, volcanic islands, and lagoons within a historical framework. As a botanist, he wrote about climbing, twining, flowering, and insectivorous plants. He was particularly interested in the reproductive methods of orchids and experimented with buttercups, snapdragons, and sweet-peas.[14] Darwin the biologist concentrated on the remarkable activities and profound values of the earthworm, oddly enough his favorite animal.

Darwin took careful notice of the appearance of varieties in cultivated plants and domesticated animals due to human intervention or artificial selection, e.g., cabbages and cauliflowers as well as dogs, sheep, horses, cattle, and especially pigeons. Attention was also paid to ducks, goldfish, silk moths, donkeys, rabbits, canaries, bumblebees, and even the rhinoceros and the armadillo. He later devoted himself to a comparative study of the expression of the emotions in higher animals (including humans) and, lastly, even investigated the formation of vegetable mold. His extensive research contributed to ecology, ethology, psychology, and sexual biology.

Darwin spent eight years of meticulous research (1846–1854) analyzing and describing variations among barnacles (subclass *Cirripedia*). This tedious study demonstrated the variability of a species in nature and resulted in a four-volume monograph on these invertebrates: two volumes on living barnacles appeared in 1851, and two volumes on fossil barnacles appeared in 1854. This intensive and definitive work was based upon the careful dissection and comparative morphology of over 10,000 specimens of his "beloved barnacles" (as he called them)! In retrospect, after completing this arduous task, Darwin did admit that perhaps he had devoted too much time and effort on such a specialized area of scientific inquiry. One may suspect that this long-range study of barnacles was deliberately self-imposed to keep the energetic naturalist away from having to grapple with his mechanistic theory of evolution and its materialist implications. Nevertheless, this project made him an astute taxonomist (to say the least).

Concerning artificial selection, Darwin took seriously the experiments of gardeners and breeders. He became more and more aware of the fact that the appearances of varieties due to artificial selection (human choices) may perpetuate some characteristics and eliminate others as a result of the deliberate crossing of particular cultivated plants and the selection of partners in the mating of domesticated animals. It dawned on him that human beings are now active instruments in the further evolution of life on earth (including the future development of their own species). There was, for Darwin, an essential connection between such practical experiments on the subspecies level of microevolution and the scientific theory of organic evolution, i.e., the mutability of plant and animal forms on the species level as well as those higher taxonomic categories of macroevolution. In fact, he referred to a variety as an incipient species; yet, his own experiments in sexual reproduction lacked both the use of mathematics and that rigorous control required for appreciating the results in terms of the basic principles of heredity. Consequently, Darwin asked: If humans can create new varieties of plants and animals through artificial selection, then could not nature itself over vast periods of time evolve new species of flora and fauna as a result of natural selection acting on random-but-beneficial-because-adaptively-useful and inheritable chance variations within populations?

Before 1842, Darwin had shared only with the botanist Sir Joseph Dalton Hooker of the Royal Botanic Gardens at Kew in London the larger consequences of his insights into a gradualist theory of descent with modification and its explanatory mechanism of natural selection: an infinitely slow evolution of life on earth challenged the conceptual foundation of early biology, which at that time was grounded in the idea that all species are forever immutable in nature.

In 1842, Darwin wrote in pencil a 35-page abstract that briefly sketched his theory of evolution. Then, in 1844, he enlarged this preliminary outline into a 230-page manuscript. Neither work was published. Still plagued by illness and anxiety, Darwin took an intriguing precautionary measure: in his will of 5 July 1844, he left both instructions and the financial means so that his scientific theory of biological evolution would be published by his wife, Emma, should he unexpectedly die before the manuscript was seen in print. One may even argue that Darwin actually had no serious intention of publishing a multivolume work on the theory of evolution during his own lifetime.

Also in 1844, a book titled *Vestiges of the Natural History of Creation* written by an unknown author appeared in print; it was the most significant work on the theory of evolution since Lamarck's *Zoological Philosophy* (1809). *Vestiges* explained organic history in terms of two impulses: one natural for adapation and the other divine for evolution itself. Charles Darwin did not take seriously the religious and vitalist viewpoints of *Vestiges*, which he correctly guessed to have been written by the Scottish encyclopedist Robert Chambers (1802-1871). As did Lamarck, Chambers discovered evolution from his serious study of the geological column and its fossil record.

For twenty years, Darwin reflected upon the creative development of life on earth and the questions it raised. During this time, one may even speak of the evolution of his own views on organic history. It seems strange that Darwin did remain so indifferent to the philosophical and theological issues surrounding the ultimate nature of this material cosmos and the origin of life within it.[15] His speculations always remained within the realm of science and reason. Nevertheless, as a biologist, he asked: Could evolution account for the human eye and hand, insect sterility, and instinctive behavior (not to mention the giraffe, mistletoe, woodpecker, peacock's tail, flatfishes, the electric organs of fishes, and the whalebone whale)?

What about the Irish elk (an excellent example of allometric principles* at work)?

While preoccupied with biological research and his poor health, Darwin lost his appreciation for poetry and music as well as for politics and religion (he found anything artificial repugnant), but he did continue to read widely in science and literature.[16] His chronic illness remained a puzzlement.[17] He suffered from boils, eczema, nausea, dizziness, shivering, giddiness, eye trouble, violent vomiting, palpitations of the heart, attacks of stomach discomfort and nervous indigestion, and, understandably, periods of insomnia and depression. These afflictions caused physical weakness along with mental fatigue.

Darwin's illness may have been due to long seasickness or, more likely, to the bite of a poisonous Benchuca bug, *Triatoma infestans*. Perhaps these afflictions were merely psychosomatic in origin (possibly due to guilt feelings resulting from the conflict between evolutionary science and traditional religion). Darwin resorted to several methods, some bizarre, to treat his enigmatic illness: snuff, chalk, ice packs, damp sheets, frequent showers, ozonized water, the taking of some mineral acids, saline solutions of the water-cure at a hydropathy spa, and even both brass and zinc wires moistened with vinegar and wrapped around his neck and waist!

On 13 November 1847, while Charles was away for hydropathic treatment, his father died. On 24 April 1851 the Darwins' beloved young daughter Annie died of a fever at Malvern. After her death, Charles totally lost his faith in Christianity as a divine revelation.

Also in 1851, Herbert Spencer published his book, *Social Statics*, that included the explanatory phrase "survival of the fittest" (which Charles Darwin himself later used).

Early in 1856, Lyell had advised Darwin to write out his own views on evolution in book form. Finally, in May, Darwin actually began to write his projected multivolume work on organic history to be titled *Natural Selection* (clearly the theory of evolution was "in the air"). He was also being urged by close friends (notably Lyell, Hooker, and his older brother Erasmus) to publish his growing manuscript.[18] Yet anticipating rejection, persecution, and controversy, the ever-cautious Darwin continued to delay offering his unique manuscript to a publisher. Indeed, Darwin seems to have deliberately avoided the whole issue. No doubt, he feared the consequences of philosophical materialism that his own theory of organic evolution implied.

In general, Darwin may have withheld the publication of his theory

---

*Allometry is the study of those changes in the proportions of various parts of an organism as a consequence of differential growth and natural selection.

of evolution because of the scandal surrounding the appearance of *Vestiges of the Natural History of Creation* (1844), as well as the metaphysical nonsense found in Lamarck's *Zoological Philosophy* (1809). Darwin also anticipated the negative reverberations that his own disturbing ideas and their disquieting implications would have in scientific, philosophical, and theological circles. Nevertheless, his steadfast reluctance to publish anything on evolution will always remain a great puzzlement.

On 18 June 1858, Darwin received an envelope postmarked from the small island of Ternate in Malaysia; it contained both a letter and manuscript from Alfred Russel Wallace. The thunderbolt had struck. After twenty years, Darwin abruptly learned that his theory of organic evolution with its explanatory mechanism of natural selection had been independently discovered and summarized by another naturalist on the other side of the globe![19] His intimate scientific friends had even anticipated the possibility of such an event. Furthermore, the explanatory mechanism of natural selection had already been clearly glimpsed by James Hutton, William Wells, and Patrick Matthew.

Wallace first desired Darwin to comment on his 4,000-word unpublished essay of 15 pages titled "On the Tendency of Varieties to Depart Indefinitely From the Original Type" (this essay would then be sent by Darwin to Lyell, who was now recognized as the dean of natural science in Victorian England). Darwin was astonished to find that the ideas and views of Wallace were strikingly similar to his own. Clearly, this whole situation of parallel discovery in science was an incredible coincidence.

There were other remarkable similarities between Darwin and Wallace: each was an English naturalist with interests in both geology and biology who had explored South America; both had visited a unique archipelago (Galapagos and Malay, respectively); had studied orchids and collected beetles; both revered life while doing original research on the barnacle and the butterfly, respectively; both read Paley, Humboldt, Lyell, and Malthus; each enjoyed reflecting on nature in solitude; both accepted the scientific theory of biological evolution as a true framework of organic history; and each had independently arrived at the explanatory mechanism of natural selection as the key force to account for the survival, divergence, and extinction of species throughout the evolution of life on earth.

Darwin had intuited the concept of natural selection as early as 1838, but never published his views on this subject (neither the spring 1842 abstract nor the expanded summer 1844 version had appeared in print). On 5 September 1857, Darwin had referred to his theory of evolution in a letter he sent to the famous botanist Asa Gray in America. However, Darwin had not established his priority in the scientific literature. On the other hand, Wallace had already presented an introduction to his theory of evolution in two published works: *The Sarawak Law* (1855) and the *Ternate Essay* (1858).[20]

As a consequence, both Hooker and Lyell sought to establish Darwin's priority. At the Linnean Society meeting in London on the soporific evening of 1 July 1858, they delivered the two papers by Darwin and Wallace under the dry joint title "On the Tendency of Species to Form Varieties; and On the Perpetuation of Varieties and Species by Natural Means of Selection" (on 20 August, the papers appeared in the *Journal of the Proceedings of the Linnean Society*). In fact, neither Darwin nor Wallace was present. The society agreed that priority should be given to Darwin as the scientific father of the biological theory of organic evolution primarily by means of natural selection—it helps to have influential friends in high places. Although both naturalists had independently recognized the significance of natural selection as the key mechanism in biological evolution, it must be emphasized that Darwin had a far greater body of documented empirical evidence and a much wider range of both experience and experimentation than Wallace had ever accumulated or could ever claim. A meritorious and respectable scientist in his own right as well as an honest and grateful man, Alfred Russel Wallace loyally conceded the priority to his fellow naturalist Charles Robert Darwin at Down House in Kent. (Of course, Wallace never had Darwin's academic, social, and financial advantages.)

It has recently been suggested that Hooker and Lyell had plotted together against Wallace in order to establish for their friend Darwin a clear victory as regards the question of priority concerning the theory of evolution by means of natural selection.[21] Be that as it may, the Linnean Society meeting had been a nonevent. Clearly, the value and significance of this whole affair went unappreciated by the scientific community. Likewise, Wallace himself seems to have never resented the priority having been given to Darwin for discovering the key mechanism of natural selection to explain biological evolution (unfortunately, the former was always an unlucky man).

As incredibly similar as these two naturalists were in 1858, in later years their interpretations of evolution increasingly diverged from each other to the point of eventual diametrical opposition: Darwin grounded his theory in mechanistic materialism, whereas Wallace turned to teleological spiritualism. Darwin was appalled at Wallace's turn from naturalism to spiritualism; the latter argued that the essential uniqueness of the human being could not be accounted for merely through the slow accumulation of slight beneficial variations over long periods of time due only to natural selection. The shy and gentle Darwin always viewed life, including the human animal, from a strictly naturalist attitude; yet, he never defended his own theory of organic evolution in public or in print.

On 24 November 1859, Charles Darwin's major work *On the Origin of Species* appeared in print; all 1,250 copies of the first edition were sold on the day of their publication.[22] It should be noted that its great naturalist author never did finish a detailed multivolume treatise on organic evolution. This "premature" book is the so-called abstract version of his proposed "big book" and it focused on that mystery of mysteries, the appearance of new forms of life on earth. Its general thesis is that species are not ("It is like confessing a murder") immutable types in material nature: structures and functions are not fixed within the flux of organic history. Either a population adapts and survives and reproduces in its changing environment, or extinction is the inevitable outcome. Briefly, the evolutionary imperative is change or perish!

Thomas Henry Huxley wrote a favorable review of the *Origin* volume in the influential London *Times* (December 26, 1859), thereby helping to ensure that the significance of Darwin's scientific theory of organic evolution was not lost in the flood of severe opposition. Even so, evolutionism has never escaped controversy from both inside and outside of the scientific community.

The *Origin* shocked the world, generating a great storm of bitter controversy.[23] It is one long, impressive argument for the theory of evolution, although originally intended to be merely a preliminary sketch of this scientific framework. Actually, this book is the theoretical core of Darwin's ten major works. What the geologist Lyell had done for inorganic nature (expanding and surpassing Hutton), the biologist Darwin now did for the living world (modifying and improving Lamarck).

There even appeared an essay in Hebrew, whose author attempted to show that the theory of evolution is, after all, already contained in the Old Testament! Of course, scientific Darwinism is incompatible with any strict and literal interpretation put upon the Holy Bible. In sharp contrast to the creation myth in Genesis, Darwin's *Origin* is an evidential masterpiece. This volume is a remarkable accomplishment in science and logic, free from theology and metaphysics. Yet, even today, outrage and furor over the theory of evolution continue in some important and influential religious circles as well as philosophical schools of myopic thought.

The publication of *On the Origin of Species* was a historical event; it has been judged by many to be the most important single scientific book of the last century. Its comprehensive and intelligible (although incomplete) explanation for the origin of new species through the process of organic evolution rests upon a logical arrangement of the documented facts and relationships in the relevant natural sciences. Darwin supported his theory with an impressive body of empirical evidence: geopaleontology, biogeography (especially isolation), ecology, taxonomy, comparative embryology and morphology (e.g., homologies), vestigial features or rudimentary organs (e.g.,

"Darwin's point" of the human ear, hair on the body, the nictitating membrane, wisdom teeth, various muscles, male mammilary glands or mammae, veriform appendix, and the immovable os coccyx vertebra),[24] and ongoing experiments in the artificial selection by human intervention of cultivated plants and domesticated animals. Darwin paid special attention to the appearance of varieties among domesticated dogs, sheep, fowl, cattle, horses, and especially pigeons. He was also aware of correlated growth and variation in animals.

Darwin argued that varieties are incipient species. He reasoned that if the accumulation of slight favorable-because-useful variations under domestication as the result of artificial selection (human intervention) produces new varieties, then chance variations in nature over vast periods of time will eventually produce new species and even higher taxa. In nature, a series of slight differences over countless generations in one species may add up to a major change resulting in a new species in its own right. As such, an endless adaptive radiation of life pervades the complex web of this organic world. The severe competition among organisms for survival in changing environments gives rise to natural selection. Today, this survival of the fittest is referred to as differential adaptation, survival, and reproduction (fecundity being an essential factor).

Darwin envisioned the ever-branching and ever-growing tree or coral of life emerging from a trunk of common origin (recognizing that although the birth of species and higher taxa throughout organic history is replete with creativity, in reality the extinction of species and higher taxa is the rule from a geobiological perspective of events).

It was clear to Darwin that there is a complex historical relationship between that vast array of forms from one or several first types of life on our primitive earth to the existing prosimians, monkeys, apes, and humans. He argued that all plants and all animals throughout planetary space and time are related in groups subordinate to groups as a result of the branching tree of organic history. The human mind reels at the unimaginable expanse of time represented by the creative evolution of all life on our planet.

Darwin went further than any other evolutionist before him. Unlike Lamarck and Chambers and Wallace, and all the other evolutionists before the publication of the *Origin*, Darwin had amassed an overwhelming body of empirical evidence to provide factual support for his theory of evolution. Even as early as October 1838, he had conceived of the materialist explanatory mechanism of natural selection to explain in part the ongoing process of descent with modification (although this major causal principle to account for organic evolution is not a totally adequate interpretation of ongoing biological history). Despite all this irresistible evidence and convincing logic, the Darwinian theory of organic evolution is a scientific framework of history above and beyond direct perceptual experience and, as a result, generates questions concerning its assumptions and implications (especially interpretations regarding the place of humankind within natural history). Anyone critically reading the *Origin* will easily see that this evolutionary theory must be extended to account also for the emergence and history of our own human zoological group.

Darwin's overwhelming achievement in the *Origin* was the drawing of true statements having general application after having meticulously analyzed a vast amount of natural facts and a wide range of personal experiences. The importance of these empirical inferences is beyond dispute. Even in light of scientific advancements since its publication, the *Origin* still presents an astoundingly accurate general picture of the evolutionary process as applied to life on earth. Modifications due primarily to modern genetics, population studies, new discoveries in paleontology, and far more sophisticated dating techniques have not discredited its overall validity. With the towering genius of Darwin, evolution ceased to be merely a rational speculation on nature or a philosophical pronouncement on the scheme of things. Instead, evolution became an all-embracing naturalistic theory and scientific fact of organic history with glorious descriptive, explanatory, predictive, and exploratory powers. Not since Galileo had a scientist so altered our view of this material world and the place of humankind within it.

Within the sweeping structure of his theory of evolution, Charles Darwin had not as yet included the human animal. Conspicuously absent in the *Origin* is any application of his theory and its mechanism to account for the emergence and history of humankind. In this volume, he made very few references to apes or monkeys and only one cryptic sentence concerning our own species. The first edition merely contained what must be the scientific understatement of all time: "Light will be thrown on the origin of man and his history."

Although Darwin had been strangely reluctant to discuss human evolution in print, it was obvious to any intelligent reader that the same theory and mechanism that had been applied to all the rest of organic history could easily be extended to include our own zoological group as well. In fact, it was the implications that evolution held for the human species in natural history that actually caused the bitter controversy that quickly surrounded this scientific viewpoint. Darwin was attacked and ridiculed by rigid religionists as well as many scientists and philosophers. He was also contemptuously criticized and bitterly opposed in the press. There were even deliberate attempts maliciously to discredit Darwinian evolution through absurd misrepresentations of its facts and ideas, particularly as they apply to the glaring similarities among man and the five apes (especially the chimpanzee and gorilla of Africa).

Interestingly enough, Darwin seems to have paid little attention to his scientific and philosophical predecessors.[25] Not until April of 1861 did he include "An Historical Sketch" to the revised third edition of his *Origin* volume, which went through six editions in all. The author was always unkind to Lamarck and gave no credit to Edward Blyth, who had actually used the mechanism of natural selection but to account only for the appearance of varieties within seemingly fixed species. As a result, Blyth's influence on Darwin remains a moot point.

In 1862, treating our planet as a cooling body, the mathematician and physicist William Thompson (later Lord Kelvin) calculated the age of our earth to be probably about 20 million years. If true, this time framework was far too short to allow for the slow and gradual evolution of all life by means of natural selection. Darwin himself thought our planet to be at least 600 million years old (modern science has determined its age to be about 5 billion years). Therefore, to account for the creative advance of organic history in this relatively short geological time span, the great biologist reluctantly developed and incorporated a Lamarckian explanation of inheritance: this was his provisional hypothesis of pangenesis with its gemmules, use and disuse, environmental causality, inheritance of acquired characteristics, and "blending" effect.

Although condemned from pulpit and press as a dangerous discovery in conflict with entrenched natural philosophy and traditional sacred theology, Darwin's *Origin* was without a doubt a popular success. The envious anatomist Richard Owen and the spiteful religionist Samuel Wilberforce ("Soapy Sam") were plotting to trounce the Darwinian theory of organic evolution both in public and in print.

Clearly, the crucial turning point in favor of biological evolution came as a direct result of the Huxley-Wilberforce open discussion on 30 June 1860. This epic confrontation occurred at the annual meeting of the British Association for the Advancement of Science held in Oxford's University Museum Library (nearly one thousand people were present). Professor John Stevens Henslow presided over this four-hour gathering (Charles Darwin remained at Down House). The vertebrate paleontologist and natural philosopher Thomas Henry Huxley defended the theory of evolution against the uninformed and irrelevant arguments of Samuel Wilberforce, Bishop of Oxford. Huxley's polite but pointed rebuttal to this religionist exposed the theoretically biased and scientifically faulty opinions of the strict biblical literalist. Huxley was always delighted to have an opportunity to defend Darwin's scientific theory of organic history. His learned and eloquent defense of biological evolution at Oxford, based on facts and logic, gave Darwin's theory its first major victory over biblical dogmatism and intellectual duplicity.

After being a member of the British Parliament and then the governor of New Zealand (both unsuccessful positions), a distressed Rear Admiral Robert FitzRoy had attended this public meeting to help support the traditional religionist sector. However, he failed miserably in his attempt to denounce Darwinism by appealing to the authority of the Holy Bible (FitzRoy certainly regretted having welcomed the young Darwin aboard the HMS *Beagle* as its unpaid naturalist and his traveling companion). It was now evident that the gentle genius of Down House would triumph after all—a clear victory for science and reason over dogmatism and superstition.

Darwin was at first greatly abused, being held to have both ridiculed traditional religion with heretical ideas and dishonored natural philosophy with fanciful assumptions.[26] His *Origin* had startled almost everyone who read it, and its theory of evolution was bitterly attacked by theologians as well as scientists, e.g., Owen in England and Agassiz in America. Nevertheless, Darwin's intellectual reputation remained secure and his materialist worldview weathered the storm.

As "Darwin's bulldog" in England, naturalist Thomas Henry Huxley remained the evolutionist's ardent defender, along with support from Joseph Dalton Hooker, Charles Kingsley, Sir John Lubbock, Alfred Russel Wallace, and W. B. Carpenter. In Germany, scientist and philosopher Ernst Haeckel did not hesitate to defend the Darwinian worldview (this monist-pantheist was referred to as "Germany's Darwin" while, in turn, he himself referred to Darwin as the "Newton of Biology"). The American botanist and taxonomist Asa Gray was an early supporter of Darwin, whose ideas were also popularized and defended by the paleontologist Vladimir Kovalevskii in Russia and, soon afterwards, by Pierre Paul Broca and Ernest Renan in Europe as well as Charles S. Peirce and Chauncey Wright in America. It must be noted, however, that Herbert Spencer had adhered to the idea of evolution since 1852.

To Darwin, Sir Charles Lyell's *Geological Evidences of the Antiquity of Man* (1863) was a disappointment. Lyell vacillated between accepting the mutability of species and rejecting the theory of organic evolution. He was even reluctant to acknowledge the growing fossil evidence for human evolution. In fact, Lyell, a strong theist, encouraged Darwin to refer to God as the creator of life; a concession made by Darwin in the last paragraph of the later editions of his *Origin* in order to diminish the controversy. Lyell even claimed that questions about the origin of our earth are outside the scope of geology.

For Huxley, the "question of questions" concerned the origin and natural history of our own zoological group. In his book *Evidence as to Man's Place in Nature* (1863), he was quick to extend Darwin's theory of evolution to account for the emergence and development of the human animal. Actually, Huxley now found the fact of evolution so obvious that he remarked how extremely stupid it was that he himself had not thought of the explanatory mechanism of natural selection before Darwin and Wallace had done so. As an agnostic (he coined the term) in science, Huxley differentiated between nature "red in tooth and claw" and the realm of human ethics, a sharp distinction that the evolutionist Peter A. Kropotkin rejected.

On 30 November 1864, Charles Darwin was awarded the Copley Medal from the Royal Society of London, England's highest honor for outstanding achievement in science (Darwin was a fellow of this distinguished group).

Five months later, on 30 April 1865, the schizophrenic Vice Admiral Robert FitzRoy, still a depressed believer, committed suicide by slashing his throat. This tragic act could have been due, at least in part, to the guilt FitzRoy felt in his having played a direct role in providing the young Charles Darwin with that unique opportunity to develop his theory of evolution. No doubt, the caring Darwin probably felt somewhat responsible for this unfortunate turn of events.

Foreshadowed in the scientific studies of plant sexuality and experiments in plant hybridization by the German botanist Joseph Gottlieb Koelreuter (1722–1806) but superseding them, the Czech monk-priest Gregor Johann Mendel (1822–1884) worked obscurely in a garden of the provincial Augustinian monastery in Brno, Moravia. There, as a result of his original research involving the artificial fertilization of common garden pea plants (genus *Pisum*), Mendel discovered the four basic principles of heredity that established the general foundation for our modern science of genetics. In doing so, he may be referred to as the first mathematical biologist. Mendel sought to understand the connection between organic inheritance and physical appearance. Through his rigorously controlled experiments and the use of arithmetic symbols, Mendel observed the transmission of specific pairs of characteristics for seven different physical traits in individual plants from generation to generation. His resultant four contributions are (1) a particulate theory grounded in the existence of discrete and indivisible but invisible hereditary units in those gametes that are clearly distinct from the somatic cells of an organism; (2) the distinction between dominant and recessive characteristics for the same physical trait; (3) the principle of segregation; and (4) the principle of independent assortment.

In 1865, after eight years of research, Mendel presented his results in two lectures to the Natural History Society of Brno, and then published them in his only monograph *Experiments in Plant Hybridization* (1866). His pioneering work in mathematical genetics surpassed all those speculations of any Lamarckian hypothesis of heredity. It is now maintained that Mendel had deliberately manipulated the results of his pea experiments to obtain (or at least closely approach) the whole number ratios desired in the physical expressions of characteristics in the offspring. Be that as it may, Mendel was appointed the abbot of his monastery in the spring of 1868 and later, in 1871, he abandoned altogether his hybrid experiments because of poor health and a hectic schedule. Unfortunately, his discoveries went unrecognized by the other monks as well as unappreciated by the scientific community. Furthermore, even Mendel himself failed to comprehend the awesome consequences of his unique research; actually it paved the way for the mutation theory, the discovery of the DNA molecule, and the science of genetic engineering.

Mendel's valuable manuscript with its crucial discoveries went unnoticed until 1900, when two botanists in Europe independently discovered this work: Hugo DeVries of Holland and Karl Erich Correns of Germany, as well as, shortly afterwards, the zoologists William Bateson in Britain and Lucien Cuénot in France.

Later, important discoveries were made by T. H. Morgan, C. D. Darlington, and C. H. Waddington as well as the controversial zoologist Richard B. Goldschmidt, who presented the "hopeful monster" hypothesis in his major book titled *The Material Basis of Evolution* (1940).

One thousand miles separated Darwin from Mendel. Although the monk knew of the great naturalist and his theory of evolution, Darwin was unaware of Mendel and his contributions to understanding and appreciating heredity. Mendel had even sent a copy of his monograph to Darwin, who never opened this package to read the enclosed publication. Had Darwin done so, biology may have had the synthetic theory of organic evolution many decades earlier. Actually, with a touch of irony, the Augustinian abbot was never an evolutionist.

Darwin continued to speculate on the mechanism of sexual heredity. He paid special attention to the selective breeding of deliberately chosen domestic pigeons to produce numerous novelties that satisfied the fancies of farmers. It was his volume *The Variation of Animals and Plants Under Domestication* (1868) that contained the hypothesis of pangenesis with its conception of hereditary particles referred to as gemmules and its reluctant reliance on that explanatory mechanism of the inheritance of acquired characteristics as a result of use and disuse (habit). The great naturalist felt a need to incorporate a Larmarckian viewpoint into his own theory of evolution: if gemmules produced from organs constitute the hereditary material, then the modification of any organ through use or disuse would result in a corresponding modification of the gemmules produced by that organ. He claimed that such modifications are inherited. In fact, Darwin never abandoned his idea that the physical environment is capable of inducing the favorable variations in organisms that are necessary for the adaptation and survival and subsequent reproduction of organisms in changing habitats. In general, he held to the "blending" sexual inheritance of unlimited, fortuitous variations.

Although Darwin taught the slow and gradual accumulation of continuous but imperceptible modifications (while Huxley argued for the sudden appearance of saltations), it disturbed him that he could not explain the emergence and transmission of such slight fortuitous variations as the raw material of natural selection. Their occurrence always remained an inexplicable puzzle to him. Yet, unlike Darwin's hypothesis of pangenesis, the particulate views of both Gregor Mendel and August Weismann clearly separated all body cells from the sex cells with their hereditary material. Before this century, only a few thinkers had ever rejected outright the supposed inheritance of acquired characteristics (e.g., Lucretius, Immanuel Kant, and Charles Bonnet).

The anthropologist and psychologist Sir Francis Galton (1822–1911), Darwin's cousin, was greatly influenced by the publication of On the Origin of Species; the reading of this volume was a major turning point in his scientific career. After studying biology and medicine, and then devoting time to geography and meteorology, Galton focused on the importance of heredity in determining the psychophysical variations, abilities, and characteristics in the human animal. He formulated, in sharp contrast to euthenics (environmental control), the main principles of eugenics in his book Hereditary Genius (1869). Galton stressed the role of nature over the influences of nurture, as did Darwin (Darwin's son Leonard was an advocate of eugenics).[27]

The Russian philosopher and naturalist humanist Peter A. Kropotkin (1842–1921) devoted his scientific curiosity to an inquiry into the geography, botany, zoology, and anthropology of eastern Siberia and northern Manchuria. Although greatly influenced by the Darwinian theory of organic evolution, Kropotkin's own studies investigated the roles that mutual aid and mutual support (or group cooperation) play in the living world as important progressive factors throughout biological history (both nonhuman and human). Kropotkin stressed the significance of these factors in his major volume Mutual Aid: A Factor of Evolution (1902), pointing out that this interpretation of life did not contradict but merely supplemented the explanatory mechanism of natural selection in Darwinism.[28]

As a devoted but critical evolutionist, Kropotkin rejected Huxley's sharp separation of the human realm from the rest of the organic world in terms of behavior. This Russian thinker explored human conduct within the frameworks of both biological evolution and the natural environment. He maintained that the ethical and social progress of our own species is the result of instincts and feelings transformed into morality and love. In his posthumously published book Ethics: Origin and Development (1922), this philosophical anarchist advocated both human solidarity and human sociability.[29] Like Darwin, Kropotkin saw the human animal as being totally within the evolving matter of concrete reality.

On 24 February 1871, after three years of writing, Darwin's *The Descent of Man* appeared in print; Huxley and Haeckel had already written about human evolution,[30] but Lyell abhorred the idea that our own species is linked with the other animals through organic evolution. In this, his second major work on evolution, the ever-cautious Darwin as geobiologist and quasi-anthropologist had finally extended his own theory of evolution to account for the origin and history of humankind. He gave special attention to the mechanism of sexual selection in regard to those physical characteristics and behavioral patterns of reproductive significance that influence male combat and female choice. All in all, his explanations for organic evolution included five mechanisms: natural selection, artifical selection, and sexual selection as well as the provisional hypothesis of pangenesis and environmental influences.

Evolution is a fact encompassing the entire organic world from life's minute beginnings to the emergence of our own species. That "question of questions" is the origin and nature of the human zoological group. Early in the nineteenth century, the prehistorian Boucher de Perthes had discovered hominid fossils and paleolithic artifacts together deep in sedimentary layers. This clearly demonstrated that the human species once lived thousands of years ago as a toolmaking animal that shared its environment with now extinct mammals.

Darwin's major thesis is that the human animal and two of the great apes (chimpanzee and gorilla) have descended through organic evolution from a common but remote ancestral group of hominoids in prehistoric Africa. He held that those terrestrial ancestors of our own species slowly changed from quadrupeds into erect bipeds; this was accompanied by language, toolmaking, and both the quantitative and qualitative increase in the central nervous system and brain. In body, mind, and behavior our human species unmistakably bears an indelible stamp of its lowly origin among the earlier primates. Likewise, Darwin further argued that all the higher animals exhibit those mental traits usually attributed only to humans: reason, emotion, attention, memory, imitation, curiosity, association, imagination, and vocal communication. Like Huxley and Haeckel before him, Darwin now claimed that there is no structure or function in the human brain that is peculiar to it alone, i.e., there is nothing in the human brain that is not already found to some degree in an ape brain.

For Darwin, the superiority and essence of the human animal resides in its moral faculties; it is this moral dimension alone that distinguishes but does not separate our own species from the great apes (unfortunately, this great biologist did cling to elements of racism and sexism). Like Huxley and Haeckel before him, Darwin emphasized the striking and undeniable similarities between the human animal and the three pongids (orangutan, chimpanzee, and gorilla). He claimed that our own species is closer to the three great apes than these are to the two lesser apes (gibbon and siamang). However, it is necessary to point out that Darwin never suggested that humankind had evolved from an ape: instead, both our own species and the three pongids have descended from a generalized hominoid ancestor of the remote past in Africa. (Haeckel had thought Asia to be the birthplace of the alleged "missing link" between apes and man.)[31] The latest comparative studies in primate biology and ethology clearly demonstrate that, in terms of genetic makeup and behavioral patterns, the human animal is closer to the three great apes (orangutan, chimpanzee, and gorilla) than even Huxley, Haeckel, or Darwin himself had thought in the last century. From this scientific perspective, our human species can no longer be considered the ultimate center of a divine creation or the final goal of cosmic evolution.

Science affects human conduct and human values. This being the case, following biological Darwinism came Social Darwinism. Darwin himself always restricted his explanatory mechanism of natural selection to the nonhuman biological realm. However, Spencer extended his "survival of the fittest" principle to include human society and its progressive development. Yet, unlike Spencer and Haeckel, Darwin did not extend his theory of evolution to embrace the whole cosmos. Likewise, he did not delve into the crucial ethical questions and moral problems generated by his controversial framework as Thomas Henry Huxley, Peter A. Kropotkin, C. H. Waddington, Antony Flew, and Marvin Farber (among others) have done. In fact, Darwin never seriously thought about the origin of life on earth or speculated on the emergence and history of matter in this universe. He simply concluded that the mysteries surrounding both the beginning and the end of all things are beyond the range of our human intellect. Furthermore, he had no explanation for the existence of excessive evil in this world.

Darwin had never been a devout believer. Yet, this gentle and tolerant scientist was never mocking or cynical in his attitude toward religion. Nevertheless, one may make a strong case that something deeply bothered him and that this problem was always to remain unsolved. For the most part, he was silent about his own inner life. Over the years, however, Emma feared that Charles was moving further and further away from any belief in a personal God and the teachings of Christianity.

For Darwin, religion is not a matter of spiritual transcendence but, instead, merely a human phenomenon of psychological, ethical, and moral significance; religious beliefs and practices (including the concept of a personal God) have evolved as aspects of sociocultural development. Darwin saw the complexity of religious devotion consisting of love, fear, submission, dependence, reverence, hope, and awe. Needless to say, the theory of evolution in the last century established the foundation for scientific naturalism and rational humanism in this century.

As a solitary scientist, Darwin continued to investigate new areas of biological research. His next book was The Expression of the Emotions in Man and Animals (1872);[32] the sixth and last edition of Darwin's Origin appeared in the same year. Among his last works are The Effects of Cross and Self Fertilization in the Vegetable Kingdom (1876), a preliminary notice in Ernst Krause's Erasmus Darwin (1879), and The Formation of Vegetable Mould, Through the Action of Worms, With Observations on Their Habits (1881). Darwin's last scientific papers were on chlorophyll bodies and roots.

Although the scientific community was finally accepting Darwinism, several important naturalists had rejected the theory of evolution in part or whole, e.g., Alexander Agassiz, Georges Cuvier, John Stevens Henslow, St. George Jackson Mivart, Richard Owen, Adam Sedgwick, Rudolph Virchow, and William Whewell.[33] Yet, their attacks were merely dust in the wind that soon settled before the power of facts and reason. Lyell presented his final position on evolution in the twelfth and last edition of Principles of Geology (1875). Of course, evolutionists themselves may disagree over mechanisms and interpretations, e.g., the debate between mutationists and selectionists early in this century (with the emphasis placed on the influence of heredity or the environment, respectively); the controversy over chance or design in organic nature; and those ongoing philosophical and theological differences, e.g., the differences among materialists, vitalists, spiritualists, and biblical fundamentalists, not to mention those opposing views between the gradualists and the punctuationists in the current evolution literature.

Darwin was a devoted husband to Emma, a loving father to their seven children (several distinguished themselves in science), a loyal friend to several naturalists, and an appreciative correspondent with those scientists elsewhere who championed his theory of evolution (notably Ernst Haeckel in Germany and Asa Gray in America). Over the years at Down House, his circle of intimate friends included Henslow, Lyell, Hooker, Wallace, and Huxley, as well as Sir John Lubbock, and George J. Romanes.

Urged by Emma to record his life in general for the family and their grandchildren, Charles did write an autobiography during the year 1876.[34] His Autobiography first appeared posthumously in 1887 (a complete edition of the original manuscript was not published until 1958). Yet, the real Darwin still eludes us. He seems never to have resolved those metaphysical issues surrounding his scientific theory of organic evolution. His innermost thoughts belonged only to the sandwalk ("my thinking path"). With the gradual acceptance of his controversial ideas by important members of the scientific community, his poor health began to improve, suggesting that there was at least a psychosomatic component to his chronic illness.

Darwin had been one of the summer vice-presidents of the British Association for the Advancement of Science. He was awarded the Royal and the Wollaston Medals of British science, received honorary doctorates from five unversities (including both Cambridge and Oxford), and was given corresponding or honorary membership in fifty-seven foreign societies (including the French Academy of Sciences in 1878). International recognition also included his being awarded the highly coveted Prussian *Ordre pour le mérite* and elected to the Imperial Academy of Science in Saint Petersburg (now Leningrad).

Karl Marx venerated Charles Darwin. This sociopolitical theorist was so impressed with *On the Origin of Species* that he expressed the desire in a letter to dedicate the English first volume of his *Das Kapital* (1882) to the esteemed evolutionist. Darwin courteously refused this sincere request, not wanting his family associated with Marx's communistic and atheistic movement. In 1905, Alexander Eric Kohts founded the Darwin Museum in Moscow. Later in this century, Russian biology suffered serious stagnation under a Lamarckian theory of evolutionary genetics dogmatically upheld by Trofim Denisovich Lysenko (1898-1976) for vested sociopolitical interests.

In his intuitive grasp of both the awesome immensity of geological time and the endless creativity of biological evolution, Charles Darwin stood alone in the middle of the last century. Yet, despite all his scientific accomplishments, Darwin was never knighted by Queen Victoria or officially honored by the British government. Unfortunately, orthodox England thought organic evolution (which argued for humans and the apes sharing a common ancestry) to be too repugnant a discovery to rank a Sir Charles Darwin among its greatest countrymen.

After living the last forty years of his life in the serenity and seclusion of Down House, Charles Darwin died of a heart attack at the age of seventy-three on 19 April 1882. Public sentiment decreed his final resting spot. On 26 April, in response to an appeal from twenty members of Parliament, this great naturalist and once-perceived "scourge of orthodoxy" was buried at Westminster Abbey near the grave of Sir Isaac Newton.[35]

In 1889, Wallace published his volume *Darwinism* and subsequently received the first Darwin Medal (1890).

Darwin's own work in organic evolution did not deal with entropy, philosophy, theology, or the cosmic perspective as such. His empirical investigations into and explanations for the theory of biological evolution were hindered by the limits of empirical knowledge and scientific technology during his lifetime. These shortcomings of a Darwinian worldview in the last century included: greatly underestimating the age of our earth; not accounting for the origin of life on this planet; the incompleteness of both the geological column and the paleontological record; the absence of transitional fossil forms in rock strata; the lack of intermediate living types in the biosphere; a slow, gradual, and uniform interpretation of the process of organic evolution; the inability to observe directly the origin of new species; the incorporation of Lamarck's two principles of animal evolution; not understanding the appearance and transmission of heritable variations; an overemphasis on the individual organism; a lack of population genetics; disregarding the possibility of celestial influences; underestimating the crucial role of extinction; and the mistaken notion that the external conditions of the physical environment directly stimulate the appearance of biological variations necessary for the adaptation, survival, and subsequent reproduction of organisms. Today, these scientific shortcomings or errors in evolutionary thought have been (or are being) corrected. The modern evolutionary framework encompasses both the origin and entire history of this material universe and, as such, one may now speak of cosmic evolution.

In this century, neither Darwinism nor neo-Darwinism (the synthetic theory of organic evolution) is synonymous with evolutionism: there is a crucial distinction between the fact of evolution and the explanations or interpretations to account for it. Differences of opinion involve the various rates, levels, distances, mechanisms, relationships, and perspectives throughout organic evolution.[36] Modern biology grounds organic evolution primarily in random genetic variability as a result of chance major or minor mutations (along with genetic recombinations) and the necessity of ongoing natural selection: refer to the writings of Sergei S. Chetverikov, Theodosius Dobzhansky, R. A. Fisher, J. B. S. Haldane, Julian Huxley, H. B. D. Kettlewell, Ernst Mayr, Bernard Rensch, George Gaylord Simpson, and Sewall Wright (among others).

The fact of evolution has also generated the following areas of ongoing inquiry: cosmic evolution (including astrochemistry and astrophysics), dynamic philosophy, and process theology. New areas of research include: plate tectonics, biochemistry, genetic engineering, systematics, biogeography, ecology, ethology, altruism, sociobiology, celestial causality, and the emerging science of exobiology (which suggests that life forms, intelligent beings, and civilizations exist elsewhere in this evolving universe).

Today, no competent person should have any doubts about the truth of species mutability and human evolution. From the writings of Sir Edward Burnett Tylor and Lewis Henry Morgan in the last century to the books by Leslie A. White and V. Gordon Childe in this one (including the works of Charles A. Ellwood, Marvin Harris, Leonard T. Hobhouse, Albert G. Keller, Maxim M. Kovalevsky, Robert M. MacIver, Raoul Naroll, and Pifirim

A. Sarokin), both anthropologists and sociologists have extended the evolutionary framework to incorporate sociocultural development. Furthermore, one must not forget the pioneering work of Sigmund Freud (1856–1939) on the natural evolution of the human psyche from earlier animals.[37]

The continuing controversy between biblical fundamentalist creationism and rational scientific evolutionism has an interesting history. In the last century, religious creationism upheld the argument from divine design in a natural theology grounded in the benevolent intelligence of a personal God held responsible for the assumed order, beauty, harmony, and efficiency throughout the inorganic cosmos and organic world. It supported the belief in an instantaneous special creation of this physical universe (including all life) from nothingness. It also maintained that the age of this planet is less than ten thousand years, that plant and animal kinds or types were created all at once and are eternally fixed in nature, and that the human animal is a special creature unrelated to the apes. Periodic global catastrophes (e.g., the Noachian Deluge) are held to account for the existence of rock strata with their sequence of fossils and artifacts. There are even attempts to extend natural theology into evolutionary science and descriptive science into process theology; of special significance are the seminal writings of the Jesuit priest and geopaleontologist Pierre Teilhard de Chardin (1881–1955).[38]

The ongoing creation/evolution controversy was foreshadowed in the disagreement beteen Cuvier and Saint-Hilaire as well as in the personal conflict between Darwin and FitzRoy (and later in the differences of opinion between Charles and Emma). Philip Gosse, struggling with the evident contradiction between evolutionary science and fundamentalist religion, had offered a bizarre compromise that satisfied neither the religionists nor the evolutionists. The public disagreement between Huxley and Wilberforce, as well as those later arguments offered against evolution by Agassiz and for Darwinism by Gray, provided a rich background for both the fundamentalists and naturalists involved at the John Scopes "Monkey Trial" in 1925; this religion-versus-science controversy had become a legal issue on the state level.[39] Today, this same controversy is of national, if not international, concern.

Special creationism is not science or natural philosophy or even modern theology.[40] In fact, it is essentially a conservative sociopolitical movement rather than a traditional religious commitment. Myopic biblical fundamentalism in its most severe form is a clear and deliberate threat to science, reason, tolerance, education, and free inquiry. Evolutionists must now defend the right to think: truth is a necessary direction in overcoming hate, prejudice, bigotry, ignorance, and fanaticism.

Charles Darwin is exemplary of the creative and courageous genius who is dedicated to both science and reason in that pursuit of truth and wisdom wherever it leads and whatever its consequences. His mind was always free, open, rational, and methodical. This celebrated naturalist had something very decisive to say about space, time, life, change, and death. Obviously, he contributed critical insights into those organic workings of the material world around us. His vision has, in fact, opened up vast fields for further research. In doing so, Darwin revolutionized the scientific study of natural history. Few men have so profoundly influenced human thought.

As collector and observer and theorist, the ingenious naturalist Charles Darwin convinced the world of the fact of organic evolution; it is now the cornerstone of modern biology.

Today, Darwin's scientific theory based on evidence and logic (with those needed modifications in terms of mechanisms) remains the only plausible explanation to account for the history of life on earth. It survives the essential tests of observation and experimentation. The truth of organic evolution is incontrovertible to any enlightened mind, and Darwin's contributions to evolution remain indispensable.

No doubt, the name of Charles Darwin will be remembered through the ages; his theory and its influence will prevail. He had caused an upheaval in the biological sciences, ushered in an intellectual revolution in natural philosophy, and undermined traditional theological beliefs. As such, Darwin was one of the greatest forerunners of modernity (along with Karl Marx, Friedrich Nietzsche, Sigmund Freud, and Albert Einstein). Modest and aloof, with integrity and simplicity, possessed of wealth and with ample leisure time for reflection and research, the wise Darwin as mature theorist had presented one of the most significant scientific frameworks in Western intellectual history. His life and work are there to influence and inspire all those who really support science, reason, and free inquiry.

The need for a continuous adaptation to new facts, concepts, and perspectives undoubtedly applies to the theory of evolution itself, which must keep developing ever further beyond Charles Robert Darwin but certainly never without him.

ENDNOTES AND SELECTED REFERENCES

1. Cf. Philip Appleman, ed., *Darwin*, 3d ed. (New York: W. W. Norton, 1990); H. James Birx, "Charles Darwin: A Centennial Tribute," *Creation/Evolution* 3(2):1–10, and "The Theory of Evolution," *Collections* 60(1):22–33; Peter Brent, *Charles Darwin: A Man of Englarged Curiosity* (New York: Harper & Row, 1981); John Chancellor, *Charles Darwin* (New York: Taplinger, 1976); C. D. Darlington, *Darwin's Place in History* (Oxford,

England: Basil Blackwell, 1960); Sir Gavin de Beer, *Charles Darwin* (London: Oxford University Press, 1958), and *Charles Darwin: A Scientific Biography* (Garden City, N.Y.: Anchor Books, 1965); Loren C. Eiseley, "Charles Darwin," *Scientific American* 194(2):62–70, 72, and *Darwin's Century: Evolution and the Men Who Discovered It* (Garden City, N.Y.: Anchor Books, 1961); Benjamin Farrington. *What Darwin Really Said* (New York: Schocken Books, 1982); F. D. Fletcher, *Darwin: An Illustrated Life of Charles Darwin, 1809–1882* (Bucks, U.K.: Shire Publications Ltd., 1975, Lifelines 34); Roy A. Gallant, *Charles Darwin: The Making of a Scientist* (Garden City, N.Y.: Doubleday, 1972); Wilma George, *Darwin* (U.K.: Fontana, 1982); John C. Greene, *Darwin and the Modern World View* (Baton Rouge: Louisiana State University Press, 1981); Gertrude Himmelfarb, *Darwin and the Darwinian Revolution* (New York: W. W. Norton, 1968); Jonathan Howard, *Darwin* (New York: Hill and Wang, 1982); Julian Huxley and H. B. D. Kettlewell, *Charles Darwin and His World* (New York: Viking Press, 1965); Stanley Edgar Hyman, *The Tangled Bank* (New York: Atheneum, 1962), pp. 9–78; William Irvine, *Apes, Angels, & Victorians: Darwin, Huxley, & Evolution* (New York: McGraw-Hill, 1972); Walter Karp, *Charles Darwin and the Origin of Species* (New York: Harper & Row, 1968); Jonathan Miller, *Darwin for Beginners* (New York: Pantheon Books, 1982); Christopher Ralling, ed., *The Voyage of Charles Darwin* (New York: Mayflower Books, 1979); Paul B. Sears, *Charles Darwin: The Naturalist as a Cultural Force* (New York: Charles Scribner's Sons, 1950); L. Robert Stevens, *Charles Darwin* (Boston: Twayne, 1978); and Irving Stone, *The Origin: A Biographical Novel of Charles Darwin* (Garden City, N.Y.: 1980).

2. Cf. Donald M. Hassler, *Erasmus Darwin* (New York: Twayne, 1973), esp. pp. 69–71; and Desmond King-Hele, *Erasmus Darwin* (London: Macmillan, 1963).

3. Cf. Richard W. Burkhardt, Jr., *The Spirit of System: Lamarck and Evolutionary Biology* (Cambridge, Mass.: Harvard University Press, 1977); and Stephen Jay Gould, "Shades of Lamarck," *Natural History* 88(8):22, 24, 26, 28. Also refer to Arthur Koestler, *The Case of the Midwife Toad* (New York: Vintage Books, 1973).

4. Cf. Bern Dibner, *Darwin of the Beagle* (Norwalk, Conn.: Burndy Library, 1960); and Alan Moorehead, *Darwin and the Beagle* (New York: Harper & Row, 1969). Also refer to Victor Wolfgang von Hagen, *South America Called Them* (New York: Alfred A. Knopf, 1945), pp. 86–168 (Humboldt) and 169–229 (Darwin); and Robert S. Hopkins, *Darwin's South America* (New York: John Day, 1969).

5. Cf. Charles Coulston Gillispie, *Genesis and Geology: A Study in the Relations of Scientific Thought, Natural Theology, and Social Opinion in Great Britain, 1790–1850* (New York: Harper Torchbooks, 1959).

6. Cf. Loren C. Eiseley, "Charles Lyell," *Scientific American* 210(2):98–106, 168; and Leonard G. Wilson, ed., *Sir Charles Lyell's Scientific Journals on the Species Question* (New Haven, Conn.: Yale University Press, 1970).

7. Cf. Richard L. Schoenwald, ed., *Nineteenth-Century Thought: The Discovery of Change* (Englewood Cliffs, N.J.: Prentice-Hall, 1965), esp. pp. 89–128.

8. Cf. Benjamin Farrington, *Greek Science: Its Meaning for Us*, rev. ed. (Baltimore, Md.: Penguin Books, 1961), pp. 112–33, 156–59.

9. Cf. H. James Birx, "The Galapagos Islands," *Collections* 62(1):12–16; Robert C. Eckhardt, "Introduced Plants and Animals in the Galapagos Islands," *BioScience* 22(10):585–90; *Galapagos: The Flow of Wildness*, 2 vols. (New York: Sierra Club & Ballantine Books, 1970); Roger Lewin and Sally Anne Thompson, eds., *Darwin's Forgotten World* (Los Angeles: Reed Books, 1978); Tui De Roy Moore, *Galapagos: Islands Lost in Time* (New York: Viking Press, 1980); Alan Moorehead, *Darwin and the Beagle* (New York: Harper & Row, 1969), esp. pp. 186–209; Ian Thornton, *Darwin's Islands: A Natural History of the Galapagos* (Garden City, N.Y.: Doubleday, 1971); and Alan White and Bruce Epler, *Galapagos Guide*, 3d ed. (Quito, Ecuador: Libri Mundi, Liberia Internacional, 1978). Concerning Darwin's finches in particular, refer to David Lack, *Darwin's Finches: An Essay on the General Biological Theory of Evolution* (New York: Harper Torchbooks, 1961); Suh Y. Yang and James L. Patton, "Genetic Variability and Differentiation in the Galapagos Finches," *The Auk* 98(2):230–42; and Frank J. Sulloway, "Darwin and His Finches: The Evolution of a Legend," *Journal of the History of Biology* 15(1):1–53. For another example of microevolution among oceanic islands, see the special feature "Hawaii: Showcase of Evolution" by Hampton L. Carson et al., *Natural History* 91(12): 16–18, 20–22, 24, 26, 28, 30, 32, 34–44, 48–72. Of literary interest see Herman Melville, "The Encantadas, or Enchanted Isles" in *Herman Melville: Four Short Stories* (New York: Bantam Books, 1959), pp. 43–103.

10. Cf. Thomas Robert Malthus, *An Essay on the Principle of Population*, ed. Philip Appleman (New York: W. W. Norton, 1976). Refer to the brilliant introduction by Philip Appleman, pp. xi–xxvii.

11. Cf. Charles Darwin, *The Voyage of the Beagle*, ed. Leonard Engel (Garden City, N.Y.: Anchor Books, 1962). First appeared as *Journal of Researches into the Geology and Natural History of the Various Countries Visited by H.M.S. Beagle* (1839) and in the second edition as *Journal of Researches into the Natural History and Geology of the Countries Visited During the Voyage of H.M.S. Beagle* (1845). Note the preference given to natural history over geology in the title of the second edition of this volume. Also refer to R. Alan Richardson, "Biogeography and the Genesis of Darwin's Ideas on Transmutation," *Journal of the History of Biology* 14(1):1–41.

12. Cf. Sir Hedley Atkins KBE, *Down, the Home of the Darwins: The Story of a House and the People who Lived There* (Chichester, England: Phillimore, 1976); and Elizabeth Lambert, "Historic Houses: Charles Darwin's Home in Kent," *Architectural Digest* (February 1983): 134–40, 150.

13. Darwin's three works in geology are *The Structure and Distribution of Coral Reefs* (1842; 2d ed., 1874), *Geological Observations on the Volcanic Islands Visited During the Voyage of H.M.S. Beagle* (1844), and *Geological Observations on South America* (1846). As a petrologist, two of Darwin's major contributions to geology were his own deformation theory for the origin of metamorphic rocks that he first formulated at Tierra del Fuego as well as his distinguishing between cleavage and sedimentary bedding. During his visit to Chile, he discovered the connections among earthquakes, volcanic activity, and both geological elevation and subsidence as local phenomena. After the voyage of the *Beagle*, he suggested the use of echo-sounding methods for measuring the depths of the oceans. Darwin's two errors in geology were his account of icebergs carrying erratic boulders from the Alps to the Jura mountains (they were actually transported over great distances by glaciers), and also his assuming that the so-called three sets of parallel beaches on the slopes of the valleys of Glen Roy and Glen Gluoy were due to the action of marine waves (they were actually formed by a glacier-caused inland lake during the last ice age). Learning of these mistakes, he never again invoked the principle of exclusion in scientific matters.

14. Darwin's works in botany include: *On the Various Contrivances by Which British and Foreign Orchids are Fertilized by Insects, and on the Good Effects of Intercrossing* (1862; 2d ed., 1877), *The Variation of Animals and Plants Under Domestication* (1868; 2d ed., 1875), *Insectivorous Plants* (1875), *Climbing Plants* (1875), *The Effects of Cross and Self Fertilization in the Vegetable Kingdom* (1876), *The Different Forms of Flowers on Plants of the Same Species* (1877), and *The Power of Movement in Plants* (with the assistance of his son Francis Darwin, 1880). Also refer to Mea Allan, *Darwin and His Flowers: The Key to Natural Selection* (New York: Taplinger, 1977).

15. Cf. Charles Darwin, *Metaphysics, Materialism, and the Evolution of Mind: Early Writings of Charles Darwin* (Chicago: The University of Chicago Press, Phoenix Edition, 1980); and Neal C. Gillespie, *Charles Darwin and the Problem of Creation* (Chicago: The University of Chicago Press, Phoenix Edition, 1979).

16. Cf. Philip Appleman, "Darwin and Literature," *Free Inquiry* 2(3):25–28.

17. Cf. Ralph Colp, Jr., M.D., *To Be an Invalid: The Illness of Charles Darwin* (Chicago: The University of Chicago Press, 1977).

18. Cf. R. C. Stauffer, ed., *Charles Darwin's Natural Selection: Being the Second Part of His Big Species Book Written From 1856 to 1858* (Cambridge,

England: Cambridge University Press, 1975). Also refer to Dov Ospovat, *The Development of Darwin's Theory: Natural History, Natural Theology, and Natural Selection, 1838-1859* (Cambridge, England: Cambridge University Press, 1981). Between 1837 and 1859, one may speak of the evolution of Charles Darwin's own views on biological evolution.

19. Cf. Bert James Loewenberg, *Darwin, Wallace, and the Theory of Natural Selection* (Cambridge: Arlington Books, 1959). This work contains the Linnean Society Papers and a list of books pertinent to the text. Also refer to H. Lewis McKinney, *Wallace and Natural Selection* (New Haven, Conn.: Yale University Press, 1972).

20. Cf. Arnold C. Brackman, *A Delicate Arrangement: The Strange Case of Charles Darwin and Alfred Russel Wallace* (New York: Times Books, 1980), reviewed by David Kohn, "On the Origin of the Principle of Diversity," *Science* 213(4512):1105-8. Also refer to Stephen Jay Gould, "On Original Ideas," *Natural History* 92(1):26, 28-30, 32-33.

21. Ibid.

22. Cf. Charles Darwin, *The Origin of Species* (New York: Penguin Books, 1968). This copy of the 1859 first edition is edited by J. W. Burrow. Also refer to *The Origin of Species by Means of Natural Selection: or, The Preservation of Favored Races in the Struggle for Life* and *The Descent of Man and Selection in Relation to Sex* (New York: Modern Library, 1936), as well as *The Origin of Species* (New York: Mentor Books, 1958). To support his theory of evolution empirically, Charles Darwin relied upon the following: fossils ("the succession of types"), zoogeography (similar "representative species" of different local environments descended from a common ancestor), biological isolation (especially on oceanic islands), and the principle of divergence (particularly the adaptive radiation of life at the Galapagos Islands), as well as the inexorable pressure of natural selection resulting in differential reproduction (survival or extinction). For a current view of this classic volume, see Richard E. F. Leakey, ed., *The Illustrated Origin of Species by Charles Darwin* (New York: Hill and Wang, 1979).

23. Cf. Julian S. Huxley et al., *A Book That Shook the World: Anniversary Essays on Charles Darwin's Origin of Species* (Pittsburgh: University of Pittsburgh Press, 1958).

24. These vestigial features or rudimentary organs are discussed in Charles Darwin's *The Descent of Man* (1871; 2d ed., 1874), pt. 1, chap. 1.

25. Cf. Bentley Glass, Owsei Temkin, and William L. Straus, Jr., eds., *Forerunners of Darwin: 1745-1859* (Baltimore, Md.: The Johns Hopkins Press, 1968). Refer to "An Historical Sketch" (of the progress of opinion on the origin of species previous to the publication of the first edition of this work) in *The Origin of Species* (New York: Penguin Books, 1968), pp. 53-63, or *The Origin of Species* (New York: Mentor Books, 1958), pp. 17-25:

Darwin wrote, "I may add, that of the thirty-four authors named in this Historical Sketch, who believe in the modification of species, or at least disbelieve in separate acts of creation, twenty-seven have written on special branches of natural history or geology" (p. 61 or p. 23, respectively).

26. Cf. David L. Hull, *Darwin and His Critics: The Reception of Darwin's Theory of Evolution by the Scientific Community* (Cambridge, Mass.: Harvard University Press, 1973); and Peter J. Vorzimmer, *Charles Darwin: The Years of Controversy, The Origin of Species and its Critics 1859–1882* (Philadelphia: Temple University Press, 1970). Also refer to Michael Ruse, *The Darwinian Revolution: Science Red in Tooth and Claw* (Chicago: The University of Chicago Press, 1979).

27. Cf. Robert I. Watson, Sr., "Galton: Developmentalism, Quantitativism, and Individual Differences," in *The Great Psychologists* (New York: J. B. Lippincott, 1978), chap. 14, pp. 319–45.

28. Cf. Pëtr Kropotkin, *Mutual Aid: A Factor of Evolution* (Boston: Extending Horizons Books, 1914). Also refer to Ashley Montagu, *Darwin: Competition and Cooperation* (New York: Henry Schman, 1952).

29. Cf. Pëtr Kropotkin, *Ethics: Origin and Development* (New York: Tudor, 1947).

30. Cf. Howard E. Gruber, *Darwin on Man: A Psychological Study of Scientific Creativity* (New York: E. P. Dutton, 1974), esp. pp. 177–200.

31. Cf. Alfred Kelly, *The Descent of Darwin: The Popularization of Darwinism in Germany, 1860–1914* (Chapel Hill: The University of North Carolina Press, 1981).

32. Cf. Charles Darwin, *The Expression of Emotions in Man and Animals*, preface by Konrad Lorenz (Chicago: The University of Chicago Press, Phoenix Books, 1963).

33. Cf. *Gists From Agassiz: or, Passages on the Intelligence Working in Nature* (Hawthorne, Calif.: Omni Publications, 1973). This work also contains Agassiz's essay on "Evolution and Permanence of Type" (1874), pp. 97–115.

34. Cf. Nora Barlow, ed., *The Autobiography of Charles Darwin, 1809–1882* (London: Collins, 1958); and Sir Francis Darwin, ed., *Charles Darwin's Autobiography* (New York: Collier Books, 1961). Also refer to Francis Darwin, ed., *The Life and Letters of Charles Darwin*, 2 vols. (New York: Basic Books, 1959), esp. the autobiography in vol. 1, pp. 25–86.

35. Recent tributes to Charles Darwin include: Roger Bingham, "On the Life of Mr. Darwin," *Science 82* 3(3):40–45; H. James Birx, "Charles Darwin: A Centennial Tribute," *Creation/Evolution* 3(2):1–10; Stephen Jay Gould, "In Praise of Charles Darwin," *Discover* 3(2):20–25; Stephen Jay Gould, "The Importance of Trifles," *Natural History* 91(4):16, 18, 20–23; Francis Hitching, "Was Darwin Wrong?" *Life* 5(4):48–52; William Van

der Kloot et al., "Evolution: 100 Years After Darwin," *BioScience* 32(6):479, 495–533; and Paul Kurtz et al., "Science, The Bible, and Darwin," *Free Inquiry* 2(3)(special issue):3, 5–23, 25–49, 51–61, 63–70.

36. Cf. Francisco J. Ayala and James W. Valentine, *Evolving: The Theory and Processes of Organic Evolution* (Menlo Park, Calif.: Benjamin/Cummings, 1979); Theodosius Dobzhansky, *Genetics of the Evolutionary Process* (New York: Columbia University Press, 1970); Douglas J. Futuyma, *Evolutionary Biology* (Sunderland, Mass.: Sinauer Associates, 1979); J. B. S. Haldane, *The Causes of Evolution* (Ithaca, N.Y.: Cornell University Press, Cornell Paperbacks, 1966); Mar K. Hecht and William C. Steere, eds., *Essays in Evolution and Genetics in Honor of Theodosius Dobzhansky: A Supplement to Evolutionary Biology* (New York: Appleton-Century-Crofts, 1970); Julian Huxley, *Evolution: The Modern Synthesis* (New York: John Wiley & Sons, Science Editions, 1964); Ernst Mayr, *Evolution and the Diversity of Life: Selected Essays* (Cambridge, Mass.: The Belknap Press of Harvard University Press, 1976); Ernst Mayr and William B. Provine, eds., *The Evolutionary Synthesis: Perspectives in the Unification of Biology* (Cambridge, Mass.: Harvard University Press, 1980); Roger Milkman, ed., *Perspectives on Evolution* (Sunderland, Mass.: Sinauer Associates, 1982); Bernard Rensch, *Evolution Above the Species Level* (New York: Columbia University Press, 1959); George Gaylord Simpson, *Tempo and Mode in Evolution* (New York: Hafner, 1965); Steven M. Stanley, *Macroevolution: Pattern and Process* (San Francisco: W. H. Freeman, 1979); and Steven M. Stanley, *The New Evolutionary Timetable: Fossils, Genes, and the Origin of Species* (New York: Basic Books, 1981). Also refer to Ernst Mayr et al., "Evolution," *Scientific American* 239(3)(special issue): passim.

37. Cf. Sir Edward Burnett Tylor, *Researchers into the Early History of Mankind and the Development of Civilization* (Chicago: Phoenix Books, 1964; first published in 1865; 3d ed. rev., 1878), *Primitive Culture*, 2 vols. (New York: Harper Torchbooks, 1958; first published in 1871), and *Anthropology* (Ann Arbor: The University of Michigan Press, 1960; first published in 1881); Lewis Henry Morgan, *Ancient Society: or, Researches in the Lines of Human Progress from Savagery through Barbarism to Civilization* (New York: Meridian Books, 1963; first published in 1877); Leslie A. White, *The Science of Culture: A Study of Man and Civilization* (New York: Grove Press, 1949), and *The Evolution of Culture: The Development of Civilization to the Fall of Rome* (New York: McGraw-Hill, 1959); V. Gordon Childe, *Social Evolution* (New York: Meridian Books, 1963; first published in 1951).

38. Cf. Pierre Teilhard de Chardin, *The Phenomenon of Man*, 2d ed. (New York: Harper Torchbooks, 1965).

39. Cf. Ray Ginger, *Six Days or Forever? Tennessee v. John Thomas Scopes* (New York: Signet Books, 1960); and Mary Lee Settle, *The Scopes Trial: The State of Tennessee v. John Thomas Scopes* (New York: Franklin Watts, 1972).

40. Cf. H. James Birx, "The Creation/Evolution Controversy," *Free Inquiry* 1(1):24–26; Niles Eldredge, *The Monkey Business: A Scientist Looks at Creationism* (New York: Pocket Books, 1982); Douglas J. Futuyma, *Science on Trial: The Case For Evolution* (New York: Pantheon Books, 1983); Laurie R. Godfrey, ed., *Scientists Confront Creationism* (New York: W. W. Norton, 1983); Philip Kitcher, *Abusing Science: The Case Against Creationism* (Cambridge, Mass.: The MIT Press, 1982); Dorothy Nelkin, *The Creation Controversy: Science or Scripture in the Schools* (New York: W. W. Norton, 1983); Norman D. Newell, *Creation and Evolution: Myth or Reality?* (New York: Columbia University Press, 1982); Robert Root-Berstein and Donald L. McEachron, "Teaching Theories: The Evolution-Creation Controversy," *The American Biology Teacher* 44(7):405, 413–420; Michael Ruse, *Darwinism Defended: A Guide to the Evolution Controversies* (Reading, Mass.: Addison-Wesley, 1982); and J. Peter Zetterberg, ed., *Evolution versus Creationism: The Public Education Controversy* (Phoenix, Ariz.: Oryx Press, 1982). Also refer to articles in the *Creation/Evolution* journal, ed. Frederick Edwords (P.O. Box 146, Amherst Branch, Buffalo, N.Y. 14226). Of general interest are Ian G. Barbour, *Issues in Science and Religion* (Englewood Cliffs, N.J.: Prentice-Hall, 1966), esp. pp. 365–418; U. J. Jensen and R. Harre, eds., *The Philosophy of Evolution* (New York: St. Martin's Press, 1981); Mary Long, "Visions of a New Faith," *Science Digest* 89(10):36–43; E. C. Olson, *The Evolution of Life* (New York: Mentor Books, 1966), esp. pp. 265–78; Karl E. Peters, "Religion and an Evolutionary Theory of Knowledge," *Zygon: Journal of Religion and Science* 17(4):385–415; W. Widick Schroeder, "Evolution, Human Values and Religious Experience: A Process Perspective," *Zygon: Journal of Religion and Science* 17(3):267–91; and Eugenia Shanklin, "Darwin vs. Religion," *Science Digest* 90(4):64–69, 116.

FURTHER READINGS

Alland, Jr., Alexander. *Human Nature: Darwin's View.* New York: Columbia University Press, 1985.

Appleman, Philip, ed. *Darwin.* 3d ed. New York: W. W. Norton, 1990.

Barlow, Nora, ed. *The Autobiography of Charles Darwin, 1809–1882.* New York: W. W. Norton, 1969.

Barrett, Paul H., ed. *Metaphysics, Materialism, and the Evolution of Mind: Early Writings of Charles Darwin.* Chicago: University of Chicago Press, 1980.

———. *The Collected Papers of Charles Darwin.* Chicago: University of Chicago Press, 1977.

Beebe, William. *Galapagos: World's End.* New York: Dover, 1988.

Birx, H. James. "Darwin & Teilhard: Some Final Thoughts." *Proteus: A Journal of Ideas* 6 (Fall 1989): 38–46.

Boorstin, Daniel J. *The Discoverers*. New York: Random House, 1983, esp. pp. 420–76.

Bowler, Peter J. *Evolution: The History of an Idea*, rev. ed. Berkeley: University of California Press, 1989.

———. *The Eclipse of Darwinism: Anti-Darwinian Evolution Theories in the Decades Around 1900*. Baltimore, Md.: Johns Hopkins University Press, 1983.

Burkhart, Frederick, and Sydney Smith, eds. *The Correspondence of Charles Darwin*. 4 vols. Cambridge, England: Cambridge University Press, 1985–1988.

Darlington, C. D. *Darwin's Place in History*. Oxford, England: Basil Blackwell, 1960.

Denton, Michael. *Evolution: A Theory in Crisis*. Bethesda, Md.: Adler & Adler, 1986.

Engel, Leonard, ed. *The Voyage of the Beagle* (Charles Darwin). Garden City, N. Y.: Anchor Books, 1962.

Clark, Ronald W. *The Survival of Charles Darwin: A Biography of a Man and an Idea*. New York: Random House, 1984.

Cohen, I. Bernard. *Revolution in Science*. Cambridge, Mass.: Belknap Press of Harvard University Press, 1985, esp. pp. 283–300.

Colp, Jr., Ralph. *To Be an Invalid: The Illness of Charles Darwin*. Chicago: University of Chicago Press, 1977.

de Beer, Gavin. *Charles Darwin: A Scientific Biography*. Garden City, N. Y.: Anchor Books, 1965.

Edelman, Gerald M. *Neural Darwinism: The Theory of Neuronal Group Selection*. New York: Basic Books, 1987.

Eiseley, Loren. *Darwin's Century: Evolution and the Men Who Discovered It*. Garden City, N. Y.: Anchor Books, 1958.

Farb, Nathan. *The Galapagos*. New York: Rizzoli International, 1988.

Ghiselin, Michael T. *The Triumph of the Darwinian Method*. Chicago: University of Chicago Press, 1984.

Gillespie, Neal C. *Charles Darwin and the Problem of Creation*. Chicago: University of Chicago Press, 1979.

Godfrey, Laurie Rohde, ed. *What Darwin Began: Modern Darwinian and Non-Darwinian Perspectives on Evolution*. Boston: Allyn and Bacon, 1985.

Gould, Stephen Jay. *Ever Since Darwin: Reflections in Natural History*. New York: W. W. Norton, 1977, esp. pp. 21–45.

———. "In Praise of Charles Darwin." *Proteus: A Journal of Ideas* 6 (Fall 1989): 1–4.

Gruber, Howard E. *Darwin on Man: A Psychological Study of Scientific Creativity.* New York: E. P. Dutton, 1974.

Harris, Marvin. *The Rise of Anthropological Theory: A History of Theories of Culture.* New York: Thomas Y. Crowell, 1968, esp. pp. 142–216, 634–53.

Himmelfarb, Gertrude. *Darwin and the Darwinian Revolution.* New York: W. W. Norton, 1968.

Humann, Paul. *Galapagos: A Terrestrial and Marine Phenomenon.* Santiago, Chile: Editorial Kactus, 1988.

Huxley, Julian, and H. B. D. Kettlewell. *Charles Darwin and His World.* New York: Viking Press, 1965.

Irvine, William. *Apes, Angels, and Victorians: Darwin, Huxley, and Evolution.* New York: McGraw-Hill, 1972.

Kohn, David, ed. *The Darwinian Heritage.* Princeton, N.J.: Princeton University Press, 1985, esp. pp. 755–812.

Mayr, Ernst. *The Growth of Biological Thought: Diversity, Evolutionism, and Inheritance.* Cambridge, Mass.: Belknap Press of Harvard University Press, 1982, esp. pp. 394–534.

Meadows, Jack. *The Great Scientists.* New York: Oxford University Press, 1987, esp. pp. 9–28, 149–168.

Ospovat, Dov. *The Development of Darwin's Theory: Natural History, Natural Theology, and Natural Selection, 1838–1859.* Cambridge, England: Cambridge University Press, 1981.

Ruse, Michael. *Darwinism Defended: A Guide to the Evolution Controversies.* Reading, Mass.: Addison-Wesley, 1982.

———. *The Darwinian Paradigm: Essays on Its History, Philosophy, and Religious Implications.* New York: Routledge, 1989.

———. *The Darwinian Revolution.* Chicago: University of Chicago Press, 1979.

———. *Taking Darwin Seriously: A Naturalistic Approach to Philosophy.* New York: Basil Blackwell, 1987.

Salwen, Peter. *Galapagos: The Lost Paradise.* New York: Mallard Press, 1989.

Scopes, John Thomas. *Center of the Storm.* New York: Holt, Rinehart, and Winston, 1967.

Steadman, David W., and Steven Zousmer. *Galapagos: Discovery on Darwin's Islands.* Washington, D.C.: Smithsonian Institute Press, 1989.

Wallace, Alfred Russel. *Darwinism: An Exposition on the Theory of Natural Selection with Some of Its Applications.* London: Macmillan, 1899.

Wichler, G. *Charles Darwin: The Founder of the Theory of Evolution and Natural Selection.* New York: Pergamon Press, 1961.

4

Henri Bergson: Creative Evolution

In the historical development of evolutionary interpretations, the critical thoughts and philosophical writings of vitalist Henri Bergson (1859–1941) represent both a chronological and an intellectual link between the earlier Darwinian materialism and the later Teilhardian spiritualism.

Bergson became an internationally known French philosopher of Polish-Jewish ancestry. He was a professor at the Collège de France (1900–1921) and spent his life teaching, lecturing, and writing on his novel philosophical worldview. In 1927, he received the Nobel Prize in literature for his major work *Creative Evolution* (1907).

Bergson rejected outright Democritus's materialist atomism, Plato's eternally fixed forms, Aristotle's immutable species, and Kant's rigid epistemological categories. Although greatly influenced by the theories of Darwin and Spencer, he opposed mechanism and materialism. His own system is grounded in an application of intuition to both the process of evolution throughout organic history and human consciousness.

Bergson's conceptual framework is ultimately dualistic, distinguishing between the special sciences and metaphysics; intellect and intuition; scientific or mathematical time and experiential or "pure" time; mechanistic determinism and the spontaneous freedom of the whole personality; the material body and the spiritual mind united through perception in real duration; the sensorimotor or "habit" memory of animals and the "pure" spiritual memory of man; materialist evolution and creative evolution; and a closed society and an open one.

According to Bergson, the following dualistic scheme prevails: science is concerned with mathematics and logic, appearances, space, matter (extension), fixity, mechanism, static nature, the human body, relative knowledge, and a closed society; whereas metaphysics is concerned with intuition and mysticism, reality, time, spirit (life), change, vitalism, creative evolution, the human mind (spirit or consciousness), absolute knowledge, and an open society.

Bergson was a seminal thinker concerned with space, time, matter, life, evolution, consciousness, creativity, free will, intuition, and mysticism. He presented original insights into the phenomenon of real duration, continuous becoming, and creative evolution as a planetary process. For Bergson, God is love and the object of love. In fact, the evolutionary process has a divine purpose (this was his final undeveloped interpretation). As a systematic thinker, he is impossible to understand and appreciate without referring to the whole spectrum of his thought. Nevertheless, his arguments for the significance of intuition can easily be taken as the central point of his entire philosophical structure. His views have led to the development of existentialism in one direction and process philosophy in another.

Ultimately, Bergson's rejection of reason in favor of intuition must be critically evaluated in light of rapid advancements in the special sciences and logical deliberation. In the last analysis, his dualistic orientation is grounded in religious beliefs and metaphysical assumptions not subject to empirical verification or rational inquiry.

Bergsonian metaphysics is incomplete without taking into account knowledge from biology and psychology. However, it is not mathematically oriented but rather time oriented: science and reason (mathematics and logic) are replaced by intuition. Actually, intuition is seen as that essential bridge between science and metaphysics: metaphysics is knowledge of things-in-themselves or absolute knowledge of reality. Therefore, where Kant's work resulted in a critique of "pure" reason, Bergson's philosophy is a critique of "pure" intuition. As such, it rejected mechanism and materialism but established a metaphysics (however incomplete) embracing planetary evolution, diverging creativity, emerging novelty, and human freedom. According to Bergson, only intuition can disclose the inner workings of reality as the creative evolution of progressive consciousness.

Bergson focused on human freedom, consciousness, and the place of our own species within earth history. His view is at the same time scientific and mystical, resulting in unavoidable ambiguities. His attempt is a bold one: the reconciliation of subjectivity and mysticism with rational thought and objective science. However, priority was clearly given to intuition rather than logic. Major problems do result from this basic presupposition. For Bergson, there is a sharp distinction between space and time as well as two categorically distinct realms, the inorganic and the organic. What results is a metaphysical dualism between matter (stuff) and life (spirit). Matter is passive and static while life is active and dynamic. When matter is infused with life, the result is creative evolution. Unlike Darwinian materialism, which offered natural selection as the primary explanatory mechanism for comprehending biological evolution, Bergsonian vitalism is an attempt at a systematic approach to the creative aspect of all evolution that had been allegedly ignored by all prior naturalists.

What method did Bergson use to become aware of this creative flux of reality? He did not appeal to science (mathematics) or reason (logic). His metaphysics rests on intuition. For Bergson, intuition not only gives one an awareness of time or duration (metaphysics) but also illuminates the process of memory (epistemology). His emphasis on subjective evolution moved from a rejection of metaphysics based upon science and reason to an exaltation of intuition and mysticism as the true source of ultimate knowledge: only intuition grasps creative reality as time and change or pure duration.

In *Time and Free Will* (1889), Bergson's philosophy asks one to turn from those fragments of science and immerse oneself in a living stream of things as the flux of reality.[1] As already noted, Bergson makes a sharp distinction between space and time. Science, being objective, is only concerned with space and arrives at its concepts through an analysis of the quantitative multiplicity of space. But for Bergson, this analysis distorts the continuous flux of reality. Therefore intuition, being subjective, is concerned with time and arrives at the concept of *durée réelle* through sympathetic insight into that qualitative multiplicity of consciousness in continuous creative flux. Science, using mathematics, can never obtain absolute reality because absolute reality (metaphysics) is known only through intuition. Number belongs to space and therefore mathematics is a tool of all the sciences. Bergson had already asserted the belief that space cannot contain or reveal absolute reality; therefore neither can mathematics or logic. According to Bergson, one is misled if one believes that time is homogeneous. Time is not homogeneous since it is distinct from space. As pure duration (*durée réelle*), it is time that holds the key to absolute reality and consequently creative evolution.

Time and duration can be known only intuitively, never through the application of science or logic or mathematics within space. Time is intuitively grasped by the flux of consciousness. Pure duration, that which endures, exists only in our mind. Unlike pure space where there are merely simultaneities, the human mind with its memory contains pure duration. In memory, things endure. Space is external and homogeneous, while duration is internal and heterogeneous. Bergson established in this early work (1) the limitations of science (mathematics and logic) and (2) the significance of intuition. Thus in Bergson's first major volume, he clearly favored intuition over science. It is immediate intuition that allows one to see time and motion. Real duration is intuitively known within the enduring memory of the flux of consciousness.

In *Matter and Memory* (1896), Bergson intuitively distinguished among matter, perception, and memory.[2] For him, immediate intuition continues to be the method by which one may discover reality. Memory adds to intuition simply because memory is the collection of previous intuitions (pure memory is the unconscious). Intuition shows the distinction between body and memory as well as their union. Although Bergson was neither a materialist nor an idealist, he still desired to make a distinction between the body (matter) in space and the soul (memory or spirit) in time. He tried to do this by interjecting perception as a third and intermediate principle. Thus, he did avoid both an extreme form of ontological dualism and spiritual monism. Briefly, it is perception that links space with duration and matter with memory.

Bergson's *An Introduction to Metaphysics* (1903) is nothing short of a critique of intuition.[3] It is a significant work in the development of Bergsonian philosophy. A crucial distinction is made between the intellect and intuition. Intellect, that method of those special sciences, can never give us a feeling of absolute reality. The latter is obtained only through that proper method of metaphysics (intuition as "intellectual sympathy").

What is discovered when one applies intuition to those inner human experiences? One discovers the flowing of personality through time: the ego is a ceaselessly changing yet continuously ongoing process. Duration is in sharp contrast to that mathematically "spatialized" time of the intellect. Duration is a heterogeneous flux of becoming always irreversibly flowing toward the future and continuously creating newness or novelty that is intrinsically unpredictable; an inexhaustible source of freedom whose living reality can never be communicated by concepts or images but rather must be directly intuited. Since both time (duration) and motion are apprehended intuitively, reality can only be known through intuition. However, metaphysics does not oppose the special sciences but rather complements them.

Bergson insists that logic, mathematics, and the special sciences (e.g., physics, chemistry, biology, psychology, anthropology, and astronomy) give us only relative knowledge as they necessarily demand the use of symbols. Yet metaphysics, by definition, is a search for absolute knowledge that does not require symbols. If intuition alone gives us absolute knowledge, then it is clearly the method of metaphysics. Bergson believed that "intellectual sympathy" is not difficult: it is a simple act to intuit space, time, motion, matter, and memory. They seem to have always been "given" to us as data of perception as well as absolute metaphysical certainties (again, such knowledge is said to require no symbolization).

For Bergson, there is an absolute reality that we seize only through intuition and not by a simple analysis or the use of symbols: it is our own personalities flowing through time. As such, each of us is immediately aware that the self endures and consequently experiences pure duration (*durée réelle*). Intuition allows the metaphysical investigation of what is essential and unique in an object: duration is essential and unique to the human self. This pure duration excludes all ideas of extension, juxtaposition, and reciprocal externality. Likewise, it is impossible to represent the self by concepts or symbols, i.e., by abstract, general, or simple ideas.

Bergson strongly believed that both empiricism and rationalism fail to obtain absolute knowledge of the ego. Only intuition reveals an absolute knowledge of the ego as pure duration. Pure duration is within the ego, because only the human ego has extensive and specific memory. Space, being exterior to the human ego, has no memory and therefore it does not endure in the same sense that memory endures. Space has no duration but only instantaneity. Only time as memory has duration. In the final analysis, intuition attains the absolute while bringing science and metaphysics together through perception.

Philosophy cannot ignore the knowledge given to it by the special sciences. Bergson himself, greatly interested in biology and psychology, became an avid evolutionist (an emergent evolutionist to be exact). What he was reacting against was the alleged misconception that the methods of science, which symbolize and spatialize, also give us absolute knowledge. For Bergson, science offers empirical facts, but it is the intuitive method of methaphysics that reveals the ultimate reality of things. That is to say, only through intuition are we aware of the flux of consciousness in time as well as, by analogy, the flux of things in the external world. In short, reality is the flow of time (whether internal or external) and this is only apprehended through intuition, which does not spatialize or symbolize.

Thus far in the development of his thought, Bergson's use of intuition has established two aspects of time (objective and subjective) and a mind-body (memory-matter) distinction. His intuitive method is a deliberate reaction against the use of understanding (science and reason) in the attainment of absolute knowledge, since it is claimed that the methods of understanding distort the ultimate flux of things by symbolizing and spatializing them. Reality is not space but rather time, and there are two "types" of time: (1) objective time (scientific time), which is an abstract mathematical conception, extended, homogeneous, and passively separated into units (it is, in essence, an illusion as the product of rational thought); and (2) subjective time (conscious time), which is an active, creative, flowing, endless, irreversible, indivisible, accumulative, immediately experienced (intuited) process. In essence, it is heterogeneous real time or "pure time" (*durée réelle*) that is grasped intuitively within the continuous flux of our own consciousness.

Once again, the real time (*durée réelle*) of consciousness and the abstract time of science are said to be united through perception. Our mind is real duration. It is continuous and irreversible. It is a heterogeneous flux of ongoing activity creating novelty and manifesting the intuition that gives man his freedom. Through perceptions mind is united with matter, which is relatively stable in comparison to the flux of consciousness. Mind and body are related by a convergence in time through perception. The human intellect uses symbols and thereby conceptualizes, giving relative external knowledge. In Bergson's view, such static abstractions of time and motion falsify real becoming because, for him, the special sciences, logic, and mathematics can never give an adequate and complete account of this dynamic universe. Intuition is immediate, nonconceptual, and nonsymbolic (intellectual sympathy); it yields absolute knowledge of the creative becoming of reality itself.

Creative Evolution (1907) is Bergson's masterpiece.[4] In it, he rebels against the mechanism and materialism of Darwinian evolution and those fixities and rigidities of all logical and mathematical interpretations of reality. He further develops his conception of intuition, illustrating its significance for a true understanding of and proper appreciation for emergent evolution on earth. However, his philosophy of evolution does not include exobiology or exoevolution.

Bergson is still time oriented and *Creative Evolution* gives him an opportunity to present his interests in matter, spirit, intuition, and evolution within a dynamic metaphysical system. To be sure, he never entertained a possiblity that the evolutionary process could be reduced to mathematical predictability.

Bergson rejects both radical finalism and radical mechanism, claiming that each distorts the free creativity of the evolutionary process. It is because of this use of intellect that the essence of emergent evolution is missed. If the intellect cannot give a true view of the evolutionary process, then intuition can. Just as intuition is able to show the absolute reality of one's own internal ego (i.e., make one aware of subjective duration or *durée réelle*), an intellectual sympathy with the evolution of life will reveal the objective duration of external nature as creative evolution.

By going "counter to the natural bent of the intellect" (i.e., by using intuition) one discovers the irreducible, irreversible, divergent, and durative nature of the evolutionary process. Evolution is creative: this universe is in a continuous flux of emergent novelty. To give an adequate philosophical account of this intuitive fact of reality, Bergson established a metaphysical concept: the *élan vital* or vital principle as that pervasive force, impetus, or "spirit" within otherwise static matter. It causes the creative divergence throughout earth history, the evidence of which science explores and examines. But alas, this *élan vital* is ultimately a metaphysical conception that is not scientifically verifiable. According to Bergson, however, this conception is metaphysically sufficient to account for the creative, divergent, holistic, and functional coadaptation of an organism through successive alterations of form toward greater and greater complexity, consciousness, and emergent novelty. It is also the *élan vital* that gives evolution its continuation and duration. Although it is continuous, this impetus manifests itself in three divergent lines: plants, insects, and animals.

For Bergson, metaphysics supplements the findings of biology through the use of intuition. Intuition makes us aware of this vital impetus as an *élan vital* within the evolutionary process itself. Consequently, intuition reveals with absolute certainty the cosmic process of becoming. Bergson argues that the *élan vital* is responsible for (1) a current of cumulative and irreversible consciousness; (2) evolutionary continuity; (3) emergent novelty as those creative "leaps" in evolution; (4) an increased complexity through time in the evolutionary process; (5) "spirit" attempting to transcend matter; (6) vitalism as a force that is the original impetus from which life springs as well as the metaphysical principle that maintains the continuity of evolution and its continuous emergent novelty; and (7) God as unceasing life, action, and freedom. Hence, as mind is the spirit of our material body, so the *élan vital* is that spirit within the matter of the cosmos. In the process of divergent evolution plants mainfest fixity and torpor, insects are locked into instincts, but animals display increasing degrees of consciousness culminating in human self-consciousness. Our own species alone displays reason (intellect), reflection, and intuition; it is the apex of creative evolution on earth.

From Bergson's point of view, both intelligence and instinct have innate knowledge. Even at the extremities of these two principal lines of evolution (the insects and the vertebrates), instict and intelligence do not occur in pure form. In general, Bergson believed that instinct is innate knowlege of things (matter) while intelligence is innate knowledge of relations (form). Instinct is sympathy for external material things as is illustrated especially in the Hymenoptera and Lepidoptera. Likewise, intellect is knowledge of static relations or states and thus cannot grasp mobility or continuity or the essence of evolution as creative becoming. However, intuition is "purified" instinct (remembering that the sympathy of instinct is free from any use of symbols, but the intellect is not) that accompanies human intellect and immediately grasps the continuity of a change as pure mobility. Under the influence of intellect, instinct has become disinterested in material things. It reflects on the interpenetration of subjective duration as a stream of consciousness and objective life as the flux of reality. There is, then, an ongoing unity between mind with its duration and matter that is in change. Somehow, there must be within this unity of the cosmos a relationship between pure duration and creative evolution (i.e., between *durée réelle* and the *élan vital*). This ongoing relationship is grasped through the use of intuition.

A continual coming and going is possible between mind and nature because all spirit in nature as the *élan vital* emerges in man as intuition. Intuition is *élan vital* conscious of itself: it has consciously become aware of itself through intuition within the self-consciousness of man. Clearly, intuition is that pivotal concept in Bergsonian philosophy, for it alone can link the consciousness of mind (*durée réelle*) with that stream of consciousness (*élan vital*) in matter. This is knowledge that the special sciences, mathematics, and logic can never provide. Only through intuition is absolute reality known as time, whether subjective (*durée réelle*) or objective (*élan vital*).

Bergson deplores the tendency in science and philosophy to mistake its conceptualizations (products of the understanding) for absolute reality. Time is ruled out by intelligence. Looking for fixity, intelligence masks the flow of time by conceiving it as a juxtaposition of "instants" on a line. Time is not oriented to the special sciences, mathematics, or logic, because it is an indivisible flow and therefore has no measurable parts. We may think we are measuring time, but in reality we are merely measuring space. True time is only the intuitive experience of duration. Therefore, one must distinguish between real lived time on the one hand and its spatialization through symbolization into those external objects, events, and relationships of scientific inquiry on the other.

Duration and Simultaneity: With Reference to Einstein's Theory (1922) is an attempt to refute the existence of multiple real times, a thesis found in the theory of relativity.[5] Bergson advanced arguments to save the absolute character of time from that problem of "asymmetrical aging" known as the clock paradox of identical twins; this theory of multiple times challenged Bergson's major philosophical intuition about the absolute unity of true time. Against Einstein's thesis, Bergson defended his intuitive view of duration as a single, indivisible flow of true time. For Bergson, there is no dichotomy between time external and time internal: the duration of this universe includes both external time or the *élan vital* and internal time or *durée réelle*. It follows that there is only one continuous time in this cosmos and only one stream of consciousness. However, the epistemological leap from *durée réelle* to the *élan vital* is unclear. One is told that it is done intuitively, of course, involving memory and perception.

There is no reason to limit duration to consciousness. According to

Bergson, evolution is only intelligible when it is extended to spirit. Duration exteriorizes itself as spatialized time: it is space, rather than time, that is measurable. Real duration is experienced and as such contains the very stuff of all existence. Time remains absolute reality and is learned as well as retained only in the realm of human intuition. Furthermore, there is only one continuous time in this dynamic universe.

Finally, Bergson's philosophical inquiry is ultimately a mystical commitment. *The Two Sources of Morality and Religion* (1932) argues that the discovery of God is made only by that sort of intuition that is the mystical experience of exceptional persons.[6] This form of intuition leads to intense activity to ensure humankind's further evolution toward a mystically grounded open society instead of the merely intellectually oriented closed society. By extending intuition to mysticism, Bergson hoped that in the future more and more mystics would change our earth into an open society, i.e., a society with unlimited possibilities, progressive as well as flexible in its moral and religious beliefs, and mystical in nature.[7] The intellectually mechanical and routine society is authoritarian, conservative, self-centered, and closed both morally (static and absolute) and religiously (ritualistic and dogmatic) and as such negates change, freedom (free will), and spontaneity (intuition).

Bergson's far-reaching influence is especially noticeable in the evolutionary synthesis of Pierre Teilhard de Chardin, who, consciously or not, never explicitly admitted an obvious and great indebtedness to his philosophical predecessor. Despite this curious point, the major positive contribution of both is their recognition that, one way or another, the fact of evolution must be taken seriously.

ENDNOTES AND SELECTED REFERENCES

1. Cf. Henri Bergson, *Time and Free Will: An Essay on the Immediate Data of Consciousness* (New York: Harper Torchbooks, 1960).

2. Cf. Henri Bergson, *Matter and Memory* (London: George Allen and Unwin, 1962).

3. Cf. Henri Bergson, *An Introduction to Metaphysics*, 2d rev. ed. (New York: Bobbs-Merrill, 1955).

4. Cf. Henri Bergson, *Creative Evolution* (New York: Modern Library, 1944). In chapter one, Bergson writes: "The 'vital principle' may indeed not explain much, but it is at least a sort of label affixed to our ignorance so as to remind us of this occasionally, while mechanism invites us to ignore that ignorance" (p. 48).

5. Cf. Henri Bergson, *Duration and Simultaneity: With Reference to Einstein's Theory* (New York: Bobbs-Merrill, 1965).

6. Cf. Henri Bergson, *The Two Sources of Morality and Religion* (Garden City, N.Y.: Doubleday, 1935). Also refer to Henri Bergson, *An Introduction to Metaphysics: The Creative Mind* (Totowa, N.J.: Littlefield/Adams, 1965). This work was first copyrighted in 1946 by the Philosophical Library, Inc.

7. Cf. Karl R. Popper, *The Open Society and Its Enemies*, 2 vols. (New York: Harper Torchbooks, 1963), vol. 1, pp. 202, 294, 314; vol. 2, pp. 229, 258, 307, 315–16, 361.

FURTHER READINGS

Bergson, Henri. *Creative Evolution*. New York: Modern Library, 1944.

———. *Duration and Simultaneity: With Reference to Einstein's Theory*. New York: Bobbs-Merrill, 1965.

———. *An Introduction to Metaphysics*, rev. 2d ed. New York: Bobbs-Merrill, 1955.

———. *Matter and Memory*. London: George Allen & Unwin, 1911.

———. *A Study in Metaphysics: The Creative Mind*. Totowa, N.J.: Littlefield/Adams, 1965.

———. *Time and Free Will: An Essay on the Immediate Data of Consciousness*. New York: Harper Torchbooks, 1960.

———. *The Two Sources of Morality and Religion*. Garden City, N.Y.: Anchor Books, 1935.

Einstein, Albert. *Relativity: Special and General*. New York: Random House, 1961.

Schubert-Soldern, Ranier. *Mechanism and Vitalism: Philosophical Aspects of Biology*. Notre Dame, Ind.: University of Notre Dame Press, 1962.

Whitehead, Alfred North. *The Concept of Nature* (1920). Cambridge, England: Cambridge University Press, 1964.

Pierre Teilhard de Chardin

PIERRE TEILHARD DE CHARDIN (1881–1955). As geopaleontologist and Jesuit priest, Teilhard sought to reconcile scientific evolutionism with religious mysticism. He interpreted this dynamic universe in terms of four basic concepts: spiritual monism, the law of complexity-consciousness, critical thresholds, and a future omega point. His controversial cosmology is actually an earth-bound and human-centered personal vision of planetary history. Over the years, this panentheisitic worldview was developed in prolific writings that included *The Divine Milieu* (1927), *The Phenomenon of Man* (1940), and *Man's Place in Nature* (1949). *Courtesy of American Museum of Natural History.*

5

Pierre Teilhard de Chardin:
The Phenomenon of Man

Pierre Teilhard de Chardin (1881–1955) was a remarkable human being indeed. He dedicated his life to both scientific research and the service of a personal God. His poetic and inspiring writings, especially *The Phenomenon of Man* (1940), are an attempt to reconcile within an evolutionary perspective the special sciences, process philosophy, and a mystical interpretation of Roman Catholic theology.

Teilhard's worldview is profoundly abstract. It emphasizes the essential unity of matter and spirit, thought and action, personalism and collectivism, plurality and unity. In propounding it, he went far beyond traditional theism and even pantheism by advocating a process panentheism, i.e., God is simultaneously both immanent as the evolving spiritual universe and transcendent as an eternal and loving personal being.

Teilhard attempted to synthesize the empirically documented facts, far-reaching philosophical implications, and inescapable religious consequences of planetary evolution with certain elements of Christian supernaturalism and cosmic mysticism.

As a result of Teilhard's courageous but unsuccessful request for the publication of his own slightly revised edition of *The Phenomenon of Man* in 1948, Pope Pius XII issued an encyclical letter, *Humani generis* (12 August 1950), in which he gave preferential priority to a Thomistic interpretation of divine revelation as contained in the Holy Scriptures over the evidence of the empirical sciences and the arguments of logic. In this document, the pope warns that opinions on the theory of evolution, the truth of which he was willing to leave an open question, may be erroneous, i.e., fictitious or conjectural. This authoritative papal pronouncement holds that evolutionism results from a desire to be novel; therefore, the theory of evolution is merely a question of hypotheses or possibly even a false science. It warns that evolutionism leads to the formless and unstable tenets of a new philosophy and sterile speculation, claiming the major error to be too free an interpretation of those historical books of the Old Testament. This encyclical letter held the doctrine of evolution to be not only plainly at variance with Holy Scriptures but false by human experience as well.

It is obvious that such a narrow view was a direct attack against Teilhard's new philosophy of evolution. In fact, it was against such dogmatic claims as expressed by Pius XII that the Jesuit scientist had attempted to establish a better Christianity, i.e., a meta-Christianity within a cosmic, spiritual, and ultimately mystical evolutionary framework. It is worth noting that although his scientific writings on evolution were duly sealed in a supernaturalist envelope, they never received the *Nihil Obstat* and *Imprimatur*, which declare that a book or pamphlet is considered to be free from doctrinal or moral error in the eyes of the Roman Catholic Church. Moreover, not only was the publication of the papal encyclical letter an implicit attack on Teilhard's less-than-orthodox thoughts, but also a *Monitum* decree (March 1962) issued by the Holy Office on this priest's works went so far as to warn bishops and heads of seminaries of those doctrinal errors said to be inherent in his interpretation of humankind within an evolving nature.

It is painfully embarrassing and profoundly regrettable that in the middle of this century a gentle, devout, and brilliant intellect like Teilhard de Chardin should have had to suffer silencing by the Roman Catholic Church because of his bold curiosity and daring originality in areas clearly lying beyond the Church authorities' scientific competence, let alone their lack of expertise in evolutionary thought.

Nevertheless, the sudden public enthusiasm elicited by Teilhard's posthumously published works was nothing less than phenomenal. Yet, a sound analysis and proper evaluation of his unique synthesis was not immediately forthcoming. Scientists referred to it as primarily religious and mystical, theologians objected to its glaring unorthodoxy, and naturalistic philosophers for the most part chose to ignore it. The truth is that this French scholar went far beyond any other philosopher of evolution in trying seriously to reconcile the relevant particular sciences, dynamic philosophy, and mystical religion within a holistic view of humankind's emergence and destiny in a divinely generated and God-embraced universe.

It is unfortunate that, during his lifetime, Teilhard never had an opportunity to submit his writings for professional criticism, which could have eliminated at least some of their many ambiguities. More than three decades after his death, Teilhard's thought and vision still remain both attractive and controversial.[1] Strongly negative assessments include those made from a very narrow viewpoint (e.g., those reviews by George Gaylord Simpson, P. B. Medawar, and Jacques Maritain). At the other extreme are those favorable assessments made by Theodosius Dobzhansky and Sir Julian Huxley; despite certain reservations, they highly praised Teilhard's bold attempt at an evolutionary synthesis.

Simpson did refer to Teilhard as a phenomenal man, an accomplished geopaleontologist and profound theologian who uniquely combined in one well-integrated personality an evolutionary scientist and a religious mystic. However, Maritain (one of the leading Thomists of this century) held that this Jesuit priest had, in fact, given to science a dazzling primacy and thereby committed an unforgivable sin against the human intellect; as such, Maritain referred to Teilhard's thought-system as theological fiction. Yet, the strongest rejection of Teilhard's worldview came from Medawar. He held it to be an antiscientific, unintelligible, and impractical philosophical fiction containing an abominably expressed feeble argument. So much for open-minded inquiry!

Briefly, Teilhard's work was referred to as a false science beyond demonstration, a subjectively contrived piece of theology-fiction, and even sheer nonsense in both science and theology. Nevertheless, there were a few scientists who were favorable to it. Sir Julian Huxley wrote a sympathetic and admiring introduction to the English translation of The Phenomenon of Man, thus giving the work scientific credibility and scholarly prestige. Huxley himself had also been attempting such an evolutionary synthesis, and therefore he welcomed Teilhard's bold and holistic conception of the place of humankind within natural history.

Similar to several earlier systematic thinkers (e.g., Aristotle, Spinoza, Leibniz, and Hegel), Teilhard had desired to present an overview of humankind's position in the universe. Where some others before him had merely been concerned with change and development, Teilhard, as did Spencer and Haeckel, saw the whole cosmos (including the human species) in evolution. Because Teilhard was at once a scientist, philosopher, theologian, and poetic mystic, his unique achievement is very difficult to place, assess, interpret, and evaluate. Yet, an attempt must be made to understand and appreciate the Teilhardian worldview in terms of scientific naturalism and rational humanism.

As will become very evident, Teilhard's shortcomings are due to the planetary and theological orientations he needed to take for his daring (but unsuccessful) effort to achieve a reconciliation between the natural universe and a supernatural God within a personalizing, converging, and involuting dynamic panentheism of mystical evolution.

It is safe to say that Teilhard's Roman Catholic philosophy of mystical evolution rests upon four basic interrelated concepts: (1) an evolutionary panpsychism or spiritual monism, (2) the accelerating cosmic law of increasing centro-complexity/consciousness, (3) the critical thresholds at transitional phases of terrestrial evolution, and (4) the future omega point.

Teilhard was not restricted to a single method or field of inquiry. What resulted from his holistic approach is a Christian natural theology favoring vitalism, teleology, and spiritualism. As such, it is still awaiting the sort of comprehensive and critical assessment that it undoubtedly merits in light of rigorous reflection and the established facts of the special sciences. For better or worse, a modern evolutionary materialist may reinterpret Teilhard in terms of scientific naturalism just as Marx had reinterpreted Hegel in light of dialectical materialism.[2]

For the scientific naturalist, the essential contribution of Pierre Teilhard de Chardin is to be found in his unequivocal commitment to the evolutionary framework in natural theology with all its implications, including almost certainly some to which he never gave explicit thought or critical attention.

Teilhard wrote only three books, but each is a major contribution: The Divine Milieu (1927), The Phenomenon of Man (1940, with a postscript and appendix in 1948), and Man's Place in Nature: The Human Zoological Group (1950).[3] His ideas are also expressed in voluminous collections of expert essays, scholarly articles, and both insightful and revealing letters to family, friends, and professional associates.[4]

Human inquiry is never free from conscious or unconscious predilection and consequent preferential judgments: all scientists, philosophers, theologians, artists, statesmen, and politicians have their unavoidable vested interests. This is why no ideology, fact-system, or belief-system springs forth unaffected by its author's personality or outside of the economic, sociopolitical, and cultural conditions of the environment as well as its author's general place in the process of history. Teilhard's worldview is no exception. To remove this Jesuit scientist and his own system from such pervasive influences would amount to an injustice not only to him personally, but also to those facts of and ideas about cosmic evolution. Therefore, in order to understand and appreciate soundly and properly the development of Teilhard's philosophy of evolution from his early World War I writings and his first book, The Divine Milieu, to his major systematic work, The Phenomenon of Man, and the following scientifically oriented Man's Place in Nature: The Human Zoological Group, it is necessary to consider the significant events in his life that determined the growth and development of his thought as recorded in his books, essays, articles, and letters. As with Darwin, Teilhard is a product of his own time.

On 1 May 1881, Marie-Joseph-Pierre Teilhard de Chardin was born at Sarcenat in the province of Auvergne, near Orcines and Clermont-Ferrand, France.[5] He was the fourth of eleven children, and his ancestry could boast of such luminaries as Pascal and Voltaire. The influence of his mother, Berthe Adèle de Dompierre, kindled in him a deep lifelong love of and wholehearted dedication to Christian mysticism. His father, Emmanuel Teilhard de Chardin, encouraged an active and ardent interest in natural history. Therefore, even as a child Teilhard was equally and simultaneously immersed in science and religion.

When a youngster, Teilhard found and hid a plowshare (his "genie of iron"). The boy believed it to be that unchangeable object his inquisitive mind had been searching for. Sometime later after a rainstorm, however, Teilhard discovered that this most precious possession had actually rusted! He then threw himself on the ground and cried the bitterest tears of his life.

Teilhard always had an irresistible need for one essential thing, e.g., a supernatural entity that would be simultaneously necessary and sufficient in itself. As a result, his developing enthusiasm for astronomy, mineralogy, biology, and entomology were emotionally (though not intellectually) secondary to a profound personal yearning for an all-embracing, supercosmic absolute.

Leaving behind his most precious possessions (a boyhood collection of rocks and pebbles), Teilhard, at the age of eleven, entered the Jesuit secondary school of Notre-Dame de Mongre in 1892. He studied Greek, Latin, German, and the natural sciences and developed a concern for philosophy. On 20 March 1899, he entered the Jesuit novitiate in Aix-en-Provence, for at the age of seventeen his desire to be "most perfect" determined his choice of future vocation as a Jesuit priest. While a junior, he continued his studies in geology and philosophy on the channel island of Jersey (1902–1905). After finishing his scholasticate in Jersey, he was then sent to teach chemistry and physics at the Jesuit College of the Holy Family in Cairo, Egypt (1905–1908). When free from teaching the natural and social sciences, which included botany, the young instructor preferred to do research in the deserts.

Teilhard's first publication was an article titled "A Week at Fayum" (1908). It was soon followed by an early professional publication under the title "The Eocene Strata of the Minieh Region" (1908). Yet despite continuing accomplishments in both geology and paleontology, his religious beliefs still remained orthodox and free from any scientific influence. Teilhard was next sent to Ore Place, the Jesuit house in Hastings on the Sussex coast, England. There, he completed his theological studies (1908–1912). While remaining interested in numerous scientific subjects (rocks, fossils, plants, insects, and birds), Teilhard's religious orientation was always present in his thoughts and writings as well as his teaching and correspondence.

On 24 August 1911, Teilhard was ordained a priest in the chapel at Ore Place. The following year, he successfully passed his final exam in theology, which resulted in his earning the equivalent of a doctorate in that discipline. Teilhard then pursued scientific studies at the Museum of Natural History in Paris under Marcellin Boule, a noted professor of paleontology. Teilhard joined the Geological Society of France and established a lifelong friendship with the Abbé Henri Breuil, a distinguished specialist in prehistory remembered for his outstanding contributions to comprehending the significance of paleolithic cave art.

Sometime during this period, Teilhard had read Henri Bergson's major philosophical work, *Creative Evolution* (1907). This seminal book had a profound influence on the Jesuit scientist's own incipient thoughts. No longer able to hold to a strict and literal interpretation of the traditional scriptural account of divine creation as recorded in the sacred myth of Genesis, Teilhard now easily shifted from dogmatic orthodoxy to an evolutionary perspective. Within a scientific and religious framework, he now viewed the entire universe as a dynamic process; as such, he even referred to the universe as a cosmogenesis. Unlike Bergson, however, who saw planetary evolution as an essentially creative but diverging movement of life, Teilhard formulated a Catholic interpretation of terrestrial evolution that stressed the converging and involuting aspects of spirit as it moves to unite a collective humanity with a personal God at the future omega point.

Bergson's view of terrestrial evolution had been founded upon an ontological dualism between inert matter and animated spirit as consciousness or the *élan vital*. He held the latter life force to be responsible for the irreversibility, historical continuity, and increasing novelty or creativity as well as complexity, diversity, and consciousness manifested throughout planetary evolution. Unlike Lamarck, Bergson offered a sophisticated vitalism.

Teilhard, however, adopted a monistic ontology giving a privileged position to consciousness or spirit. In brief, Teilhardian evolution is vitalistic, cumulative, irreversible, accelerating, personalizing, teleological, and converging or involuting toward ever-greater improbabilities, complexity/consciousness, value, freedom, and unity as a creative equilibrium or spiritual synthesis, with the perfection of being itself occurring at the future omega point in terms of human evolution reaching its final goal on earth.

For Teilhard, the theory of evolution provided that necessary framework within which he could now synthesize his many interests in science and philosophy with his deepest commitments to theology and mysticism. This dynamic orientation could also do justice to his cosmic interpretation and eschatological vision of Christ (he was already publicly advocating the validity and even necessity of an evolutionary perspective for grasping the essence of Christianity).

Teilhard's first encounter with human paleontology (physical anthropology) was an unfortunate one, indeed. In 1909, he had met the amateur geologist Charles Dawson. In 1912, joined by Sir Arthur Smith-Woodward of the British Museum of Natural History, they visited Piltdown, England, where Dawson unearthed a new fragment of a human skull and Teilhard found an elephant molar. The three returned to this same site in 1913. This time, Teilhard discovered an additional canine tooth attributed to the lower jaw of the now-infamous Piltdown Man (*Eoanthropus dawsoni*). However, this incident of dubious value is nevertheless of personal significance because it kindled in Teilhard a lasting interest in paleoanthropology. Fortunately, this cautious young scientist remained skeptical of the skull fragments and teeth then attributed to Piltdown Man since this fossil find presented a structural anomaly in the then emerging picture of human evolution.

In 1953, Dr. Kenneth P. Oakley, using the then newly developed fluorine relative dating method, revealed that this specimen was, in fact, a fraudulent miscombination of a far more recent human cranium with an orangutan mandible![6] Teilhard remarked that "Nothing seemed to 'fit' together-it's better that it all fell through."

Recently, the geopaleontologist Stephen Jay Gould of Harvard University has made an attempt at reconstructing this whole unfortunate footnote to the history of evolutionary science.[7] However, his own interpretation of the event seems to be more imaginative than convincing, more assumptive than sound. If guilty at all, the highly ethical Teilhard could be so only by circumstantial association and perhaps omission, but not because of a deliberate conspiracy and/or conscious commission.

Be that as it may, Teilhard's scientific research now took him to every major human paleontological site known during his lifetime. In fact, the removal of Piltdown Man from the legitimate record of fossil evidence for hominid evolution actually favored this Jesuit priest's preferred interpretation by supporting his own hypothesis of a continuous evolution of the human brain and its cultural outcome.

Charles Darwin had erroneously maintained that the modern human cranial capacity had evolved before our own species became a toolmaker; one may now safely assume that the human animal was an erect toolmaker over two million years ago with only about a third of its present cranial capacity. Human intelligence and cultural development have evolved together in a creative feedback process between ongoing biological and psychosocial causes and effects.

In 1913, Teilhard joined the Abbé Breuil on an expedition to the prehistoric sites in the Pyrenees.[8] Early mammalian fossils from Quercy and the Rheims areas in the southwest of France provided Teilhard with the empirical evidence for his dissertation in geopaleontology. He began work on his doctoral thesis at the Institute of Human Paleontology in the Museum of Natural History, Paris.

Teilhard's research met sudden and unusual interruption when he was drafted in December 1914 and subsequently assigned to the thirteenth division of the medical corps. He served at the front as a stretcher bearer (1914–1919) and was cited three times for distinguished service: he was made Chevalier of the Légion d'Honneur and held both the Croix de Guerre and the coveted Médaille Militaire. Teilhard had shown great courage and humility, and his letters to his cousin Marguerite Teillard-Chambon (Claude Aragonnès) illustrate his continued optimism and developing mysticism even during the grimmest years of the war.

Teilhard's experiences during this tragic conflict did not deaden his spirit.[9] On the contrary, he emerged more dedicated to the two objects of his passion: a personal God and this evolving universe. His belief in a spiritual cosmogenesis prevented him from adopting the intellectually tempting but emotionally unsatisfying position of cosmic pantheism. His own mystical interpretation of a converging and involuting universe allowed for the future culminating union of a spiritualized evolving planet with an all-enclosing personal God. Although he had temporarily and discreetly replaced traditional theism with a dynamic but provisional pantheism, he now reconciled and transcended both of these antithetical viewpoints in a synthetic but process panentheism. Teilhard focused on this last position and dedicated his life to a philosophical clarification of and scientific demonstration for his belief in the validity and necessity of what was for him the definitive viewpoint concerning the survival and fulfillment of humankind on earth. For his unique theological perspective and spiritual inspiration, this Jesuit priest relied upon the writings of Saints Paul and John of Patmos to substantiate the seemingly Roman Catholic orthodoxy of his cosmic and mystical view of a finalistic panentheism in terms of God, humankind, and planet earth.

After the war, with renewed strength and hope in the future progress of our species, Teilhard took his solemn vows on 26 May 1918 at Sainte-Foy-Les-Lon, France. His early writings during the war had been primarily devoted to an explication of the immanence and transcendence of a personal God, i.e., the converging and involuting evolution of the spirit of this world toward a final creative union with an Absolute through the attracting cosmic force of love energy. Now, this Jesuit priest returned to his scientific research. He taught as an associate professor of geology at the Institut Catholique (1920–1923) while resuming work on his doctoral thesis about the mammifers of the Lower Eocene in France and the layers of soil in which they are found. However, his profound religious commitment and pervasive global optimism endured throughout this scientific research and beyond.

In 1922, Teilhard received the title of doctor with distinction at the

Sorbonne for his thesis on *The Mammals of the Lower Eocene Period in France* (he was by now a formally recognized professional geopaleontologist).

Père Emile Licent, a fellow Jesuit, had built a museum and laboratory at Tientsin in China for the purpose of promoting the study of Chinese geology, mineralogy, paleontology, archeology, and biology. As the director of this project, Licent invited Teilhard to join the French Paleontological Mission. Thus, on 6 April 1923, a very enthusiastic Teilhard left Paris for Tientsin. He accompanied Licent on an expedition to inner Mongolia and the Ordos Desert. It was during his journey into the interior of this desolate region that, on a Sunday and having no means to celebrate Mass on the Feast of the Transfiguration, Teilhard finished his religious and mystical poem entitled "The Mass on the World" (1923).[10] For Teilhard, cosmogenesis as evolutionary convergence focused on this earth will lead to an ultimate mystical unification of our humanized and spiritualized planet with the supracosmic and trans-spatiotemporal personal God at a future omega point as the fulfillment of Christogenesis.

On 13 September 1924, Teilhard left China and returned to France. During the winter months of 1925–1926, he gave four lectures on the theory of evolution and originated his concept of the noosphere, an idea that represents the hominization of this earth considered from both the biosocial and psychocultural perspectives. Considered spiritually, the noosphere represents an accumulation of all persons as reflective monads or centers-of-consciousness within an earth-bound layer or terrestrial envelope of minds and their thoughts. Through continued human evolution, Teilhard held these sensate units of psychic force are collectively converging and involuting toward a personal God as the supreme center of a process reality.

Because of his evolutionary orientation and largely unorthodox interpretation of original sin, the Jesuit superiors now felt compelled to withdraw Teilhard from the Institut Catholique; his lectures on evolution were already challenging the traditional Catholic view of a more or less static and complete planet with its hierarchy of eternally fixed forms. Moveover, Teilhard was now confined to scientific research in descriptive geopaleontology, forbidden to publish or teach in the areas of philosophy and theology, and exiled from France back to China in 1926 (thereby removing him from the European intellectual community).

During his return stay in Tientsin, from November 1926 to March 1927, Teilhard wrote his first book entitled *The Divine Milieu*, a spiritual essay on the life of an inward vision that attempts to reconcile the adoration of God with a love for nature. It advocates the divinization of human activities and passivities while supporting a ubiquitous application of the theory of evolution. Critical of Christian asceticism, it teaches that the goal of human evolution will be reached only through a collective effort of our species.

Like some before him (e.g., Saint Bonaventure, Nicholas of Cusa, and Duns Scotus), Teilhard's philosophy of man within nature is not only clearly geocentric and anthropocentric but also, and more importantly, essentially Christocentric. His panentheistic perspective of reality is now very evident in his writings: a personal God is at once both immanent to and transcendent of this spiritual world.

For Teilhard, sidereal evolution is spiritualizing and therefore purifying itself. Cosmic history, which at first reveals itself as a multiplicity of separate evolutions, is actually the unfolding of a single great mystery through the immanent force of Christ as the ultimate source of meaning and purpose to the whole process of universal and directional development. As a process philosopher, Teilhard was convinced that through a concerted effort by humankind a new and better earth is slowly being engendered. As a result, the evolutionary fulfillment of planetary Christogenesis at the point of Parousia (the Second Coming of Christ) will bring about the synthetic and mystical union of a collective humanity with a personal God, jointly forming the Pleroma (the divine end of human evolution in the fullness of time). It is evident that Teilhard's teleological position rests upon profound religious faith rather than certain scientific knowledge. In any event, he was refused the official Church permission needed for the publication of his first book; surely, it was a theologically unorthodox interpretation of the human situation in terms of a cosmic and mystical evolution.

Nevertheless, Teilhard continued to see Christ in universal history and a personal God as the beginning and end of cosmic evolution. He was never to abandon this highly personal and very controversial position. In fact, he would later attempt to give his vision both scientific support and theological justification in his second and major work titled *The Phenomenon of Man*.

In 1929, Teilhard was appointed scientific advisor to the National Chinese Geological Survey as well as advisor and collaborator of the Cenozoic Research Laboratory established by Dr. Davidson Black, then director of the Peking Union Medical College.[11] Teilhard also participated in both the Central Mongolian Expedition (1930) and the Yellow Expedition into central Asia (1931–1932). During his twenty years in China, he managed to take part in geopaleontological field trips to India, Burma, Java, Ethiopia, and the United States. As a natural scientist, Teilhard was primarily a geologist, secondarily a paleontologist specializing in mammals, and only in the third place a paleoanthropologist. Even so, he is now best remembered for his association with the fossil hominid discoveries from the Western Hills of Zhoukoudian near Beijing, China.

It is ironic, indeed, that Teilhard was relocated in China because of his evolutionary convictions. Only twenty miles southwest of Beijing is Zhoukoudian, an area that was soon to become and remain one of the world's most significant human paleontological sites. In 1928, the remains of a fossil hominid jawbone had been found and, in the following year, the largest part of a fossil hominid cranium was also unearthed at this same site. The jawbone and cranium from Zhoukoudian were held to belong to Peking Man (*Sinanthropus pekinensis*), a hominid estimated to be at least 350,000 years old.[12] Teilhard received worldwide recognition for his scientific articles that popularized the series of findings from this site in the Western Hills, although he neither discovered this fossil evidence nor analyzed it. However, Teilhard did contribute to an understanding of and appreciation for the geological and paleontological contexts of these important finds through his own numerous and detailed scientific reports. Likewise, he always remained aware of their broader philosophical and theological implications. Unfortunately, this Jesuit priest was not allowed to express any of these metascientific insights in his published writings (although he continued to record his private thoughts and personal visions for posterity, hoping that one day these ideas and speculations would be publicly available in print).

The discovery of this *Sinanthropus* material (1928–1934) reinforced Teilhard's evolutionary perspective as well as his own interpretation of human evolution, which needed additional scientific support.[13] For this singular man of insightful genius, evolution is applicable not only to geology and biology in general but also, and more particularly, to the origin as well as sociocultural history and psychic destiny of our own species on earth.

Teilhard's involvement with the *Sinanthropus* material was supplemented by scientific field trips to central Asia with George Barbour (1934), to India and then Burma with Helmut de Terra (1935, 1937–1938), and also twice to Java to investigate the pithecanthropine evidence at the invitation of G. H. R. von Koenigswald (1935, 1938).

In 1937, after years of scientific research and rigorous reflection, Teilhard, now a famous geopaleontologist, started his preliminary notes for a grand synthesis that would interpret evolving humankind within natural history from a comprehensive perspective. Work progressed steadily, with the Jesuit priest writing one or two paragraphs each day. Finally, after two years (from June 1938 to June 1940), he completed his major manuscript and titled it *The Phenomenon of Man*. On 6 August 1944, however, the author learned that ecclesiastical permission to publish his controversial volume had been flatly refused. Teilhard remained in Beijing until after the war; in 1946, he left for France and was never to return to China again.

In 1947, Pierre Teilhard de Chardin was elected to the French Academy of Sciences for his outstanding work in geopaleontology throughout a long and distinguished career as natural scientist. He had already held a position as corresponding member of mineralogy in this prestigious society.

Although Teilhard suffered his first heart attack in 1947, the following year he journeyed to Rome personally to request official permission for the publication of both *The Divine Milieu* and *The Phenomenon of Man* (the latter had been slightly altered for that purpose), as well as to seek permission to succeed the late Abbé Breuil as professor of prehistory at the Collège de France. Unfortunately, his earnest requests were rejected outright.

Discouraged but still optimistic, Teilhard now wrote his third book titled *Man's Place in Nature: The Human Zoological Group* (1950). Although it is a somewhat more scientific and lucid restatement of the arguments found in his major volume, this work was not to be published during his lifetime. Nevertheless, he continued to express his original thoughts and unconventional beliefs in essays privately circulated among close friends as well as through prolific correspondence around the world. Fortunately, at least his important contributions to science were publicly recognized and professionally appreciated.

In 1951, Teilhard made his seventh visit to New York City. He now accepted a research post at the Wenner-Gren Foundation for Anthropological Research (Viking Fund). At the invitation of C. Van Riet Lowe, this organization sponsored Teilhard's two trips to the australopithecine sites in South Africa (1951, 1953). In fact, during his life, this Jesuit priest had managed to examine the major australopithecine, pithecanthropine, and neanderthal fossil evidence.

Earlier in the year, Teilhard had expressed a wish to die on the Feast of the Resurrection. Ironically, at the age of 74, he died of a sudden stroke on 10 April 1955 in New York City. With a certain poignant appropriateness, his death actually occurred on Easter Sunday evening! He was buried at Saint Andrew on the Hudson, in the cemetery of the Jesuit novitiate for the New York Province. Encouraged by others to do so, this brilliant but controversial thinker had wisely given all his unpublished manuscripts to the care of Jeanne Mortier, a loyal and trustworthy friend (Teilhard's complete bibliography amounts to over five hundred titles). By the fall of 1955, the first edition of *The Phenomenon of Man* was published in France in its author's native language.

The Phenomenon of Man is a truly remarkable synthesis of science, philosophy, theology, and poetic mysticism within a liberal Roman Catholic interpretation of cosmic evolution and earth history. In its preface, Teilhard writes that he is offering a scientific treatise dealing with the whole phenomenon of humankind, i.e., an introduction to his explanation for the place of our own species considered solely as a unique phenomenon within a dynamic universe. Yet, this book does abound with assumptive reasoning and is ultimately grounded in personal faith, process theology, and a pervasive mysticism. Why, then, did Teilhard claim scientific status for this treatise? First, because he wanted to gain the attention of the scientific community.

Second, because he actually believed his vision to be scientifically sound. Lastly, in sharp contrast to dogmatic theology, his own work did recognize open science and free inquiry.

In his forward, Teilhard holds that either one must adequately see the phenomenon of man in its entirety or face the prospect of probable human extinction. With an employment of seven categories (space, time, quantity, quality, proportion, motion, and organic unity) for the illumination of one's vision, he taught that the result of a scientific investigation of man's place in nature would reveal that our own species is still the structural and spiritual center of this evolving universe. His frank geocentrism and bold anthropocentrism are steadfast and unequivocal .

Within an evolutionary framework, Teilhard saw the sudden emergence of three fundamentally unique events: the appearance of prelife, then life, and finally thought. He gave human thought a privileged position in this universe, clearly wanting to supplement previous analytical studies of the external structure of so-called material evolution with a much needed complementary, but more important phenomenological, investigation of the internal emergence of spirit (emphasizing the historical convergence of psychic evolution). Teilhard claimed neither to have given a final explanation of things nor to have contructed an original metaphysical system. As such, *The Phenomenon of Man* is a largely tentative but personal, suggestive but incomplete essay on humankind within nature founded upon a mystical presupposition: fuller being means closer union with a personal God.

To understand and appreciate his message adequately, one must consider each of Teilhard's four fundamental but interrelated assumptions that give coherence to his evolutionary scheme of things: (1) spiritual monism, (2) law of complexity/consciousness, (3) critical thresholds, and (4) the future omega point.[14]

Teilhard's Catholic philosophy of cosmic evolution rests upon an ontological monism. As a result of giving a privileged position to mind or consciousness over matter and energy, he held that the universe is ultimately psychic or spiritual in nature; it is developing toward greater unity and eventual perfection. His position of philosophical idealism as evolutionary panpsychism is religious in conception, inspiration, and orientation. Although it is more implicit than explicit in his writings, this universal spiritualism is essential for comprehending the Teilhardian cosmology and planetology.

Teilhard desired to take account of the meaning and purpose of consciousness in the cosmos. He started from both objective and subjective considerations of our own species as a recent biological, sociocultural, and spiritual event in nature. Teilhard extrapolated the necessary conditions of this universe that are required to bring our human phylum into existence and also guarantee its survival and fulfillment. In an early essay titled "Cosmic

Life" (1916), he presented an embryonic form of his intellectual and spiritual testament.

Although Teilhard had adopted a position of panpsychism, he retained an unclarified distinction between matter and spirit. What is clear, however, is that he held cosmic evolution to be a process of an ever-increasing spiritualization of matter. In "The Mystical Milieu" (1917), he committed himself to an ontological monism, maintaining that there is but a single substance created to sustain the successive growths of consciousness in this evolving universe.

In "Creative Union" (1917), Teilhard sounded in some ways very much like Gottfried Wilhelm von Leibniz, who was, however, though a teleologist and a spiritualist, neither a finalist nor an evolutionist. In "The Eternal Feminine" (1918), Teilhard taught that matter is a tendency or direction: it is the side of spirit that we meet as we fall back. He never doubted the spiritual evolution or ultimate unity of this entire cosmos, claiming that both plurality and matter are merely temporal aspects of reality. In "The Universal Element" (1919), he wrote that "strictly speaking, there is in the universe only one single individual (a single monad), that of the whole conceived in its organized plurality."

Written during World War I, Teilhard's early essays reflect a deeply religious mind struggling to reconcile the material world of a scientist with the spiritual orientation of a theologian. In his "Hymn of the Universe" (1916), he had even tried to describe his mystical experiences as a soldier-priest at the front lines of mortal combat.

Unlike René Descartes, whose subjective methodology unfortunately reintroduced into the history of philosophy an ontological dualism between mind and matter (in fact, Descartes held that there are three separate substances: thought, extension, and God), Teilhard developed a monistic metaphysics. It is evident that his thoughts were always inclined to give a preference to spiritual monism rather than dualism, pluralism, or pervasive materialism. In "My Universe" (1924), he gave a summary of his theistic and spiritualistic conception of the world.

Teilhard maintained that the order of interacting monads is not a continuum but, instead, manifests significant levels differing in quality and kind. Each successive ascending level of monadic existence is held to be both higher in evolution and closer to perfection than that preceding lower level, i.e., more conscious and therefore more spiritual (like Leibniz's monads, all those fragmentary elements of the Teilhardian universe mirror God imperfectly and differently). Leibniz had held that this world is an unbroken series of an infinite number of developing monads, each mirroring the absolute monad by degree in such a way that no two monads could ever be psychically identical at the same time. In his essay "The Phenomenon of Man" (1928),

Teilhard emphasized that sidereal evolution is not only spiritual but also irreversible. Like Leibniz, he held that God is responsible for a pervasive order and direction throughout the monadic development of this evolving cosmos.

During his years in China, free from public appearances and teaching responsibilities, Teilhard had ample time to reflect upon his scientific research and the far-reaching implications of the theory of evolution from a Christian perspective. Although immersed in both geological and paleontological investigations, this preoccupation with so-called matter did not divert him from his primary concern for the spiritual fulfillment of the human species. Teilhard's participation in the *Sinanthropus* excavations as well as the field trips to other major fossil hominid sites in India and Java strongly reinforced his view that the human animal is the most recent product of an organic evolution that had its origin billions of years ago. Although covering a span of serveral million years, human existence on earth represents, nevertheless, a recent and unique psychic event in terms of cosmic evolution. Through the emergence of humankind, this spiritual universe has become at least once aware of itself. Less concerned with the origin and emergence of our human phylum than he had been in the past, Teilhard now concentrated on developing a view conducive to a suitable outcome for the future spiritual-ization and global convergence of a collective humanity. In his popularizations of the *Sinanthropus* finds, Teilhard could only hint at the theological im-plications he was discerning from the design he believed to be manifested throughout cosmic and planetary evolution.

Teilhard did continue to write philosophical and theological essays, hoping that someday his superiors would grant him the permission needed to see them in print. There was always present in him the desire to write a book in which he would synthesize science, philosophy, and theology to give a proper account of that significance our own species holds in the cosmic flux of a spiritual universe.

Teilhard held that the existence of a cosmic development of pervasive spirit is the greatest discovery made by modern science. He also argued that our own species is actually an end product of universal evolution: if the essence of a human being is that it is a person, then this unfolding universe is ultimately a process of cosmic personalization. In the continuation of this argument, he also assumed that planetary evolution on earth will terminate in the formation of a supreme person at the future omega point.

Teilhard's clearest position on the nature of this evolving universe is presented in his major work, *The Phenomenon of Man*. Plurality, energy, and unity are held to be the three basic characteristics of this cosmos, i.e., the boundless world viewed as a whole manifests a system, a quantum, and a totum respectively. Matter is ultimately composed of homogeneous units of psychic energy that are dynamic, interdependent, and infinitesimal. These cosmic units of matter/energy/spirit reveal themselves in an indefinite development within duration as a converging space-time continuum. This philosophical atomism is reminiscent of the seeds of Anaxagoras and the atoms of Leucippus, Democritus, and Lucretius as well as the monads of Bruno and Leibniz. It should also be noted that although Teilhard sometimes refers to this dynamic universe as being boundless, one may safely assume that he actually held it to be finite (although of an indeterminately enormous size).

For Teilhard, spirit evolves from a homogeneous base to a future heterogeneous limit, i.e., the omega point as that supreme upper pole of this finite earth at the end of its planetary convergence. Structurally, he held that the evolution of spirit represents a cone or pyramid. This spiraling nature of evolution is absolutely crucial to his terrestrial model, as is the finite sphere of the globe (both must constantly be kept in mind). Briefly, Teilhard's converging spiritual evolution is in sharp contrast to Bergson's diverging creative evolution.

Teilhard considered the ontological status of a cosmic embryogenesis through which the potentialities of the universe are being actualized within an involuting space-time continuum. He held that universal evolution is divinely self-enclosed and self-sufficient, there being no addition of external spirit to replace that psychic energy that is irrecoverably lost through entropy as the necessary cost of creative syntheses. In short, cosmogenesis is subject to the two principles of the conservation and dissipation of energy as spirit. However, Teilhard actually held that the evolution of love energy is in a direction antithetical to cosmic entropy: while the so-called physical energy of this universe is diverging and dispersing to a state of equilibrium, the psychic energy of our cosmos is evolving to and will converge upon a future end point of dynamic concentration and final fulfillment.[15]

At first, Teilhard attempted to unite the two positions of materialism and spiritualism by merely making a conceptual distinction between determined tangential energy (the without-of-things as a materialistic horizontal component of nature) and free radial energy (the within-of-things as that vitalistic vertical component of evolution). These two forms of energy are interrelated so that everything in the dynamic universe appears to have both a physical aspect and a psychic pole.

Teilhard held that most scientists are concerned only with the structures of matter while most Marxists are preoccupied only with dialectical materialism as economic determinism. As such, they both fail to do justice to the emergence of human consciousness and its spiritual significance. Teilhard also held that the phenomenologists and existentialists do not take seriously enough the theory of evolution with its far-reaching consequences. Likewise, naturalists and humanists do not sufficiently appreciate the significance of Christianity for the survival and fulfillment of our own species on a finite earth.

Teilhard's own approach is a bold and unique attempt to do justice to both psychic and physical evolution, although preference is clearly given to spirit over matter (consciousness over energy). However, Teilhard did not retain this distinction between radial and tangential energies when he left the phenomenal level of investigation to consider the ultimate ontological status of this universe. In a crucial passage in *The Phenomenon of Man* he not only gives radial energy a privileged position in this cosmos but clearly commits himself to an ontological monism grounded in pervasive spirit. For him there is but a single energy operating within the whole world, and it is psychic in nature. (In the first English translation of *The Phenomenon of Man*, this one fundamental energy had been mistranslated as being physical rather than psychic in nature. As a direct result, the glaring error has been responsible for many unfortunate ambiguities concerning Teilhard's true ontological position of pervasive spiritualism.)

Surprisingly, this fundamental and crucial aspect of Teilhard's philosophy of evolution has been either entirely overlooked or largely misunderstood. Teilhard did continue to use the term *matter* in his later writings, while equating progress with the growth of consciousness. However, this ambiguity may be resolved by making the philosophical distinction between appearance and reality, i.e., the distinction between epistemology and ontology. From the epistemological perspective, one may distinguish between external objects (i.e., that which appears to be independent of human experience and is referred to as matter) and one's own internal consciousness that perceives such material objects. In contrast to this distinction, Teilhard as religious ontologist preferred to interpret all reality as being ultimately spiritual in nature: instead of spirit being merely a temporary emanation of matter, matter is only a temporary perception of alienated spirit.

At this point, a crucial distinction is called for between objective and subjective methodologies in terms of the actual status of this concrete world. If it is to be intelligible at all within his own view of things, then Teilhard's position must be acknowledged to support pervasive spiritualism. He also held that the beginning of cosmogenesis was prior to and independent of human experience. Philosophically, he even claimed that consciousness causes the further evolution of itself. Theologically, moreover, Teilhard went so far as to maintain that the immanent nature of God is manifested as radial or spiritual energy in the process of becoming (while the transcendent nature of God is that supreme center of love as being itself). By equating God's immanent nature with a cosmic interpretation of the Christ, Teilhard believed that this divine presence throughout the universe allows God's transcendent nature to make things make themselves. Teleology and vitalism are both important elements in his dynamic philosophy and process theology. In brief, universal evolution is ultimately the cosmic Christ unfolding and fulfilling himself toward a future omega point. To my knowledge, Teilhard is the only evolutionist who had dared to equate the Christ with this evolving cosmos (or at least our evolving earth).

Teilhard held that the universe has an internal structure not accessible to objective investigation. Therefore, it is necessary but not sufficient to deal with the external structure or complexity of things. Teilhard taught that the internal realm is more significant than the external one, as its consideration provides that means for comprehending the purpose and goal of planetary evolution. The emergent evolutionists had seen consciousness only as a recent quality. Unlike Teilhard, they did not trace the origin of consciousness as such back to the outset of cosmic evolution.

Teilhard held that all the objects of nature have both an interior as well as an exterior. Bergson had acknowledged this duality when he postulated that a stream of consciousness causes and runs through the creative evolution of the otherwise inert matter of this universe (therefore he formulated an unwarranted vitalistic dualism). Teilhard's unique phenomenology of cosmic evolution attempts to account for the development of both matter and mind. Yet, to do justice to the evolving interiority of this universe, Teilhard did an injustice to its material foundation by giving ontological preference to spirit over energy.

As a scientist and Jesuit priest, Teilhard sought to reconcile naturalism with theism; he could not satisfy himself with mere agnosticism or tempting pantheism. As a devout Christian, Teilhard rejected both atheism and materialism in favor of theistic personalism and a spiritualist metaphysics. Since Christian eschatology was always foremost in his thought, there was no compromising as far as ontology is concerned. To allow for the future convergence of humankind with God, his faith took preference over science. Unlike the orthodox conception of theism, Teilhard established a process metaphysics supporting dynamic panentheism. For a scientific naturalist, the Kantian separation of the phenomenal realm from the noumenal world is untenable, since it leads to a superfluous and unnecessary ontological dualism. One might look at the universe from the position of perspectives: there are an infinite number of possible perspectives of the material cosmos, and our human perspective is only one of these.

Teilhard had erroneously overextended the ontological status and evolutionary importance of mental activity. His spiritualist monism is colored by religious bias; the material universe is seldom in accordance with one's needs, wishes, hopes, and desires.

The Jesuit priest held that, from its beginning to its end, the evolving cosmos is nothing other than the growth of consciousness. However, the emergence of consciousness on earth as a natural activity is, as a matter of fact, a relatively recent event in cosmic history. It is anthropocentric in the extreme to claim that the fleeting phenomenon of human consciousness is of primary importance in this material universe.

Ideally, philosophical concepts should supplement scientific facts and rational speculations, thereby making the kaleidoscopic process of this complex world intelligible. On the realistic assumption that the material universe itself is the foundation of life and thought, science and reason ought to be sufficient to account for evolutionary creativity. Within a naturalistic attitude, cosmic evolution appears to be an endless series of interrelated objects and events and relationships grounded in the underlying and pervasive continuity of matter and energy. A more rigorously scientific philosophy of evolution acknowledges the primacy of matter or energy and sees the human mind as merely one material form of natural activity resulting from biological development. Specific questions concerning the historical, structural, and functional aspects of matter should be resolved by further scientific inquiry and logical reflection.

As a natural scientist, Teilhard devoted his life to descriptive geopaleontology. He was an expert on Chinese geology and specialized in vertebrate paleontology, concentrating on mammalian evolution. In later years, his interest shifted to primate paleontology in general and human evolution in particular. His work was both analytic and synthetic, for every fact was quickly incorporated into an evolutionary framework. Teilhard taught that science and religion supplement each other, never failing to extend the implications of his vision to the future evolution of our own species on earth. Although he had been permitted to publish his strictly scientific articles, he was repeatedly warned not to delve into philosophy or theology. Fortunately, his published popularizations of the *Sinanthropus* material brought him worldwide fame and academic recognition.

Like Aristarchus in antiquity, Nicolaus Copernicus had held that the geocentric model of this cosmos could be replaced by his own heliocentric theory of a sun-centered universe. Later, Galileo Galilei gave this heliocentric theory considerable scientific support through his remarkable discoveries in descriptive astronomy. Furthermore, Giordano Bruno even argued that reality is an eternal and infinite universe with its center everywhere and its circumference nowhere. As a result of these new facts and perspectives, man's terrestrial habitat no longer occupied a central place or special position in this cosmos. The earth was now thought of as being merely one heavenly body among innumerable other similar celestial objects. As anticipated by Nicholas of Cusa, Giordano Bruno, and Johannes Kepler (among others), it is even probable that life, including sentient beings, exist in other solar systems elsewhere in this universe. Consequently, both geocentrism and anthropocentrism have taken a fatal blow from science and reason.

In spite of these impressive developments, however, Teilhard still argued that humankind is in fact the center of this universe. He held that our own species does occupy both a unique and crowning place in the history of planetary evolution as well as cosmic development. According to Teilhard, humanity is capable of investigating both the psychic (internal) and physical (external) structures of the cosmos. He taught that since our own species is the highest and latest product of organic evolution, it has to be the most important in terms of scientific significance. Furthermore, as the only self-conscious animal in nature (and therefore qualitatively far above all other species), humankind has to be of singular religious and philosophical importance as well.

To Teilhard, this universe is a divinely created but self-unfolding and ever-progressing process of evolution. All objects, events, and relationships are viewed correctly if and only if they are seen as parts of this universal becoming. The old conception of a geostatic cosmology had misrepresented the dynamic history and spiritual unity of this evolving cosmos.

Teilhard's new vision called for many neologisms. He wrote of a cosmogenesis that included geogenesis, biogenesis, and noogenesis. The entire history of the universe, from stars to persons, represents more or less a continuous process of progressive development. Evolution is not merely a provisional and useful hypothesis. Rather, cosmic evolution governs and influences all inorganic, organic, and psychic phenomena.

Teilhard even held that he had discovered a law of evolution that gives meaning and purpose to the multiplicity of cosmic objects, events, and relationships: a universal law that not only explains the past history of the cosmos but, if extended to its final logical conclusion, can also foretell the condition of an ultimate earth as well as the end of human development upon it. Teilhard referred to this regularity in evolution as the law of complexity/consciousness. To be more exact by taking into account the converging and involuting nature of this evolving universe, he referred to it as the accelerating cosmic law of increasing centro-complexity/consciousness.

As a theologian, Teilhard sought to reconcile planetary evolution with the existence of a personal God as well as the personal immortality of the human soul, the responsible freedom of the human will, and a divine destiny for a collective humankind. He taught that science, philosophy, and theology are merely three distinct but not separate levels of intellectual investigation. Each level presents one major perspective of a single cosmic process. Likewise, each level supplements the incompleteness of the other two in giving a comprehensive and intelligible overview of the place of humankind within the universe.

As a mystic, Teilhard experienced the evolving unity of a spiritual universe as a cosmogenesis that is progressively developing toward a personal center of supreme consciousness. Yet, he never neglected the sciences. In fact, it is though a contemplation of evolving nature that Teilhard arrived at his mystical vision of a God-surrounded and God-imbued world. At the end of terrestrial evolution, and in terms of a collective human consciousness, he taught that dynamic panentheism will finally be transcended by a true pantheism in which God shall be all in all at the future emergence of an omega point.

Teilhard is concerned with the whole phenomenon of humankind, especially in terms of its unique appearance within cosmic evolution and its final goal at the end of planetary history. Actually, one of the major difficulties in giving a definitive assessment of the overall value of this attempt at an evolutionary synthesis is the regrettable fact that one cannot say without hesitation whether its perspective is essentially cosmic in scope or merely planetary in orientation. Despite some cursory references to other possible processes of evolution elsewhere in this universe, Teilhard leaves this crucial point unsettled and indeterminant; his cosmology is really a planetology.

Every descriptive methodology necessarily implies an ontology (there is no human inquiry entirely free from presuppositions that are conscious or unconscious, stated or implied). Although Teilhard claims to be giving merely a scientific description of the process of evolution, one is not surprised when this Jesuit priest ultimately grounds his philosophy of humankind within nature in a process theology as mystical panentheism.

What Teilhard does give us, at least in part, is a phenomenological analysis of planetary evolution. His orientation was objective and subjective to do justice to both external (material) and internal (conscious) evolution, respectively. The result is a description of the essential structures, functions, relationships, and essences within the phenomena of the universe. Without doing an injustice to established facts, Teilhard wanted to reduce all phenomena to their basic information to reveal a fundamental law that would disclose the meaning, purpose, direction, and goal of cosmic evolution as well as the place and destiny of our own species on earth. It is unfortunate, however, that his phenomenology of evolution did not manage to bracket out the methodologically dispensable religious presuppositions he believed necessary for a true interpretation of both celestial and terrestrial history.

For Teilhard, this entire universe manifests a cosmic design that gives meaning and a single direction to evolution in general as well as a purpose and ultimate goal to the human phylum in particular. In addition to Pascal's two abysses (the infinitely great and the infinitely small), this Jesuit scientist recognized and emphasized the evolutionary significance of a third and no less perilous abyss in the universe: that abyss of the infinite complexity of elemental arrangements.

Teilhard viewed this dynamic universe from two major perspectives: a horizontal or synchronic perspective concerned with the size of things, and a vertical or diachronic one concerned with the evolution of things. The former recognizes that tremendous range of size in cosmic objects, from the subnuclear elemental parts of atoms in the microcosm to the immense stars and comets of the macrocosm. In this breathtaking range, our own species has been held to occupy a position midway between the infinitely small and the infinitely great. From such a horizontal viewpoint, humanity's place in the cosmic scheme of things may seem to be at first both insignificant and inconsequential. However, the Jesuit scientist argued that this conclusion is, in fact, erroneous when one considers the evolutionary viewpoint in terms of internal complexity.

Teilhard taught that if one takes a vertical view of the universe, i.e., if one considers the position of things within the historical growth and development of the cosmos, then our own species becomes the most significant object in the totality of terrestrial nature (if not in the entire universe). This follows from acknowledging the degree of interior complexity and the resultant manifestation of radial energy found in the human animal.

According to Teilhard the process philosopher, there are three discernible stages of planetary evolution thus far: prelife, life, and thought. Within each stage there is a great spectrum of diversity and degrees of consciousness. Yet, while ascending the atoms of Mendeleev's periodic table and following through organic and psychosocial emergence, one finds that each successive manifestation of evolution is always more complex in its internal configuration of elements, and at the same time each displays increasing degrees of radial energy. The simplest living things, e.g., bacteria and, under some conditions, viruses, are infinitely more complex in their interior structure than the composition of mere stars and geological formations. Similarly, the metazoans are infinitely more complex than the protozoans, while the vertebrates are even more complex than the invertebrates. While moving through the fishes, amphibians, reptiles, and mammals (especially the primates) toward the appearance of the human animal, one notices that this natural tendency toward ever greater interior complexification and therefore consciousness continues, but at an accelerating rate.

In Teilhard's own interpretation of this tendency, the degree of consciousness is directly proportional to that degree of interior complexity and underlying concentration of the structural elements in the central nervous system and brain of an organism. The position of any object in the evolutionary scale of terrestrial nature is determined by its degree of complexity/consciousness. Teilhard further argued that our own species is the most significant creature now known in this universe because it is the most complex and therefore the most conscious of all. Unfortunately, this geocentrism and anthropocentrism is pervasive and persistent in his thought (reminding one of the Aristotelian worldview). The Jesuit priest would have been on less controversial ground had he clearly limited his plausible claim to this planet alone instead of boldly suggesting that it might apply to the whole universe.

Teilhard maintained that to write the true natural history of the world, one would have to be able to follow it from within. Therefore, to correlate this evolutionary increase in complexity or interiorization of matter with the resultant increase in psychic activity, he held that there must be a certain law on which reality is based (a pervasive law of increasing complexity and consciousness in advancing creative unity). Teilhard referred to that qualitative curve of the universe as the accelerating cosmic law of increasing centro-complexity/consciousness. The law provided him with a means for viewing geological, biological, and psychosocial evolution as three distinct stages of one continuous process converging and involuting toward an ultimate spiritual end of this planet in terms of humankind on the earth.

Teilhard saw this law of complexity/consciousness manifested in those successive developments of planetary evolution: units of psychic energy (the monadic stuff of the universe), elementary corpuscles of the cosmos (electrons, protons, neutrons, etc.), systems of atoms (molecules and megamolecules), cells (protozoans and metazoans), and persons. In organic evolution, the acceleration of this law becomes obvious as one moves up through the fishes, amphibians, reptiles, and mammals to the subhuman primates and finally to the appearance of our own species. The acceleration of this increasing complexity has led to a concentrated manifestation in the differentiation of the tissue in the central nervous system and especially the expansion of the brain (particularly in primate evolution).

Teilhard saw the plant, insect, and bird kingdoms as unsuccessful deviations from this general tendency throughout organic history on earth. Strangely enough, this French scholar did not see the interdependent plant, insect, and bird evolutions as supportive of and supplementary to humankind's physical survival and social development, or as essential preconditions for the spiritual maturation and fulfillment of our own species. He held that although the insects (particularly the Hymenoptera and Lepidoptera orders) and birds represent a great proliferation in complexity and diversity, nevertheless their radial energies have been arrested in their growth and consequently are now ossified as instinct. Likewise, the psychism as small amounts of radial energy in plants is far too diffuse to be of any serious consequence. Whatever the objections to Teilhard's scheme, this law provided him with an argument for direction, meaning, purpose, and even a final goal within planetary evolution. He always held that global history has a precise orientation and a privileged axis in terms of the psychic development of spiritual energy.

Teilhard emphasized the special importance of the primate order. During the nearly two million years of primate evolution, one sees the successive emergence of first the prosimians, then the monkeys of both the New World and Old World, and finally the lesser and greater apes as well as our own species. In particular, the diversified Miocene dryopithecinae complex of hominoids eventually evolved in three separate directions: the line of hylobates, pongids, and that of the hominids (the latter line led to the human animal of today).

The Jesuit priest noted that from the evolution of some early tree shrew form to *Homo sapiens sapiens*, the accelerating law of increasing centro-complexity/consciousness concentrated with rapidity on the central nervous system and brain. Although one may be impressed with the size of the dinosaurs that dominated the earth for about 140 million years during the Mesozoic era before the appearance of the primates of the following Cenozoic era, these reptiles nevertheless had relatively small brains and therefore a meager degree of consciousness.

Teilhard speaks of cephalization and cerebralization: the former refers to an increase in the size of a brain in general, while the latter refers to the increase in the size of the cerebrum in particular. The frontal lobes of a cerebrum are necessary for abstract thinking and, therefore, intelligence. Primate evolution is important because it manifests this rapid development in the central nervous system and brain. As such, the superior intelligence of our own species is due not only to the fact that its brain is three times as large as the average brain of a gorilla but also, and even more importantly, to the fact that its neural structure is infinitely more complex. Teilhard maintained that the continued accleration of this law of centro-complexity/consciousness within the ongoing evolution of our human phylum is directly responsible for the emergence of that central phenomenon referred to as reflective consciousness. He argued that from a planetary perspective, the future outcome of this unique phenomenon could only be a superconscious and supercomplex collectivity of reflective persons converging and involuting around our finite earth.

In his essay "Science and Christ" (1921), Teilhard insisted on the fundamental groping of evolution through a process of "directed chance" toward ever-higher convergent integration within a precise orientation and toward a specific target (i.e., a teleologically preordained terminal synthesis as the unity of humankind with God).[16] Although aware of the obvious influences of both mutations and changing environments on the process of biological evolution, he proposed that from a planetary perspective the whole history of organic development reveals a trend too obvious to be merely the result of cumulative chance genetic variability and necessary natural selection. By underestimating these two major aspects of evolution and holding that somehow tangential energy is determined while radial energy is free, Teilhard never satisfactorily resolved the problem of freedom versus determinism. Nevertheless, he did claim that through the cosmic evolution of the universe freedom is steadily increasing. Presumably, then, there are degrees if not kinds of emerging freedom.

This cosmic law of centro-complexity/consciousness is as crucial to Teilhard's thought as is his spiritual monism. He even made a daring leap of faith from the alleged order of and design in cosmic evolution to the existence of a supreme intelligence. From his discernment of a cosmic law in this universe, Teilhard concluded that there must be an absolute cause behind it. He never doubted that universal evolution reveals such a cosmic order and creative designer. In his article "A Defense of Orthogenesis in the Matter of Patterns of Speciation" (1955), written only three months before his death, he still clung to the assertion that there is pervasive teleology in the paleontological record.

Like many others, Teilhard saw human psychosocial development as the continuation of organic evolution. The law of centro-complexity/consciousness now manifests itself in scientific and technological development. Teilhard held that, for the most part, our own species has remained biologically stable for about the last fifty thousand years. Except for its large brain, the human animal remains (somatically speaking) relatively unspecialized. The innovation of toolmaking, however, was analogous to a major and most favorable biological mutation in the human psychosocial realm. Paleolithic culture allowed our own species to cope with an ever-wider range of environments. For over three million years, the human animal has been using tools and weapons, ranging from the crudest stone and bone tools/weapons in the deepest paleolithic times to the most sophisticated metallic and electronic instruments of our present space age science and technology. This Jesuit scientist correctly saw these tools and weapons as specialized extensions of the human body that enhance adaptation and survival. Through their use, the human species can actively modify and transform the environment to suit its own needs and desires instead of just passively submitting to its challenges and opportunities. As a direct result, during the last fifty thousand years or so, the acceleration of complexity in tools/weapons has been paralleled by and mirrored in the increasing complexity of human social structures, functions, and relationships. Although human social organization and behavior has its origin in the activities of the earlier primates, the superior intelligence of our species allows it to become ever more capable of not only influencing but also directing its own further evolution.

Teilhard always remained an optimist. He simply held that wars are merely the "growing pains" of a converging human evolution, and rejected the possibility that all of mankind could be aborted by a global catastrophe. As a result of the continuing increase in human population with a consequent rise in demographic pressure on the closed surface of the planet, accompanied by further acceleration in the advancements of science and technology, Teilhard envisioned a psychically unified earth as the ultimate goal of complexity/consciousness in terms of a global and spiritual synthesis, i.e., the eventual formation of a final creative unity of all persons as the inevitable result of this terrestrial concentration of a monadic universe that is, in the last analysis, evolving toward spiritual fulfillment. From this perspective, Teilhard reinterpreted the Aristotelian "Great Chain of Being" in terms of an evolutionary process with its progressive scale of ever-increasing complexity/consciousness ranging from the homogeneous psychic units of the universe to an ultraconscious and ultracomplex human world of the future (also like Aristotle, he saw planet earth as structured in terms of spheres).

Lastly, this law of involuting complexity/consciousness may be understood and appreciated as a dialectical process as long as one remains concerned only with the phenomena of cosmic evolution. In Teilhard, there is a direct relationship between evolving matter and psychic energy throughout the history of the universe. Like Hegel and Marx, this Jesuit priest believed that throughout cosmic evolution increased changes in quantity do eventually result in sudden changes in quality. Whereas Hegel claimed this dialectical process to be eternal, Teilhard viewed it as finalistic; terrestrial evolution is teleological with an eschatological end in terms of human spirit.

All of the philosophers of evolution have been aware that universal history displays a chain of products ranging, by and large, from the very simple to the ever more complex. Like Teilhard but before him, Herbert Spencer had formulated a cosmic law to account for this trend throughout universal history. In his phenomenological analysis of cosmic evolution, Teilhard had reduced this process to the fundamental law of accelerating centro-complexity/consciousness. There are, however, serious errors in his impressive interpretation of cosmic evolution. For example, the law itself is not cosmic because Teilhard has chosen a perspective that is essentially planetary in scope. To date, there is no scientific evidence that biological and psychosocial evolution have been taking place elsewhere in the solar system, let alone anywhere else throughout the rest of the dynamic universe. It is possible, but highly improbable, that evolution as it has taken place on the earth has been occurring elsewhere on other celestial bodies in an exactly identical manner. Certainly, the astronomical number of variables involved that would have to be duplicated in precisely the same sequence staggers the human imagination (one recalls Nietzsche's cosmic idea of the eternal recurrence of the same). As with the rest of us, the only evolutionary process about which Teilhard was factually knowledgeable is the one occurring on this planet. Of course, it seems certain that organic evolution has occurred elsewhere in this vast universe but in very different ways; exobiology infers exoevolution.

Surely, it is a dogmatic assumption to hold that evolution must always result in greater complexity and increasing consciousness. Teilhard knew that various orders of animals have evolved without displaying a tendency toward reflective awareness. It has already been noted that he generally ignored the history and value of plants, insects, and birds because of his almost exclusively anthropocentric and spiritual orientation. Unlike Teilhard's, Bergson's philosophy of creative but divergent evolution is clearly less dogmatic and less anthropocentric because it recognizes the scientific importance of plants, insects, and birds as well as the human animal. Our own species is not, as Teilhard argued, indispensable to the cosmos if one assumes that humankind is merely a recent outcome of biological evolution, explained in terms of random genetic variability and necessary natural selection taking place in a seemingly eternal and infinite material universe (in fact, some plants, insects, and birds may yet inherit the earth).

Teilhard dealt with scientific descriptions. As such, his alleged a priori law of cosmic evolution is actually an a posteriori generalization of things arrived at through an ex post facto review of the available empirical evidence. It is not a universal law but rather a synthetic generalization limited in application to both celestial and terrestrial history as we now know it. One recalls Alfred North Whitehead's wise clarification of the fallacy of misplaced concreteness: a crucial distinction is necessary between human idealizations and physical reality, the latter being both independent of and prior to the human knower. The extreme complexity of nature and the probability that it, whether in part or as a whole, may evolve into a plurality of new structures and functions within seemingly limitless space and endless time make the assumption of an a priori law of cosmic evolution in a process universe very unlikely. There is no empirical guarantee that the end of our planet from a human perspective will result in the formation of a superconscious and ultracomplex spiritual entity in earth's future.

Teilhard's preferential judgments are clearly anthropocentric. According to his cosmic law of evolution, something increases in value as it increases in complexity/consciousness. Our own species is held to be the most valuable object in this universe because it is the most complex and therefore the most conscious animal on earth. However, one may value something for a great many reasons: its beauty, uniqueness, necessity, simplicity, practicality, and ephemerality (to mention just a few).

Evolution may be, but is not necessarily, synonymous with progress. Therefore, a crucial distinction between evolution and progress is called for. The perception of a general tendency throughout natural history does not constitute proof of a preestablished design in cosmic or planetary evolution. Likewise, the current situation on earth certainly demonstrates that scientific and technological advances are no guarantee of concomitant ethical and moral progress in the human realm. To infer from the presently available empirical evidence that evolution is the result of the activity of a supreme mind or ultimate personal center is to make an intellectually unconvincing and philosophically perilous leap from science to theology. Both celestial and terrestrial evolution are natural processes and, as such, they are subject to human investigation in terms of science and reason (metaphysical speculations are always tentative).

A sound understanding of and proper appreciation for the nature of humankind and its place within this universe is crucial to the ongoing quest for truth. Scientific philosophical anthropology requires an ontology and cosmology supported by facts and logic. However, it is unfortunate that many exponents of an alleged scientific philosophical anthropology have little or no empirical evidence to justify their conceptions of our own species in this dynamic universe. The limiting of rigorous inquiry to a single method or the disregarding of established facts of the special sciences when formulating an interpretation of man within nature is unacceptable indeed. Such a restricted view usually yields unwarranted assumptions resulting in an untenable evaluation of the issues under discussion. Scientific philosophical anthropology must first consider the established facts of the natural and social sciences if it is to be at all meaningful and true. Likewise, the awesome perspective of cosmic evolution is indispensable to any attempt to relate humankind to the rest of this material universe.

However, it is to Teilhard's credit that he had organized and mastered such a wide breadth of knowledge from the natural and social sciences. His twenty years of academic and social isolation in China were actually very beneficial, not only in providing opportunities for geological and paleontological research but also in giving him ample time in which to do a vast amount of reading, reflecting, and writing.

Whatever Teilhard's shortcomings in his interpretation and evaluation of the facts, his inquiry into the nature of the human species had the advantage of being, at least to a significant extent, scientific in approach. No other philosopher of evolution in the twentieth century can match his extensive and versatile scientific training, field research, and global experiences. Yet, Teilhard unfortunately separated the human animal from the rest of evolving nature in too sharp a manner by asserting that throughout planetary history there has been a series of progressively arranged critical thresholds resulting in the periodic formation of new kinds of phenomena. This assertion is theologically oriented as well, since the Jesuit priest relies upon it as an argument for the personal immortality of the human soul.

In walking through a large museum or great zoo, one becomes very impressed with the vast complexity and enormous diversity of life on earth. It is even more amazing when one realizes that this display of life represents but an infinitely small part of all the living forms that have resulted from the process of organic evolution that, to the best of our present knowledge, has been taking place on the planet for about four billion years. At the same time, one also becomes aware of the striking similarities within this awesome array of organisms. Whatever the significance of the undeniable biological differences, there is no scientific evidence justifying any ontological separations in the evolutionary process (there are merely distinctions due to natural differences). For theological reasons, however, Teilhard desired to have the human animal partake in a level of evolution higher than the rest of the animal kingdom. Once again, as with his spiritual monism and cosmic law of teleological evolution, this Jesuit priest has used a theological presupposition to distort the scientific evidence to save his claim to man's central and crowning position in dynamic nature.

For Teilhard, a point of crucial value is reached at the end of each successive stage of terrestrial evolution. At such a critical threshold (phase or level), there occurs a new and quasi-spontaneous change of state (aspect, condition, or nature). This critical point results in a creative movement forward and upward. Although he held that each critical juncture is crossed only once and instantly, Teilhard was too scientifically sophisticated to believe in haphazard spontaneity. The result of this sequence of critical thresholds has been a series of successive layers of psychic development (zones, spheres, envelopes, or orders), each one different from all previous ones not merely in degree but especially in kind. The crossing of each critical transition in evolution results in a qualitative change in psychic energy, while a graded spectrum of quantitative differences in consciousness is manifested within each level of evolution. Every living organism manifests its own particular degree of psychic development according to its own position in the evolutionary hierarchy. Human psychic energy differs in kind from all those earlier manifestations of consciousness in the biosphere, belonging as it does to a new and therefore higher level of existence (i.e., the sphere of thought).

The total evolution of planet earth has thus far generated three unique events. The first critical threshold is said to have occurred at the granulation of the stuff of this universe, resulting in the sudden formation of the elementary corpuscles in the cosmos. Further evolution resulted in the phenomena of polymerization and crystallization.

The second critical threshold distinguishes between the inorganic level or prebiosphere and the formation of a living level or biosphere; a qualitative leap had taken place from the units of preliving atoms to the units of living cells. Teilhard claimed that organic evolution is qualitatively different from, higher than, and superior to inorganic evolution. Organic evolution is homogeneous and coherent. It also manifests the emergence of new kinds of phenomena, which Teilhard refers to as elementary movements of life: con-

jugation, reproduction, renovation or diversification, aggregation or association, multiplication or duplication, ramification, and controlled additivity
as the directed increase in complexity. Throughout organic evolution, the
phenomenon of controlled additivity acts as a vertical component and is,
therefore, responsible for biological evolution in a predetermined direction.
Teilhard saw life disclosing groping profusion, constructive ingenuity, and
even indifference while structurally forming a global unit. He envisioned
the solidarity of organic evolution as a single gigantic organism developing
toward an ultimate planetary goal of spiritual unity.

The crossing of that latest critical threshold resulted in the qualitative
distinction between mere life and reflective thought. Consequently, one has
the emergence and vitalization and hominization of radial energy as consciousness. Atoms, cells, and persons represent the building blocks of the
three successive stages of planetary evolution: geogenesis, biogenesis, and
noogenesis, respectively.

For Teilhard, this appearance of the human species as a reflective animal
represents a unique event not only on a planetary scale but also, interestingly
enough, from the cosmic perspective as well. He taught that the human
animal now constitutes the axis of evolution pointing the way to that final
unification of the terrestrial world in an all-embracing sphere of thought
at the future omega point.

Also crucial to Teilhard's philosophy of evolution is the finite spherical
surface of the earth. This factor prevents the indefinite dispersion of each
distinct evolutionary layer. Prelife, life, and thought are able to converge
and involute around this finite planet. As a result of the law of succession
(which refers to the necessary qualitative leaps in evolution when energy
has reached a certain critical point of internal concentration) and the finite
spherical nature of the earth, Teilhard abstracts three superimposed but
coextensive terrestrial circles. Each successive circle differs in kind from all
the previous ones. From this perspective of a converging and involuting
evolution, one has the successive formations of the geosphere (including
the barysphere, lithosphere, hydrosphere, atmosphere, and stratosphere), the
biosphere (including the plant, insect, bird, and nonhuman animal worlds),
and the noosphere: inorganic, organic, and reflective global envelopes,
respectively. Each sphere represents a distinct stage of planetary emergence,
for the axis of geogenesis was extended through biogenesis into noogenesis.
Hence there have been three sudden leaps that represent gaps and
discontinuities in the otherwise continuous, converging, and involuting
evolution of this planet.

Teilhard stresses that as a direct result of planetary evolution having
crossed a third critical threshold, the human species differs in kind, not
merely in degree, from all other earlier animals (let alone plants, insects,

and birds). With the emergence of humankind, global evolution took a qualitative leap forward and upward. Because of the finite surface of this planet, humans have been evolving collectively around the earth. Converging human thought is forming a thinking envelope around the planet. Teilhard calls this emerging thinking envelope the noosphere (a layer of consciousness).

To give a proper evaluation of the position of the human species in the cosmos and a realistic anticipation of its planetary destiny, Teilhard held that it is necessary to consider both the "within" as well as the "without" of things. Looked at from without, the human animal has acquired an erect position, bipedal gait, and a large complex brain (particularly a structurally sophisticated and very intricate cerebrum that allows for the use of symbolic language with the subsequent rapid development of society and culture). Looked at from within, which for Teilhard actually reveals the true measure of qualitative evolution, our species represents a unique phenomenon in the universe. For the first and only time in biological evolution (and at one single stroke), the evolution of radial energy had taken a dramatic and irreversible leap forward and upward: this resulted in a chasm between all nonhuman life and human thought. Our own species not only knows, but it also knows that it knows (at best, all other animals can only know). In other words, at a certain critical level of biological evolution, consciousness ceased to be merely an adaptive device for the perceptual handling of other objects but became a conceptual instrument for systematic and rigorous reflection about itself. Thus, the noosphere began to emerge on earth.

For Teilhard, the human animal is unique in this universe because it is a person (i.e., a reflective and self-reflective center-of-consciousness or an intelligent sociocultural being). As a result, this Jesuit priest claimed that humans still occupy a special place in nature: from without, the human zoological group represents just one more quantitatively distinct biological species; but from within, our phylum now represents the advancing spiritual-ization of the cosmos. He saw all of planetary evolution as a preparation for the emergence of our own species and even held that the ultimate realization and global fulfillment of the human animal would be accomplished on earth at the future omega point.

Teilhard also argues that the immortal human soul is a natural product of biological evolution: an inference from his assumption that the radial energy of humans differs in kind rather than in degree from the radial energies of all other nonhuman animals. Since for Teilhard there was no first couple but a plurality of such couples to start our human race, the doctrine of original sin takes on a cosmic dimension. Everything in the universe is imperfect proportional to its spatiotemporal distance from God. Through evolution, however, this universe as a cosmogenesis is both spiritualizing and perfecting itself.

Teilhard argued that since the appearance of our own species is a qualitative leap in biological evolution, it therefore represents a difference in kind (not merely in degree) from all other animals. As a direct result, the human mind (i.e., human radial energy) is held to differ in kind from those degrees of radial energy manifested within the ascending hierarchy of organic complexification. What is the significance of this radical change of state in the evolution of radial energy? The answer is a theological assumption: personal immortality of the human soul.

Teilhard held that each species in the animal kingdom manifests its own particular degree of radial energy as a direct result of its own inner degree of centro-complexification. The more advanced a species is in its neural complexity, the higher its position in the scale of nature and consequently the greater its degree of consciousness. At the death of a nonhuman animal, its radial energy is transformed back into and reabsorbed by tangential energy. Teilhard argues that this is not the case, however, at the death of a human being. Since human life resulted from the crossing of that most recent critical threshold, its radial energy is unique among the psychic energies manifested by all other animals in the biosphere. Human thought as radial energy does not change its nature as a result of the physical transformation of the human body at death. Rather, human thought endures forever as a disembodied spiritual center-of-consciousness (the immortality of personality).

It should be noted that Teilhard uses a philosophical speculation to support a theological assumption (both positions are scientifically unwarranted). It is necessary to point out the unique nature of humanity as the symbolizing species. One may disregard the distinction between radial and tangential energies (Teilhard, however, ultimately maintained a position of spiritual monism). If there have been no critical thresholds and if in evolution there is no ontological distinction between the two postulated energies of the universe, then there can be no personal immortality for the human soul. In the Teilhardian view, the intensity and centration (mental focus) of the psychic centers in all animals prior to humans have not evolved to that critical point where nonhuman centers are able to maintain their identity without a physical substructure: hence, the alleged cosmic uniqueness of our own species in terms of personal immortality.

One may or may not hold that the difference in degree between the mental faculties of the three great apes and humans is so essential that this distinction may be said to amount to a difference in kind.

Such a distinction between the levels of mere consciousness and a higher reflective awareness does not, however, warrant Teilhard's assumption that there are quantum jumps or discrete discontinuities in the otherwise progressive continuity of psychic evolution in biogenesis. (There is no sudden leap from a consciousness of this material world to self-awareness during the psychic development of the human mind, but rather a gradual increase in the awareness of one's self as distinct from one's environment. Briefly, one may speak of the gradual emergence of human personality.) Even if a qualitative distinction is made between the psychic manifestations of the apes and humans, this does not infer believing in the personal immortality of the human soul. Teilhard's philosophy of humankind, as with his view of the cosmos, is religiously oriented. If he had not had to reconcile an evolutionary perspective with his own fideistic predilections and the dogmas of the Roman Catholic Church, then his own philosophical anthropology would certainly have been significantly different indeed.

With the rise of biology as a special science in the last century, some philosophers started to argue for an organismic view of the universe. The theory of evolution suggested such a turn away from the crude mechanistic views of the early materialists. However, one may still hold to a sophisticated pervasive materialism.

For centuries, it has been taught that human superior mental powers are due to participation in a transcendent world; somehow the human animal bridges the gap between a realm of being and the universe of becoming. The medieval Church had once forbidden inquiry into the nature of humans, holding that the human body is sacred and therefore must not be subject to scientific investigation. It also taught that humanity is endowed with a simple, nonmaterial, immortal soul (this position inferred an ontological dualism, which resulted in a relational problem between the material body and a spiritual soul). Only since the last century have biology, anthropology, and psychology become special areas of scientific inquiry. The results have been most productive and innovative: all empirical evidence points to the position that the human being is a relatively recent product of biological, sociocultural, and psychological development, rooted in the evolution of matter. As such, our own species is totally within this physical universe.

It is clearly egocentric and anthropocentric for the human animal to place itself on an ontological plane higher than the rest of the biological kingdom. Similarly, this world is independent of and prior to human experience. Such a perspective, grounded in the special sciences and given rational coherence by a comprehensive theory of cosmic evolution, is incompatible with any form of idealism or spiritualism. In summary, mental activity is only an organic function of those neurological structures of the material brain and therefore subject to scientific inquiry and inevitable mortality.

No single characteristic separates our own species from the great apes. However, a uniquely human orchestration of several characteristics does distinguish our species from its closest primate relatives. Likewise, the human animal is increasingly capable of choosing the direction of its own future evolution on earth and elsewhere.

Teilhard's mature thought was less interested in the geological and paleontological evidence in support of planetary evolution but much more concerned with the future survival, direction, and final goal of our human species on earth. He had always maintained that scientific naturalism is necessary but not sufficient to provide an adequate interpretation of planetary evolution. In Teilhard's own view, neo-Darwinism or the present synthetic theory of organic evolution (which adds those principles of heredity and population genetics to the explanatory mechanism of natural selection in Darwin's original teaching) is incomplete for it neglects the religious dimension of human existence in general and its psychosocial development in particular. This is why the Jesuit scientist argued that theological beliefs and philosophical concepts are necessary to supplement the purely scientific evidence for and logical conception of human evolution.

If planetary evolution is spiritual, as claimed by Teilhard, then it expectedly infers a spiritual end point. He stressed that for survival and fulfillment, a collective humankind must supplement science and technology with a belief in both progress and unity under the guidance of a personal God.

Within this organismic and spiritual framework, Teilhard focused upon the human phylum. In our own species, consciousness has been raised to the second power, i.e., awareness has involuted back upon itself so that it has become an object of its own reflection and contemplation. The material human brain folding or convoluting back upon itself has resulted in human thought having the capacity to reflect upon itself. Another planetary transmutation had taken place, resulting in the formation of a single and unbroken global tissue: a quasi-membrane consisting of a complex and converging network of reflective consciousness with all its contents (i.e., the human noosphere is emerging).

For thousands of years, the human species has remained a single biological unit; it has not undergone further adapative radiation resulting in actual speciation as has occurred over long periods of time in many plants, insects, birds, and other animals. Due to this biological uniqueness and the global finitude of the planet, our own species has begun to converge and involute upon itself. Like geogenesis and biogenesis, but even more so, noogenesis is a single and unified process (according to Teilhard, conscious reflection or personalization and global reflexion or hominid planetization are those two essential aspects of the phenomenon of humankind).

In a general sense, hominization is the progressive phyletic spiritualization of our human layer. The superhominization of this layer is occurring as a result of the noosphere closing in upon itself under the structural pressure of the finite sphericity of this earth. Human thoughts are increasingly encircling and enveloping the planet, forming a psychic membrane or noosphere.

For Teilhard, no evolutionary future awaits our own species except in a collective association on the earth. He saw sociocultural evolution as an extension of biological evolution, holding that the axis of biogenesis had culminated in a new movement toward neolife or superlife (i.e., the personalization of planetary evolution). Human evolution has manifested itself in two phases: the socialization of expansion or divergence represented by those early stages of cultural evolution followed by a socialization of compression or convergence started during the Neolithic Metamorphosis (i.e., during the last ten thousand years). As a result of this human biosocial convergence, the modern earth first experienced global migration and then ongoing global acculturation and immigration.

Always optimistic on the theological level of interpretation, Teilhard held that although the human layer will reach a physical end, it will not experience a spiritual death. Since biosocial evolution and involution are both inevitable, he envisioned an ultimately personalized earth. In short, there are two major stages of hominization: (1) socialization of divergence followed by (2) socialization of convergence. Psychological as well as intellectual and moral convergence will culminate in the planetary collectivity of humankind in terms of spirit.

Teilhard offers a quasi-Lamarckian interpretation of cultural evolution, emphasizing the cumulation of knowledge from generation to generation. As a crucial part of noogenesis, technology is accelerating and thereby complements the evolving biopsychic and symbolic convergence of our own human species (along with certain predominately progressive, optimistic, and eschatologically oriented philosophies of history). This technological accumulation and spiritual involution of the human layer is a phenomenon aided by the already-mentioned roundness of this planet. The Jesuit priest was aware that, through both euthenics and eugenics, it is feasible for our species to control and improve upon its own further evolution on the earth. Briefly, humankind is more and more directing the very process of super-hominization to its ultimate end point on this planet.

There are striking similarities but major differences between aspects of Teilhard's view and that of Marx. One way or another, both modern world-views originated from a naturalist attitude that emphasizes the humanistic values of our collective species and expresses a deep concern for the present human condition as well as its survival and fulfillment on earth. Teilhard saw the resolution of human problems in the human situation through an increase in love (the amorization of our evolving species). Of course, the Marxists as passionate atheists do not accept his theological orientation. Yet, it is ironic that certain Marxists have been more sympathetic and receptive to the humanistic core of Teilhard's thought than were most members of his own Jesuit order. In a manner reminiscent of their treatment of Hegel, what some Marxists did was to turn Teilhard's metaphysical system upside down, inverting spiritualism into materialism with appropriate other philosophical modifications as well.

For obvious reasons, Christianity played a very crucial role in Teilhard's evolutionary philosophy. He saw the "Christian phenomenon" itself as a progressive continuation of human psychosocial evolution. Through the influence of Christianity, planetary noogenesis is the intensification of a cosmic Christogenesis (universal evolution as a cosmogenesis is essentially Christocentric).[17] For Teilhard, Christianity is playing a most important role in uniting our human species through its emphasis on love and compassion as well as personality and spirit. Within the evolving noosphere, humankind continues to progress forward and upward, although this movement may be almost imperceptible from time to time.

Teilhard envisioned a "Human Front" with faith in futurism, universalism, and personalism.[18] He taught that only Christianity is capable of guiding spiritual evolution to its ultimate completion; only it is capable of saving the person (in contrast to the mere individual) within a collective humankind by acknowledging the primacy of spirit. From without, human beings are merely individuals of our particular species. From within, however, they are souls endowed with personalities. An individual is an agent of psychocultural interaction endowed with both rights and responsibilities, while a personality is the inner structure of a soul, i.e., the psychological and moral organization of personal thoughts, feelings, attitudes, and values. Although human beings may submerge their physical individuality within a collective, they nevertheless enhance their own personality as a result of sharing emotional, intellectual, and spiritual interactions with others.

Christianity replaces pessimism and the passivity of isolated individuals with optimism and the activity of collective persons. Christ is no longer merely a historical figure; he takes on a cosmic and process dimension. Teilhard's mystical vision is bold enough to identify a dynamically conceived Christ with both universal and terrestrial evolution. Through Christ, in fact, evolution is made both whole and holy. The end point of Christogenesis is the formation of a collective superhumanity manifesting hyperreflection. This consummation of the world in terms of spirit will be brought about through love energy, that highest form of radial energy.

Teilhard taught that the present anguish of humankind (e.g., fearing the possibility of collective self-destruction and total death) is caused by an acute awareness of its seemingly diminutive stature when contrasted with the enormousness of space, time, and number. The scientific discoveries of the past centuries, especially a cosmic perspective in astronomy, as well as the theory of relativity in physics and the theory of evolution in biology, have jarred and jolted the human mind from its geocentric and geostatic torpor. The meaning and purpose of human existence are being questioned and debated, while competing forms of essentialism and existentialism divide the intellectual circles. For Teilhard, the ultimate answer is still to be found in Christianity. He taught that only love is capable of uniting and completing noogenesis on earth; the survival and fulfillment of humankind requires an increase in scientific knowledge and wisdom coupled with a parallel increase in love and compassion. He taught that the cosmic energy of love, which is everywhere in this universe in some extended form, is capable of preserving and perfecting that which it unites. Christianity represents a phylum of love, and Christogenesis is the increase of radial energy (now in the form of love energy) and a concomitant decrease in tangential energy. Through this increase in love energy, cosmic evolution represents a divinizing convergence toward God-omega at the omega point as the end of planetary history in terms of humanized spirit.

For Teilhard, human survival and fulfillment demand a superior form of existence.[19] He held that this ultimate source and object of love is above and ahead of terrestrial evolution. It is not something material but rather a supreme someone: the ultimate object of love can only be a personal God as that great presence guaranteeing the teleological process of cosmic evolution and a spiritual outcome for terrestrial evolution.

It is very important to note Teilhard's distinction between God-omega and the omega point. God-omega is that transcendent focus and end of this personalizing universe (i.e., God-omega is the personal center of universal convergence), while the omega point is that final event of planetary evolution. God-omega is the transcendent ultimate monad, and both cosmic and terrestrial evolution are sustained by this prime mover ahead as the ultimate gatherer and consolidator of souls. Teilhard held that autonomy, actuality, irreversibility, and transcendence are the four attributes of God-omega. Being already in existence as provider and revealer, God-omega is directing and unifying and purifying this personalizing, converging, and involuting cosmogenesis (at least in terms of earth history). In essence, love is extending this cosmogenesis into a Christogenesis that will be fulfilled at the future omega point on earth.

For Teilhard, God-omega does not directly create matter but rather causes it to create itself by being immanent in cosmic evolution as Christ (the vital principle or force) and transcendent as the final cause of planetary evolution. The conception of an omega point is surely his most original but most vulnerable assumption. Likewise, as is now obvious, his shifting back and forth from a celestial perspective to a terrestrial framework adds more ambiguity to an already-novel but obscure worldview.

By way of a grand metaphor, the mystical body of Christ is literally that collectivity of all monadic persons that have been emerging throughout the human phase of terrestrial evolution. As such, the planet is acquiring the equivalent to both a brain and a heart. Extended to its logical conclusion, Teilhard taught that the earth will acquire the semblance of a single mind and a single soul that will become united in only one supreme center of thought and feeling. Remembering that he envisions planetary evolution as a spiraling pyramid or cone, this process is now narrowing and converging upon its apex, which is the omega point. This future union of a superconscious humankind on a global scale is to be followed by its fusion with the personal God-omega (this final creative synthesis is the theosphere, the fulcrum of human destiny and the human apogee of terrestrial evolution).

Teilhard assumes that the physical death of a person merely liberates an immortal center-of-consciousness. For thousands of years since the first reflective beings died this side of that last critical threshold between life and thought, all such liberated reflective monads have been forming a planetary layer of thought. Thus, hominization extended into a superhominization will be finalized in a global arrangement or terrestrial mega-synthesis of immortal persons forming a unified and harmonized superconsciousness around the finite spherical geometry of the earth.

For Teilhard, the omega point is an inevitable extraplanetary event, i.e., the final mystical event of human evolution that will transcend space and time. His principle of emergence is thereby extended to its very limit in this formation of a spiritual pole as the result of a supercentration of persons into an absolutely original person or single supreme center of divine consciousness. In turn, this monadified aggregate of persons (this spiritual macromonad, so to speak) converges upon and unites with God-omega, thus bringing human evolution to an end.

Teilhard taught that evolution unites like to like, that everything that rises must converge, and that closer union is fuller being. The omega point is that final, creative, and clearly differentiated meta-union of persons as a result of a converging cosmogenesis focused upon this earth. With the attainment of the omega point, Teilhard's law of the recurrence of creative unions is manifested for a last time (at least as far as human evolution on this planet is concerned). With the omega point, one has that ultimate synthesis in which the universal and the personal fuse as a result of an interior totalization of the spirit. In summary, one has a dynamic natural theology: if converging and involuting cosmic evolution is spiritualistic and teleological, then it can terminate only in a personal center of ultimate complexity and maximum consciousness.

The omega point represents Teilhard's most daring assertion. By repeatedly resorting to theologically oriented philosophical concepts within a mystical framework, this Jesuit priest has not limited himself to a strictly naturalist phenomenology of planetary evolution. Far from being a naive and shallow optimist guided by wishful thinking, Teilhard realized and did admit that there may even be an evolution of evil on earth since there seemed to be an excess of evil in the world as it is. He made the common distinction between physical and moral evil. Yet, he claimed that ultimately there is only one form of evil. Evil is matter, determinism, multiplicity, and death. It is that which retards and prevents evolution toward creative union. However, growth itself requires and necessitates evil. The evolution of this universe represents a nearly infinite number of steps from absolute evil (i.e., total matter, nothingness, or plurality) to that supreme good that is total being as spirit in unity. Evolution is that progressive movement from imperfect plurality to perfect oneness. Evil is statistically inevitable as a by-product of cosmic evolution proceeding by trial and error (directed chance) toward perfect unity as spiritual oneness.

If evil continues to increase—an alternative possibility that Teilhard realistically acknowledged—then the end of this world process will bring about a bifurcation in the noosphere. An internal schism within this human layer would result in a polarity between a love of the physical and a love of the spiritual. As a result, only that segment of the noosphere that has synthesized itself beyond all evil as well as across space and time will be united with God-omega at the future omega point.

From Teilhard's theological perspective, God's love vitalizes this spiritual cosmos from within and divinely attracts directional evolution from above. This evolving universe emanated from God at the alpha point, and humankind will return to God as the final goal of cosmogenesis at the future omega point.

For a mind that thought in terms of billions of years and held that human psychosocial progress would probably continue at an accelerated rate for thousands of years to come, it is astonishing indeed that Teilhard never allowed for the possibility of human divergence on a solar or galactic or cosmic scale. Actually, his whole system of converging and involuting evolution would collapse if one day our own species should extend its development beyond the confines of the spherical surface of this finite planet.

Teilhard did speculate that perhaps other omega points have been, are being, or will be reached on other planets elsewhere in the universe. Even so, he held that communication with other such worlds with their emerging omega points is very unlikely. In any event, the only omega point that mattered to him was our own emerging one here on earth. One may speculate on innumerable omega points being reached on other planets throughout the universe; although the cosmic convergence of such units of spirit may be imagined by a mystic, such an event is unthinkable for the materialist. Teilhard as process theologian held that a spiritual goal is necessary to motivate into action an otherwise apparently meaningless and purposeless process of human evolution. However, his own personal needs and desires are not necessarily those of the rest of our human zoological group. For example, a naturalist may accept a form of mysticism if it is simply an intense personal awareness of the spatiotemporal unity of this seemingly eternal and infinite universe without resorting to a spiritual monism or anticipating a future omega point. Religious mysticism usually supports a spiritual monism or ontological dualism, whereas a scientific mysticism may support a pervasive materialism.

Teilhard will be a subject debated decades if not centuries from now.[20] He will be remembered for having contributed not only to the advancement of science but also to the enlightenment of theology. Of course, humankind is deeply indebted to his moral fiber and intellectual integrity. Teilhard was truly a great scientist as well as a noble thinker who, despite sometimes overwhelming adversities, had the courage to attempt a synthesis of facts, concepts, and beliefs. Consequently, he saw far above and beyond the horizons of most of his contemporaries in science, philosophy, and theology.

As a natural scientist, process philosopher, and deeply religious mystic, Pierre Teilhard de Chardin offered a lofty vision. One lasting contribution is his serious introduction of the fact of evolution into modern theology. Surely, he will be most remembered by posterity neither for his impressive science nor for his inspired theology but rather for the magnificent tapestry of human thought into which he attempted to weave them both. Perhaps his philosophical synthesis will continue to elicit moral respect, aesthetic delight, and intellectual admiration as well as ongoing controversy long after our own species has finally accepted the fact of cosmic evolution.

ENDNOTES AND SELECTED REFERENCES

1. For the evaluations of Sir Julian Huxley, George Gaylord Simpson, and P. B. Medawar refer to Philip Appleman, ed., *Darwin*, 3rd ed. (New York: W. W. Norton, 1990), pp. 342-62.

2. Cf. Claude Cuénot et al., *Evolution, Marxism and Christianity: Studies in the Teilhardian Synthesis* (London: Garnstone Press, 1967).

3. Cf. Pierre Teilhard de Chardin, *The Divine Milieu*, rev. ed. (New York: Harper Torchbooks, 1968); *The Phenomenon of Man*, 2d ed. (New York: Harper Torchbooks, 1965); and *Man's Place in Nature: The Human Zoological Group* (New York: Harper & Row, 1966).

4. Cf. Pierre Teilhard de Chardin, *Letters From Egypt, 1905-1908* (New York: Herder and Herder, 1965); *Letters From Hastings, 1908-1912* (New York: Herder and Herder, 1967); *The Making of a Mind: Letters From a Soldier-Priest, 1914-1919* (New York: Harper & Row, 1965); *Letters to Léontine Zanta* (New York: Harper & Row, 1969); *Letters From a Traveller* (New

York: Harper Torchbooks, 1968); *Letters to Two Friends, 1926–1952* (New York: The New American Library, 1968); and *Letters From My Friend Teilhard de Chardin, 1948–1955* (New York: Paulist Press, 1976).

5. Cf. George B. Barbour, *In the Field with Teilhard de Chardin* (New York: Herder and Herder, 1965); Nicolas Corte, *Pierre Teilhard de Chardin: His Life and Spirit* (New York: Macmillan, 1961); Claude Cuènot, *Teilhard de Chardin: A Biographical Study* (Baltimore, Md.: Helicon, 1965); Henri de Lubac, *Teilhard de Chardin: The Man and His Meaning* (New York: Hawthorn Books, 1965); Helmut de Terra, *Memories of Teilhard de Chardin* (New York: Harper & Row, 1964); Robert T. Francoeur, ed., *The World of Teilhard* (Baltimore, Md.: Helicon, 1961); Paul Grenet, *Teilhard de Chardin: The Man and His Theories* (London: Souvenir Press, 1965); Mary Lukas and Ellen Lukas, *Teilhard* (New York: McGraw-Hill, 1981); Jeanne Mortier and Marie-Louise Auboux, *Teilhard de Chardin: Album* (New York: Harper & Row, 1966); Michael H. Murray, *The Thought of Teilhard de Chardin: An Introduction* (New York: The Seabury Press, 1966); Oliver Rabut, *Teilhard de Chardin* (New York: Sheed and Ward, 1961); Charles E. Raven, *Teilhard de Chardin: Scientist and Seer* (New York: Harper & Row, 1962); René Hague, *Teilhard de Chardin: A Guide to His Thought* (London: Collins, 1967); Robert Speaight, *Teilhard de Chardin: A Biography* (London: Collins, 1967); Claude Tresmontant, *Pierre Teilhard de Chardin: His Thought* (Baltimore, Md.: Helicon, 1959); and N. M. Wilders, *An Introduction to Teilhard de Chardin* (New York: Harper & Row, 1968).

6. Cf. Ronald Millar, *The Piltdown Men* (Boulder, Colo.: Paladin, 1974); K. P. Oakley, "Further Evidence on Piltdown," in *Man's Discovery of His Past: Literary Landmarks in Archaeology*, ed. Robert F. Heizer (Englewood Cliffs, N.J.: Prentice-Hall, 1962), pp. 37–40, as well as "Further Contributions to the Solution of the Piltdown Problem," *Bulletin of the British Museum* (Natural History), Geology, 2 (3): 244–48, 253, 256–57; J. S. Weiner, *The Piltdown Forgery* (New York: Oxford University Press, 1955); and J. S. Weiner et al., "The Solution of the Piltdown Problem," in *Man's Discovery of His Past: Literary Landmarks in Archaeology*, ed. Robert F. Heizer (Englewood Cliffs, N.J.: Prentice-Hall, 1962), pp. 30–27, and "The Solution of the Piltdown Problem," *Bulletin of the British Museum* (Natural History), Geology, 2(3): 141–46.

7. Cf. Stephen Jay Gould's "Piltdown Revisited," *Natural History* 88 (3): 86–87, 89–90, 94–97; "The Piltdown Conspiracy," *Natural History* 89 (8): 8, 10–11, 14, 16, 18, 20, 22, 25–26, 28; "Vision with a Vengeance," *Natural History* 89 (9): 18–20; and "Piltdown in Letters," *Natural History* 90 (6): 12, 14, 16, 18, 20–28, 30. Also refer to Stephen Jay Gould's "Piltdown Revisited," in *The Panda's Thumb: More Reflections in Natural History* (New York: W. W. Norton, 1980), pp. 108–24.

8. Cf. Alan Houghton Broderick, *Father of Prehistory: The Abbé Henri Breuil, His Life and Times* (New York: William Morrow, 1963), esp. pp. 191–204.

9. Cf. Pierre Teilhard de Chardin, *Writings in Time of War* (New York: Harper & Row, 1968).

10. Cf. Pierre Teilhard de Chardin, *Hymn of the Universe* (New York: Harper & Row, 1965), esp. pp. 19–37. Also refer to Thomas M. King, *Teilhard's Mysticism of Knowing* (New York: The Seabury Press, 1981).

11. Cf. Dora Hood, *Davidson Black: A Biography* (Toronto: University of Toronto Press, 1964), esp. pp. 29, 62, 72, 74, 83, 100, 105, 112, 123, 125, 127.

12. Cf. Christopher G. Janus with William Brashler, *The Search for Peking Man* (New York: Macmillan, 1975), esp. pp. 30–31, 55, 208; and Harry L. Shapiro, *Peking Man* (New York: Simon and Schuster, 1974), esp. pp. 39–41, 53, 79, 81.

13. Cf. Pierre Teilhard de Chardin, *The Appearance of Man* (New York: Harper & Row, 1965), and *The Vision of the Past* (New York: Harper & Row, 1966). Also refer to Pierre Teilhard de Chardin's "The Idea of Fossil Man," in *Anthropology Today: Selections*, ed. Sol Tax (Chicago: Phoenix Books, 1962), pp. 31–38, and Sol Tax et al., eds., *An Appraisal of Anthropology Today* (Chicago: Midway Reprint, 1976), esp. pp. 13, 44, 150, 152, 338–39.

14. Cf. H. James Birx, "Knezević and Teilhard de Chardin: Two Visions of Cosmic Evolution," *Serbian Studies* 1 (4): 53–63, and "Teilhard and Evolution: Critical Reflections," *Humboldt Journal of Social Relations* 8 (3): 151–67.

15. Cf. Pierre Teilhard de Chardin, *Activation of Energy* (New York: Harcourt Brace Jovanovich, 1971), and *Human Energy* (New York: Harcourt Brace Jovanovich, 1969).

16. Cf. Pierre Teilhard de Chardin, *Science and Christ* (New York: Harper & Row, 1968) for similar views.

17. Cf. Pierre Teilhard de Chardin, *Christianity and Evolution* (New York: Harcourt Brace Jovanovich, 1969), and *The Heart of Matter* (New York: Harcourt Brace Jovanovich, 1978).

18. Cf. Pierre Teilhard de Chardin, "The Antiquity and World Expansion of Human Culture," in *Man's Role in Changing the Face of the Earth*, ed. William L. Thomas, Jr. (Chicago: The University of Chicago Press, 1962), pp. 103–12. Also refer to Pierre Teilhard de Chardin, *Building the Earth* (Wilkes-Barre, Penn.: Dimension Books, 1965).

19. Cf. Pierre Teilhard de Chardin, *The Future of Man* (New York: Harper & Row, 1964), and *Toward the Future* (New York: Harcourt Brace Jovanovich, 1975).

20. Cf. H. James Birx, "Pierre Teilhard de Chardin: A Remembrance," *The Science Teacher*, 48 (9): 19, and "Teilhard and Evolution: Critical Reflections," *Humboldt Journal of Social Relations* 9(1): 151–67. Also refer to Stephen Jay Gould, *Hen's Teeth and Horse's Toes: Further Reflections in Natural History* (New York: W. W. Norton, 1983), esp. pp. 201–40, 245–50; Thomas M. King, S.J., and James F. Salmon, S.J., eds., *Teilhard and the Unity of Knowledge: The Georgetown University Centennial Symposium* (Ramsey, N.Y.: Paulist Press, 1983); and Mary Lukas and Ellen Lukas, "The Haunting," *Antiquity* 57 (219): 7–11.

FURTHER READINGS

Barbour, George B. *In the Field with Teilhard de Chardin*. New York: Herder and Herder, 1965.

Birx, H. James. "Teilhard and Evolution: Critical Reflections." *Humbolt Journal of Social Relations* 9(1): 151–167.

Blinderman, Charles. *The Piltdown Inquest*. Buffalo, N.Y: Prometheus Books, 1986, esp. pp. 123–43.

Brooks, Daniel and E. O. Wiley. *Evolution as Entropy: Toward a Unified Theory of Biology*. Chicago: University of Chicago Press, 1986.

Cuénot, Claude. *Teilhard de Chardin: A Biographical Study*. Baltimore, Md.: Helicon, 1965.

de Terra, Helmut. *Memories of Teilhard de Chardin*. New York: Harper & Row, 1964.

Dodson, Edward O. *The Phenomenon of Man Revisited: A Biological Viewpoint on Teilhard de Chardin*. New York: Columbia University Press, 1984.

Gribbin, John. *The Omega Point*. New York: Bantam Books, 1988.

Hanson, Anthony, ed. *Teilhard Reassessed: A Symposium of Critical Studies in the Thought of Pierre Teilhard de Chardin*. London: Darton, Longmann and Todd, 1970.

Lukas, Mary, and Ellen Lukas. *Teilhard*. New York: McGraw-Hill, 1981.

Rifkin, Jeremy, and Ted Howard. *Entropy: A New World View*. New York: Bantam Books, 1980.

Schmitz-Moormann, Nicole, and Karl Schmitz-Moormann, eds. *Pierre Teilhard de Chardin: Scientific Works*. 11 vols. Olten und Frieburg im Breisgau: Walter-Verlag, 1971.

Soleri, Paolo. *The Omega Seed: An Eschatological Hypothesis*. Garden City, N.Y.: Anchor Books/Doubleday, 1981.

Speaight, Robert. *Teilhard de Chardin: A Biography*. London: Collins, 1967.

Teilhard de Chardin, Pierre. *Activation of Energy*. New York: Harcourt Brace Jovanovich, 1970.

Teilhard de Chardin, Pierre. "The Antiquity and World Expansion of Human Culture." In *Man's Role in Changing the Face of the Earth,* edited by William L. Thomas, Jr., Chicago: University of Chicago Press, 1956, pp. 103–12.

――――. *The Appearance of Man.* New York: Harper & Row, 1965.

――――. *Building the Earth.* Wilkes-Barre, Penn.: Dimension Books, 1965.

――――. *Christianity and Evolution.* New York: Harcourt Brace Jovanovich, 1971.

――――. *The Divine Milieu,* rev. ed. New York: Harper Torchbooks, 1965.

――――. *The Future of Man.* New York: Harper & Row, 1969.

――――. *The Heart of the Matter.* New York: Harcourt Brace Jovanovich, 1978.

――――. *Human Energy.* New York: Harcourt Brace Jovanovich, 1970.

――――. *How I Believe.* New York: Harper & Row, 1969.

――――. *Hymn of the Universe.* New York: Harper & Row, 1965.

――――. "The Idea of Fossil Man." In *Anthropology Today: Selections,* edited by Sol Tax, Chicago: University of Chicago Press, 1962, pp. 31–38.

――――. *Letters from Egypt: 1905–1908.* New York: Herder and Herder, 1965.

――――. *Letters from Hastings: 1908–1912.* New York: Herder and Herder, 1968.

――――. *Letters from Paris: 1912–1914.* New York: Herder and Herder, 1967.

――――. *Letters from a Traveller.* New York: Harper & Brothers, 1968.

――――. *Letters to Léontine Zanta.* New York: Harper & Row, 1969.

――――. *Letters to Two Friends: 1926-1952.* New York: New American Library, 1968.

――――. *The Making of a Mind: Letters from a Soldier-Priest (1914-1919).* New York: Harper & Row, 1965.

――――. *Man's Place in Nature: The Human Zoological Group.* New York: Harper & Row, 1966.

――――. *The Phenomenon of Man.* 2d ed. New York: Harper Torchbooks, 1965.

――――. *Science and Christ.* New York: Harper & Row, 1968.

――――. *Toward the Future.* New York: Harcourt Brace Jovanovich, 1975.

――――. *The Vision of the Past.* New York: Harper & Row, 1966.

――――. *Writings in Time of War.* New York: Harper & Row, 1968.

Wilders, N. M. *An Introduction to Teilhard de Chardin.* New York: Harper & Row, 1968.

6

Darwin and Teilhard de Chardin: Some Final Thoughts

An understanding of and appreciation for the true place of the human species within the material universe require taking both a cosmic perspective and the evolutionary framework seriously. Darwin and Teilhard presented major but contrasting interpretations of life on earth, and each has made significant contributions to evolutionary thought.

Evolution is a fact of our expanding universe. Yet, one must distinguish between this natural process in the real cosmos and all the interpretations of evolution now in world literature.[1] Viewpoints range from mechanistic materialism through creative vitalism and finalistic spiritualism to cosmic mysticism. Emphasis may be placed on facts or concepts or beliefs or intuitions; the perspective is usually cosmic or planetary.

Since the beginning of the last century, scientists have accumulated an impressive body of empirical evidence to substantiate strongly the mutability of species within a dynamic history of the earth.[2] The geological column of rock strata and the fossil record of extinct plants and animals clearly attest to the immense age of our planet and the creative sweep of life on the globe (including the emergence of our human species).[3]

Early scientists like Nicholas Steno, Abraham Gottlob Werner, James Hutton, Sir Charles Lyell, and William Smith founded the discipline of geology.[4] This serious study of rocks within an empirical framework served to free earth history from biblical mythology. Natural forces were held to cause alterations in the geological structures of the planet (a correct interpretation glimpsed by Leonardo da Vinci in the Italian Renaissance). This convincing mechanistic and materialistic worldview challenged the entrenched religious belief that the planet and the cosmos were suddenly created by a personal God fewer than 6,000 years ago.[5]

In this century, the meteorologist Alfred Wegener boldly hypothesized that even the seemingly fixed continents have altered their size and shape from an original single land mass that existed countless millions of years ago.[6] This theory of continental drift, once ridiculed by historical geologists, has given rise to the modern science of plate tectonics (a new comprehension of the earth that helps to explain a wide range of geological features and activities as well as fossil, floral, and faunal distributions).

During the Presocratic age, Xenophanes was the first philosopher to recognize both the biological and historical significance of fossils as the remains of once-living organisms. Even the genius of Aristotle failed to comprehend the importance of fossils; the father of biology saw them as being merely chance aberrations in rock strata. Aristotle interpreted nature as a great chain of being, a terrestrial hierarchy of eternally fixed forms. Each plant and animal type has its natural place within this static ladder of life. Aristotle taught that there is no creation or evolution or extinction. As such, for two millennia, the idea of evolution has had to overcome this dogmatic Aristotelian worldview.

The big-bang theory in current cosmology along with the study of stellar history in modern astronomy offers data that support universal evolution.[7] Within a sidereal framework, one may speak of the birth, development, and death of galaxies. Furthermore, the nebular hypothesis argues for the origin of our solar system from the condensation through gravity of cosmic dust and gas clouds.

Although the idea that species are indeed mutable goes back to some rational speculations in Greek antiquity, this insight received important impetus during the Age of Enlightenment. It was Lamarck who first wrote a book for the sole purpose of defending the truth of animal evolution. His major work, The Philosophy of Zoology (1809), appeared exactly fifty years before Darwin published his pivotal volume on organic history.[8] Regardless of the fact that Lamarck at least recognized the mutability of species and freed his view of life on earth from traditional theology, he nevertheless offered four unscientific mechanisms to explain organic evolution: the spontaneous generation of the most primitive plants and animals, the two laws of use/disuse and the inheritance of acquired characteristics, and a semi-vitalistic interpretation of life to help account for the evolution of the higher or more complex animals. In light of modern science, there is no conclusive empirical evidence to verify Lamarckism.[9] Even so, this French naturalist did pave the way for Charles Robert Darwin, Henri Bergson, and Pierre Teilhard de Chardin (among others).

In the middle of the nineteenth century, the intellectual atmosphere was ripe for some courageous naturalist to present a comprehensive and intelligible interpretation of organic evolution in terms of science and reason. This formidable task fell to Darwin (1809–1882), whose epoch-making writings elevated the idea of evolution in philosophy to its present status as a powerful theory of description, explanation, prediction, and exploration in the special sciences.[10] In doing so, Darwin prepared the way for the acceptance of evolution as a pervasive fact of the universe. Today, all of the special sciences, from astronomy and chemistry to anthropology and psychology, owe an enormous debt to the Darwinian framework of biological evolution. Of course, Darwin had neither the first word nor the last on explaining the mechanisms of organic evolution, although he still remains the most important evolutionist in terms of both the theoretical development of this scientific worldview and the numerous areas of relevant research it opened up to empirical inquiry.

Admittedly, Charles Darwin's essentially mechanistic and materialistic interperetation of organic evolution is incomplete. Of lasting value is the major explanatory principle of natural selection,[11] which is supplemented by artificial selection (human intervention) and sexual selection. However, Darwin himself could not account for the sudden appearance of the physical variations that are the raw material for the selective pressures throughout the history of life on earth. Reluctantly, his own hypothesis of pangenesis advocated the inheritance of acquired characteristics through the use or disuse of gemmule-producing organs. It is unfortunate that Darwin never read Mendel's research monograph, *Experiments in Plant Hybridization* (1866).[12] Had Darwin done so, he could have substituted those sound Mendelian principles of inheritance for his own erroneous hypothesis of pangenesis. Actually, the synthetic theory of organic evolution grounded primarily in chance genetic variation and necessary natural selection did not appear until the middle of this century.[13]

In serious French literature the philosophical writings of Henri Bergson, especially his volume *Creative Evolution* (1907), represent a link between the Darwinian materialism of the last century and the Teilhardian spiritualism of this century.[14] Bergson's vitalism was an attempt to account for the ongoing emergence of novelty throughout diverging evolution by resorting to a metphysical principle, the *élan vital*, but this viewpoint did not satisfy scientists, philosophers, or theologians.

As a geopaleontologist, philosopher, and Jesuit priest, Teilhard de Chardin (1881–1955) attempted to give an admittedly personal interpretation of cosmic history in general and planetary evolution in particular.[15] He sought to understand and appreciate the emergence of humankind within a holistic view of the universe, an ultra-anthropology that recognizes the value of the special sciences, dynamic philosophy, and process theology.

Teilhard's ingenious synthesis argues for a personal God, free will, immortality of the human soul, and a divine destiny for our own species.[16] Yet it has failed to convince modern scientists that organic evolution is teleological, to persuade natural philosophers that the entire universe is spiritualistic, or to demonstrate for enlightened theologians that the basic religious beliefs of traditional Christianity are defensible in terms of empirical evidence. Furthermore, in the last analysis, Teilhard does ground his unique vision of human evolution in a cosmic mysticism that is far removed from the mechanism and materialism of Neodarwinism as well as scientific naturalism[17] and rational humanism.[18]

Only about eighty years separate the publication of Darwin's *On the Origin of Species* (1859)[19] and the writing of Teilhard's *The Phenomenon of Man* (1938–1940).[20] However, these two seminal interpretations of earth history and human evolution are radically different. Of course, there have been several crucial advances in the evolutionary sciences since the last century, notably in the areas of genetics, ecology, and dating techniques. Unfortunately, neither Darwin nor Teilhard had the advantage of knowing about DNA (the code of life or language of heredity).[21] If they had been familiar with this key molecule of inheritance and the modern principles of population genetics, then their interpretations of organic evolution would have disregarded all traces of simplistic Lamarckism. Likewise, there is no need to resort to vitalism, teleology, or mysticism in explaining the history of life on earth.

It is in fact the recent Teilhardian synthesis that is religious, whereas the earlier Darwinian worldview advocated mechanism and materialism. Why then is there such a glaring discrepancy between these two major interpretations of, and explanations for, evolution? To a significant degree, one sees that this difference is due to the intellectual influences on and vested interests of Darwin and Teilhard (irrespective of the acknowledged empirical evidence itself).

As has often been pointed out, a serious thinker and his worldview cannot be removed from historical conditions and a particular sociocultural milieu. Darwin and Teilhard are not exceptions. They were both dedicated to research, organizing empirical evidence in the special sciences, and publishing their results. As youngsters, they enjoyed nature, and their passion for earth history lasted a lifetime. Since each was born into a wealthy family, there was sufficient time to develop both breadth and depth in geology, paleontology, biology, and (to some degree) physical anthropology. Their curiosity and wonder about evolution increased even during their adult years.

Darwin and Teilhard were inspired by time spent in the country (both collected rocks, fossils, and insects). They derived pleasure and inspiration from being immersed in the awesome complexity and essential unity of nature itself. Despite this enthusiasm for scientific investigation, there was no indication of the latent creative genius in each of these adolescents during their formative years: Darwin was at best an average student, and Teilhard did not excel in his religious studies. In fact, traditional theology interested neither of them!

After less than two years of medical studies at Edinburgh University, Darwin, with glaring irony, majored in theology at Christ's College, Cambridge. Actually, he never took his religious courses seriously; he always preferred to attend lectures in the natural sciences and to participate in academic field trips into the English countryside. In later years, however, he became an agnostic (if not a silent atheist).[22] Of course, much of the controversy over Darwinian evolution is due to its obvious challenge (especially its severe threat to fundamentalist creationism)[23] to those cherished beliefs of Christianity concerning the assumed spiritual uniqueness of the human being.

Teilhard always found the traditional conception of God to be too small for his personal vision of a cosmic theology in which traditional theism and tempting pantheism merged into a dynamic panentheism. He even saw scientific inquiry as a religious activity. However, a major objection to this Teilhardian synthesis is that it attempts to give natural arguments for religious beliefs. Even near the close of this century, many Christians are still unwilling to embrace the evolutionary framework because of its scientific implications and materialistic consequences for human existence within cosmic history.

Darwin and Teilhard studied the geology, paleontology, and biology of the Western and Eastern hemispheres, respectively. One point of interest is that they both lacked mathematical skills. Nevertheless, this shortcoming did not seriously interfere with their making significant contributions to natural science. Both had a poetic bent and gave outstanding descriptions of the evolving world. Darwin's notebook from his global voyage on the HMS *Beagle* is not only of great scientific value but also remains a unique contribution to natural literature. No evolutionist should neglect to read the vivid chapter about his five-week visit to the Galapagos Islands.[24] Teilhard's writings on planetary evolution and the future of the human species are an incredible blend of empirical facts with mystical intuitions. His captivating style and positive attitude contributed to the popularity of these works, while their support of science gave them a sense of credibility.

Clearly, Darwin and Teilhard benefited greatly from their powerful imaginations. Interestingly enough, both scientists became evolutionists as a result of reading a major book and taking an extensive trip (among other influences). However, Darwin was at first reluctant to acknowledge his indebtedness to the earlier French naturalists who had glimpsed the theory of evolution (among others, including his paternal grandfather Erasmus Darwin), while Teilhard seldom referred to Darwin's science or Bergson's philosophy of evolution, although both had a major influence on his own worldview.

Before his circumnavigation of the Southern Hemisphere, Darwin accepted the pervasive doctrine that all species are forever fixed in nature. At that time, his interests in geology and biology did not encompass earth history as such. It was fortunate indeed that this unpaid naturalist and traveling companion was given the first book of Lyell's three-volume work, *Principles of Geology* (1830–1833),[25] to read at the beginning of the voyage. During this trip, Darwin eventually read the entire series with lasting benefit. In general, it was Lyell's scientific theory of geological uniformitarianism that gave to Darwin that sweeping perspective of time and change within which he would later conceive of organic evolution. Briefly, slow changes in the geological structures on the surface of the earth over vast periods of time due to natural forces suggested to him that perhaps plant and animal species are also alterable throughout earth history. Although the always cautious Darwin was certainly not quick to accept the mutability of species, there is no doubt that the writings of Lyell played an essential role in finally convincing the young naturalist that species are in fact mutable.

Teilhard was greatly influenced by reading Bergson's *Creative Evolution* (1907). It was this philosophical work that finally persuaded the Jesuit priest that the evolutionary framework is true. Surely, it was Bergson's emphasis on human consciousness and his mystical approach that appealed to Teilhard's own spiritual view of earth history. Bergson and Teilhard appreciated Darwinian evolution, but both rejected its mechanistic and materialistic interpretation of organic history. Their own views took metaphysical considerations seriously, particularly directed vitalism and the uniqueness of human reflection as well as mysticism.

Indeed, it was fortuitous that the young Darwin was offered an opportunity to sail around the world as a perceptive naturalist and ironic that the admonished Teilhard was sent to Beijing (near Zhoukoudian) because of his outspoken evolutionism.

In retrospect, Darwin's visit to the Galapagos Islands in 1835 provided the key to understanding and appreciating organic evolution. This unique archipelago is a microcosm where the naturalist experienced the creative results of evolutionary forces. Only after the voyage of the HMS *Beagle* did Darwin comprehend the evolutionary significance of the physical variations among plants and animals (especially in the finches, mockingbirds, iguanas, and tortoises) throughout these islands. Differences among species and varieties and the ongoing adaptive relationship between organisms and their environments, as well as the fossil record and comparative morphology, strongly argued for the mutability of flora and fauna forms throughout time rather than a divine intervention, a series of special creations or the external fixity of living types.

Likewise, Teilhard's travels to Egypt and Asia provided him with rich opportunities to develop his growing expertise in geopaleontology.[26] In the late 1920s, with appropriate irony, this exiled Jesuit priest actually participated in the discovery of Peking Man, *Sinanthropus pekinensis*, in the Western Hills of Zhoukoudian, China. This particular fossil hominid find reinforced Teilhard's commitment to evolution and brought him international fame. (Darwin's discovery of the fossils of giant mammals in Argentina had suggested to the young naturalist that there may be a biohistorical relationship between the now extinct animals of a remote past and the similar species now existing in South America.)

Darwin and Teilhard were both more or less isolated from the hectic mainstream of social life and professional obligations. Darwin became more and more removed from the scientific community, while Teilhard in general was ostracized from the Society of Jesus (however, both were elected secretary to a professional geological organization). As such, they each had about twenty years to reflect rigorously on the evidences for, and implications of, planetary evolution. Although prolific writers, each is primarily remembered for a single but controversial work that presented his major ideas on organic history to the intellectual world.

Interestingly enough, both Darwin and Teilhard were delayed in the publication of their major works. Darwin kept putting off completing the *Origin* manuscript because of his puzzling but debilitating illness as well as his fear of the inevitable controversy surrounding the theory of evolution. Unfortunately, Teilhard was prevented from seeing the *Phenomenon* volume in print during his own lifetime because of the tragic restrictions placed on him by the dogmatic and myopic superiors who objected to his unorthodox beliefs and evolutionary orientation. Each despaired that his most important book might never be published. The clever Darwin left provisions in his will so that his wife could publish the *Origin* volume should it not be printed before his death, while the wise Teilhard left his unpublished *Phenomenon* manuscript in the care of Jeanne Mortier, a loyal and trustworthy friend.[27]

Darwin's *On the Origin of Species* (1859) offered the theory of evolution to the scientific community, whereas Teilhard's *The Phenomenon of Man*

(1940) introduced the fact of evolution into modern theology. Each book focused on earth history and ignited severe criticism. However, it must be added that both Darwin and Teilhard were influenced by Lamarck, rejected the social theories of Spencer and Marx, but at least glimpsed the possibility of life existing elsewhere in the universe (although neither took a serious interest in astronomy).

Despite the similarities between Darwin and Teilhard, there are major differences in their interpretations of evolution. These crucial differences are due in part to historical influences. Darwin was a product of English empiricism, a scientific tradition established by John Locke and Isaac Newton (among others). Since neither his father nor paternal grandfather was religious, Darwin felt no need to accommodate theological beliefs into his own account of organic history.[28] In fact, one may argue that it is because of Darwin's lack of concern for philosophy and religion that his theory of evolution was acceptable to the scientific community in a relatively short period of time. Without metaphysical speculations, his mechanism and materialism paved the way for further scientific inquiry into the areas directly related to verifying and clarifying the claims of organic evolution (e.g., comparative studies in genetics, taxonomy, embryology, and morphology). He even anticipated our modern research in human paleontology and primate behavior.

Darwin wrote about the tree of life or coral of evolution on this planet; for him, there is no overall design or final goal to organic earth history. Teilhard, however, conceived of planetary evolution as a pyramid; life emerged out of matter and is (through human consciousness encircling this globe) advancing toward an omega point.

For Darwin, organic evolution has been a slow and continuous process of chance variation and necessary selection. Rejecting this pervasive gradualism, Teilhard maintained that there have been three sudden leaps forward and upward, i.e., the crossing of three critical thresholds, throughout cosmic history. These crucial steps have resulted in the sudden emergence of matter, then life, and finally thought: geogenesis, biogenesis, and noogenesis, respectively. Because of the geometry of the earth, involuting evolution has brought about the sequential formation of three distinct but superimposed layers: a geosphere, biosphere, and noosphere (Teilhard even envisioned a future theosphere as the final creative synthesis). This idea of qualitative leaps in organic evolution does remind us of the punctuated equilibria hypothesis in modern paleontology.[29]

In the last century Darwin, like Huxley[30] and Haeckel,[31] claimed that our own species and the three great apes shared a common ancestry.[32] Unlike most and without prejudice, Darwin held Africa to be the prehistoric birthplace of humankind. He also maintained that man and the pongids differ merely in degree rather than in kind (furthermore, the human animal is argued to be closer to the gorilla, chimpanzee, and orangutan than these three apes are to the gibbon and siamang, much less the monkeys). Briefly, for Darwin, man is a product of and totally within natural history.

In sharp contrast to Darwin's empirical outlook, Teilhard's worldview grew out of Lamarck and Bergson, an orientation favorable to vitalism, teleology, and mysticism. As such, this Jesuit priest ultimately gave preference to religion over science. His natural theology supports geocentrism, anthropocentrism, and cosmic spiritualism. Despite these differences, it is particularly important to note that both Darwin and Teilhard completely abandoned any strict and literal interpretation of Genesis; as such, their views clearly challenged the sterile ideas of a restrictive fundamentalist creationism.[33]

Teilhard's philosophy of evolution is grounded in four basic concepts: spiritual monism, the accelerating and centralizing law of complexity/consciousness, critical thresholds, and the emerging omega point. It argues that a directed cosmogenesis reveals a personalizing universe in which human beings have an immortal soul (they differ in kind from the apes). The ultimate goal of planetary evolution is the mystical unity of a collective humankind with a personal God at the end of this world. In short, Teilhard has given us an earth-bound and human-centered perspective of things.[34] Of course, his religious vision and mystical orientation are far removed from the Darwinian framework of mechanism and materialism.

In sharp contrast to Teilhard, the atheist Nietzsche, whose interpretation of cosmic evolution as the eternal recurrence of the same prophesied not the coming of a superior humanity but rather only a few superior creative individuals as the future "supermen" (Übermenschen) on earth, saw our own species as now representing the "diseased skin" of the planet.[35]

An evolutionist is painfully aware of the extinction of life. For Darwin, the death of his beloved daughter Annie was the tragic event that finally turned him from Christianity to embrace agnosticism (if not silent atheism).[36] In sharp contrast, although Teilhard was literally surrounded by death during the First World War, he overcame this tragic event by emerging even more optimistic and mystical.[37] Of course, Darwin as materialist found no comfort in the religious belief of an afterlife, while Teilhard as spiritualist envisioned human immortality enjoying a future ultra-personalized omega point.

Darwin never defended his own ideas on evolution in public or in print, while Teilhard never had an opportunity to answer the inevitable professional criticisms of his personal worldview. Darwin's theory was supported by Huxley, Haeckel, and Gray[38] (among others), while Teilhard's philosophy was favorably introduced by Sir Julian Huxley, although this great biologist could not seriously accept the Jesuit priest's entire synthesis.[39]

Had Darwin been born into a religious family, one wonders if he would have interpretred organic evolution in such a way as to save the basic beliefs of Christianity (Alfred Russel Wallace, his contemporary and a co-founder of the explanatory mechanism of natural selection, moved further and further from naturalism to finally embrace occultism and spiritualism).[40] Darwin himself never abandoned materialism even though his wife, Emma Wedgwood, was a Christian and feared that her famous husband would be remembered primarily as an atheistic evolutionist. Remaining only within the purview of science, Darwin never speculated on the problem of evil or the origin and destiny of life, or considered the ultimate nature of the dynamic universe.

Teilhard's mother, a devout Catholic, had had an enormous influence on him during his childhood years. Yet, if Teilhard had not felt a religious calling to the Jesuit order, then his own interpretation of evolution probably would have been considerably different.

Darwin and Teilhard both died suddenly in the month of April, and each was buried in a sacred spot away from his hometown and family. The English naturalist was laid to rest in Westminster Abbey, while the French mystic was interred at Saint Andrew on the Hudson, in the cemetery of the Jesuit novitiate for the New York Province (a recent attempt to return Teilhard's remains to France was unsuccessful).[41]

For his outstanding contributions to natural science and evolutionary thought, Darwin was often recognized and honored during his own lifetime. Unfortunately, Teilhard received public fame and appreciation only after his death (most of Teilhard's writings, including his three books, were published posthumously).[42] Of great importance for evolutionists is Darwin's abductive method for deriving scientific truths from the creative synthesis of empirical facts and rational concepts, as well as Teilhard's profound concern for the survival and fulfillment of our own species on this finite planet.

It must be emphasized that all interpretations of evolution are incomplete and openended in principle: there are always new facts, concepts, hypotheses, and perspectives to be included. No doubt other mechanisms may be discovered that have influenced the origin and history of life on earth and elsewhere (including the emergence of our own species). Surely, human intervention will determine more and more the future of organic evolution. But, neither Darwin nor Teilhard envisioned the cosmic study of comparative exoevolution. Of course, a total and sound comprehension of this dynamic universe remains a scientific ideal.

It is true that intellectual pioneers do stand on the shoulders of earlier thinkers. Charles Robert Darwin especially owed a great debt to the geologist Lyell, while Pierre Teilhard de Chardin was particularly influenced by the philosopher Bergson. As evolutionists, Darwin and Teilhard were a product of intellectual history. Likewise, each is a glowing testament to that enormous creativity of the human mind and the power of an idea.

ENDNOTES AND SELECTED REFERENCES

1. Cf. H. James Birx, *Theories of Evolution* (Springfield, Ill.: Charles C. Thomas, 1984), and "Evolution and Unbelief," in *Encyclopedia of Unbelief* (Buffalo, N.Y.: Prometheus Books, 1985), pp. 195–211. Also see Philip Appleman, ed., *Darwin*, 3rd ed. (New York: W. W. Norton, 1990); Bentley Glass, Owsei Temkin, and William L. Strauss, Jr., eds., *Forerunners of Darwin: 1746–1859* (Baltimore, Md.: Johns Hopkins University Press, 1968); Gertrude Himmelfarb, *Darwin and the Darwinian Revolution* (New York: W. W. Norton, 1968); and Ernst Mayr, *The Growth of Biological Thought: Diversity, Evolution, and Inheritance* (Cambridge, Mass.: The Belknap Press of Harvard University Press, 1982).

2. For a general introduction to the scientific theory of organic evolution, see Fred A. Racle, *Introduction to Evolution* (Englewood Cliffs, N.J.: Prentice-Hall, 1979); and E. Peter Volpe, *Understanding Evolution*, 5th ed. (Dubuque, Ia.: Wm. C. Brown, 1985). Also see A. Lee McAlester, *The History of Life*, 2d ed. (Englewood Cliffs, N.J.: Prentice-Hall, 1977).

3. Cf. H. James Birx, *Human Evolution* (Springfield, Ill.: Charles C. Thomas, 1988); Peter J. Bowler, *Theories of Human Evolution: A Century of Debate, 1844–1944* (Baltimore, Md.: Johns Hopkins University Press, 1986); C. Loring Brace, *The Stages of Human Evolution: Human and Cultural Origins*, 3d ed. (Englewood Cliffs, N.J.: Prentice-Hall, 1988); and S. L. Washburn and Ruth Moore, *Ape into Human: A Study of Human Evolution*, 2d ed. (Boston: Little, Brown, 1980). Also see Maitland A. Edey and Donald C. Johanson, *Solving the Mystery of Evolution* (Boston: Little, Brown, 1989); Mary D. Leakey, *Disclosing the Past: An Autobiography* (Garden City, N.Y.: Doubleday, 1984); Richard E. Leakey, *One Life: An Autobiography* (Salem, N.H.: Salem House, 1984); and Donald C. Johanson and Maitland A. Edey, *Lucy: The Beginnings of Humankind* (New York: Simon and Schuster, 1981).

4. Cf. Claude C. Albritton, Jr., "Geologic Time," *Journal of Geological Education* 32(1):29–37. Also see William B. N. Berry, *Growth of a Prehistoric Time Scale: Based on Organic Evolution*, rev. ed. (Palo Alto, Calif.: Blackwell Scientific Publications, 1987); Charles Coulston Gillispie, *Genesis and Geology: The Impact of Scientific Discoveries Upon Religious Beliefs in the Decades Before Darwin* (New York: Harper Torchbooks, 1959); Stephen Jay Gould, *Time's Arrow, Time's Cycle: Myth and Metaphor in the Discovery of Geological Time* (Cambridge, Mass.: Harvard University Press, 1987); and

Rachel Laudan, *From Mineralogy to Geology: The Foundations of a Science, 1650–1830* (Chicago: University of Chicago Press, 1987).

5. Cf. Ronald Lane Resse, Steven M. Everett, and Edwin D. Graun, "The Chronology of Archbishop James Ussher," *Sky and Telescope* 62 (5):404–5.

6. Cf. Alfred Wegener, *The Origin of Continents and Oceans,* 1915, 1920, 1922, 1929 (New York: Dover, 1966), translated by John Biram from the fourth revised German edition. Also see Martin Schwarzbach, *Alfred Wegener: The Father of Continental Drift* (Madison, Wis.: Science Tech, 1986).

7. Of particular historical importance is Georges Lemaître, *The Primeval Atom: An Essay on Cosmology* (New York: D. Van Nostrand, 1950); and George Gamow, *The Creation of the Universe* (New York: Viking Press, Compass Books, 1956). For recent speculations in cosmology, refer to Stephen W. Hawking, *A Brief History of Time: From the Big Bang to Black Holes* (New York: Bantam Books, 1988). Also see John Gribbin, *In Search of the Big Bang: Quantum Physics and Cosmology* (New York: Bantam Books, 1986); and Hubert Reeves, *Atoms of Silence: An Exploration of Cosmic Evolution* (Cambridge, Mass.: MIT Press, 1984).

8. Cf. J. B. Lamarck, *Zoological Philosophy: An Exposition With Regard to the Natural History of Animals* (New York: Hafner, 1963).

9. Early in this century, Russian biology suffered stagnation under a Lamarckian theory of evolutionary genetics dogmatically upheld by Trofim Denisovich Lysenko (1898–1976) for vested sociopolitical interests. Cf. T. D. Lysenko, *Heredity and Its Variability* (New York: King's Crown Press, 1943). Also see Loren R. Graham, *Science, Philosophy, and Human Behavior in The Soviet Union* (New York: Columbia University Press, 1987), esp. pp. 102–156; and Richard Levins and Richard Lewontin, *The Dialectical Biologist* (Cambridge, Mass.: Harvard University Press, 1985), esp. pp. 163–96.

10. Cf. H. James Birx, *Theories of Evolution* (Springfield, Ill.: Charles C. Thomas, 1984), pp. 98–113. Also see C. D. Darlington, *Darwin's Place in History* (Oxford: Basil Blackwell, 1960); Sir Gavin de Beer, *Charles Darwin: A Scientific Biography* (Garden City, N.Y.: Anchor Books, 1965); Loren Eiseley, *Darwin's Century: Evolution and the Men Who Discovered It* (Garden City, N.Y.: Anchor Books, 1961); and David Kohn, ed., *The Darwinian Heritage* (Princeton, N. J.: Princeton University Press, 1985). Of general intererst are Daniel J. Boorstin, *The Discoverers* (New York: Random House, 1983), pp. 420–76; Ronald W. Clark, *The Survival of Charles Darwin: A Biography of a Man and an Idea* (New York: Random House, 1984); I. Bernard Cohen, *Revolution in Science* (Cambridge, Mass.: The Belknap Press of Harvard University Press, 1985), esp. pp. 283–300; and Jack Meadows, *The Great Scientists* (New York: Oxford University Press, 1987), pp. 149–68.

11. For two recent books on natural selection, see J. H. Bennett, ed., *Natural Selection, Heredity, and Eugenics* (Oxford, England: Clarendon Press, 1983), and Elliott Sober, *The Nature of Selection: Evolutionary Theory in Philosophical Focus* (Cambridge, Mass.: MIT Press, 1985).

12. Cf. Gregor Johann Mendel, *Experiments in Plant Hybridization* (Cambridge, Mass.: Harvard University Press, 1967).

13. Of particular historical importance, see Julian Huxley, *Evolution: The Modern Synthesis* (New York: John Wiley & Sons, 1964). Also see Ervin Laszlo, *Evolution: The Grand Synthesis* (Boston: Shambhala, 1987).

14. Cf. Henri Bergson, *Creative Evolution* (New York: The Modern Library, 1944). Also see H. James Birx, *Theories of Evolution* (Springfield, Ill.: Charles C. Thomas, 1984), pp. 209–19. Also refer to Peter J. Bowler, *The Eclipse of Darwinism: Anti-Darwinian Evolution Theories in the Decades around 1900* (Baltimore, Md.: Johns Hopkins University Press, 1983), esp. pp. 56–57, 105, 116, 147, 150, 179.

15. Cf. H. James Birx, *Theories of Evolution* (Springfield, Ill.: Charles C. Thomas, 1984), pp. 244–93. Also see George B. Barbour, *In the Field with Teilhard de Chardin* (New York: Herder and Herder, 1965); Claude Cuénot, *Teilhard de Chardin: A Biographical Study* (Baltimore, Md.: Helicon, 1965); Helmut de Terra, *Memories of Teilhard de Chardin* (New York: Harper & Row, 1964); Mary Lukas and Ellen Lukas, *Teilhard* (New York: McGraw-Hill, 1981); Robert Speaight, *Teilhard de Chardin: A Biography* (St. James's Place, London: Collins, 1967); and N. M. Wildiers, *An Introduction to Teilhard de Chardin* (New York: Harper & Row, 1968).

16. Cf. Ludwig Feuerbach, *The Essence of Christianity* (New York: Harper Torchbooks, 1957), as well as Sigmund Freud, *Civilization and its Discontents* (New York: W. W. Norton, 1962), and *The Future of an Illusion* (Garden City, N.Y.: Doubleday, 1957). Both authors, among others, gave a psychological account for the origin of religious beliefs and behavior patterns.

17. Of particular importance are John Dewey, *The Influence of Darwin on Philosophy* (Bloomington: Indiana University Press, 1965), esp. pp. 1–19; and Marvin Farber, *Basic Issues of Philosophy* (New York: Harper Torchbooks, 1968), esp. pp. 219–22.

18. Of particular importance are Corliss Lamont, *The Philosophy of Humanism*, 6th ed. (New York: Frederick Ungar, 1982); and Paul Kurtz, *A Secular Humanist Declaration* (Buffalo, N.Y.: Prometheus Books, 1980), esp. pp. 21–22.

19. Cf. Charles Darwin, *The Origin of Species and The Descent of Man* (New York: The Modern Library, 1936), esp. pp. 3–386. Also see Charles Darwin, *The Illustrated Origin of Species*, abridged and introduced by Richard E. F. Leakey (New York: Hill and Wang, 1979).

20. Cf. Pierre Teilhard de Chardin, *The Phenomenon of Man*, 2d ed.

(New York: Harper Torchbooks, 1965). Also see Edward O. Dodson, *The Phenomenon of Man Revisited: A Biological Viewpoint on Teilhard de Chardin* (New York: Columbia University Press, 1984).

21. Cf. James D. Watson, *The Double Helix: A Personal Account of the Discovery of the Structure of DNA* (New York: Mentor Books, 1968). Also see John Gribben, *In Search of the Double Helix: Quantum Physics and Life* (New York: Bantam Books, 1987); and Horace Freeland Judson, *The Eighth Day of Creation: Makers of the Revolution in Biology* (New York: Simon and Schuster, 1979).

22. Cf. Nora Barlow, ed., *The Autobiography of Charles Darwin: 1809–1882* (New York: W. W. Norton, 1969), esp. pp. 85–96.

23. For an early but bizarre attempt to reconcile fundamentalist creationism with scientific evolutionism, see Philip Henry Gosse, *Creation (Omphalos): An Attempt to Untie the Geological Knot* (London: John Van Voorst, 1957). Also see H. James Birx, "Evolution and Unbelief," in *Encyclopedia of Unbelief* (Buffalo, N.Y.: Prometheus Books, 1985), pp. 202–3.

24. Cf. Charles Darwin, *The Voyage of the Beagle*, ed. Leonard Engel (Garden City, N.Y.: Anchor Books, 1962), esp. pp. 373–401. Also see Alan Moorehead, *Darwin and the Beagle* (New York: Harper & Row, 1969), esp. pp. 186–209.

25. Cf. Loren C. Eiseley, "Charles Lyell," *Scientific American* 20(2):90–106, 168. Also see Leonard G. Wilson, ed., *Sir Charles Lyell's Scientific Journals on the Species Question* (New Haven, Conn.: Yale University Press, 1970).

26. Of indispensable value is Pierre Teilhard de Chardin, *L'Oeuvre Scientifique* (Scientific Writings, 1905-1955), 11 vols., ed. Nicole and Karl Schmitz-Moormann (Olten and Freiburg im Breisgau, West Germany: Walter Verlag, 1971).

27. Fortunately, the stubborn Jeanne Mortier would not submit to the wishes of either Teilhard's family or the Jesuit authorities concerning the publication of his manuscripts. Unfortunately, however, there are numerous errors even in those editions of Teilhard's three books, especially *The Divine Milieu*, which was repeatedly rewritten by him during the 1920s and 1930s. In fact, Jeanne Mortier herself haphazardly chose different sections from these disparate copies to produce the manuscript that she herself wanted to see appear in print!

28. Cf. Paul H. Barrett, ed., *Metaphysics, Materialism, and the Evolution of Mind: Early Writings of Charles Darwin* (Chicago: The University of Chicago Press, 1980); and Neal C. Gillespie, *Charles Darwin and the Problem of Creation* (Chicago: The University of Chicago Press, 1979).

29. Cf. Stephen Jay Gould and Niles Eldredge, "Punctuated Equilibria: The Tempo and Mode of Evolution Reconsidered," *Paleobiology* 6(1):96–

118. Also see Niles Eldredge, *Time Frames: The Rethinking of Darwinian Evolution and the Theory of Punctuated Equilibria* (New York: Simon and Schuster, 1985); and Steven M. Stanley, *The New Evolutionary Timetable: Fossils, Genes, and the Origin of Species* (New York: Basic Books, 1981).

30. Cf. Thomas Henry Huxley, *Evidence as to Man's Place in Nature* (Ann Arbor: University of Michigan Press, 1959). Also see William Irvine, *Apes, Angels, and Victorians: The Story of Darwin, Huxley, and Evolution* (New York: McGraw-Hill, 1972).

31. Cf. Ernest Haeckel, *The Evolution of Man: A Popular Scientific Study*, 2 vols. (New York: G. P. Putnam's Sons, 1910), and *The Riddle of the Universe at the Close of the Nineteenth Century* (New York: Harper & Brothers, 1905).

32. Cf. Charles Darwin, *The Descent of Man and Selection in Relation to Sex* (New York: The Heritage Press, 1972), esp. the "Preface" by Ashley Montagu, pp. v–x.

32. Cf. Niles Eldredge, *The Monkey Business: A Scientist Looks at Creationism* (New York: Pocket Books, 1982); Laurie R. Godfrey, ed., *Scientists Confront Creationism* (New York: W. W. Norton, 1983); Philip Kitcher, *Abusing Science: The Case Against Creationism* (Cambridge, Mass.: MIT Press, 1982); Chris McGowan, *In the Beginning . . . A Scientist Shows Why Creationists Are Wrong* (Buffalo, N.Y.: Prometheus Books, 1984); Ashley Montagu, ed., *Science and Creationism* (New York: Oxford University Press, 1984); Norman D. Newell, *Creation and Evolution: Myth or Reality?* (New York: Columbia University Press, 1982); Michael Ruse, ed., *But Is It Science: The Philosophical Questions in the Creation/Evolution Controversy* (Buffalo, N.Y.: Prometheus Books, 1988); J. Peter Zetterberg, ed., *Evolution versus Creationism: The Public Education Controversy* (Phoenix, Ariz.: The Oryx Press, 1983).

34. Cf. H. James Birx, "Teilhard and Evolution: Critical Reflections," *The Humboldt Journal of Social Relations* 9(1):151–67.

35. For an introduction to Nietzsche, see H. James Birx, *Theories of Evolution* (Springfield, Ill.: Charles C. Thomas, 1984), pp. 184–98. Also see David B. Allison, ed., *The New Nietzsche: Contemporary Styles of Interpretation* (Cambridge, Mass: MIT Press, 1985), esp. p. 96.

36. Cf. Nora Barlow, ed., *The Autobiography of Charles Darwin, 1809–1882* (New York: W. W. Norton, 1969), esp. pp. 97–98.

37. Cf. Pierre Teilhard de Chardin, *The Making of a Mind: Letters from a Soldier-Priest, 1914–1919* (New York: Harper & Row, 1965), and *Writings in Time of War* (New York: Harper & Row, 1968). Teilhard's three wartime mystical experiences (France, 1916) and his "The Mass on the World" (Ordos, 1923) are found in *Hymn of the Universe* (New York: Harper & Row, 1965), pp. 19–37, and 41–55, respectively.

38. On 5 September 1857, Charles Darwin had referred to his theory

of evolution in a letter he sent to the famous botanist Asa Gray in America. Cf. Bert James Loewenberg, *Calendar of the Letters of Charles Darwin to Asa Gray* (Wilmington, Del.: Scholarly Resources, 1973), pp. 10–13.

39. Cf. Sir Julian Huxley, "Introduction," in Pierre Teilhard de Chardin's *The Phenomenon of Man*, 2d ed. (New York: Harper Torchbooks, 1965), pp. 11–28.

40. Cf. Alfred Russel Wallace, *The World of Life: A Manifestation of Creative Power, Directive Mind and Ultimate Purpose* (New York: Moffat/ Yard, 1910). Also see Loren C. Eiseley, "Alfred Russel Wallace," *Scientific American* 201(2):90–106, 168. Of special interest is John Langdon Brooks, *Just Before the Origin: Alfred Russel Wallace's Theory of Evolution* (New York: Columbia University Press, 1984).

41. I learned this fact in a conversation with Professor Dr. Karl Schmitz-Moormann, philosopher and theologian as well as co-editor of Pierre Teilhard de Chardin's *L'Oeuvre Scientifique* (Scientific Writings, 1905–1955), 11 vols. (Olten und Freiburg im Breisgau, West Germany: Walter Verlag, 1971).

42. Teilhard's three books are *The Divine Milieu*, rev. ed. (New York: Harper Torchbooks, 1968); *The Phenomenon of Man*, 2d ed. (New York: Torchbooks, 1965); and *Man's Place in Nature: The Human Zoological Group* (New York: Harper & Row, 1966).

FURTHER READINGS

Barlow, Nora, ed. *The Autobiography of Charles Darwin, 1809–1882* (1876, 1887, 1929, 1950, 1958). New York: W. W. Norton, 1969.

Berry, William B. *Growth of a Prehistoric Time Scale: Based on Organic Evolution.* Palo Alto, Calif.: Blackwell Scientific Publications, 1987.

Birx, H. James. "Teilhard and Evolution: Critical Reflections." *Humboldt Journal of Social Relations* 9(1):151–67.

Blinderman, Charles. *The Piltdown Inquest.* Buffalo, N.Y.: Prometheus Books, 1986, esp. pp. 123–43.

Cuénot, Claude. *Teilhard de Chardin: A Biographical Study.* Baltimore, Md.: Helicon, 1965.

de Beer, Gavin. *Charles Darwin: A Scientific Biography.* Garden City, N.Y.: Anchor Books, 1965.

Eiseley, Loren C. *Darwin's Century: Evolution and the Men Who Discovered It.* Garden City, N.Y.: Anchor Books, 1961.

Gillispie, Charles Coulston. *Genesis and Geology: A Study in the Relationships of Scientific Thought, Natural Theology, and Social Opinion in Great Britain, 1790–1850.* New York: Harper Torchbooks, 1959.

Laudan, Rachel. *From Mineralogy to Geology: The Foundations of a Science, 1650–1830*. Chicago: University of Chicago Press, 1987.

Schmitz-Moormann, Nicole, and Karl Schmitz-Moorman, eds. *Pierre Teilhard de Chardin: Scientific Works*. 11 vols. Olten und Freiburg im Breisgau: Walter-Verlag, 1971.

Schwarzbach, Martin. *Alfred Wegener: The Father of Continental Drift*. Madison, Wis.: Science Tech, Inc., 1986.

Wegener, Alfred. *The Origin of Continents and Oceans*. New York: Dover, 1966.

Wilders, N. M. *An Introduction to Teilhard de Chardin*. New York: Harper & Row, 1968.

7

Evolution: Science or Theology?

Evolution is not only a powerful theory but also a basic fact. Today, most scientists and natural philosophers accept the ongoing process of organic evolution as an established aspect of material reality. The amassed empirical evidence now supporting an evolutionary worldview is overwhelming, scientifically convincing, and beyond any serious rational dispute as to its essential implications. Of course, what still remain open to free and critical inquiry as well as continuing expert dialogue are the particular patterns, mechanisms, relationships, and conflicting interpretations of this pervasive natural phenomenon.

In the middle of the seventeenth century, the strict biblical literalist Archbishop James Ussher wrote a book titled *Annals of the Ancient and New Testaments* (1650). In this work, the Irish primate declared that our planet was created by a personal God in 4004 B.C.E.[1] He derived this particular year by appealing to biblical chronology, i.e., by calculating the ages of the various prophets in the Old Testament. Consequently, his own argument maintained that the earth is less than 6,000 years old.

A few years later, adding insult to injury, Dr. John B. Lightfoot (1602–1675) claimed to know the exact date and hour of this momentous event. According to the vice-chancellor of Cambridge University, the world was created in 4004 B.C.E., on October 23 at 9:00 A.M.! In fact, this exact date was added to editions of the Holy Bible.

241

Today, most religious fundamentalists still believe that a recent date for the divine creation is true. They dogmatically hold to a strict and literal interpretation of the Mosaic cosmogony as presented in Genesis.

Geological facts reveal the age of the planet to be nearly five billion years. Recently discovered paleontological remains in Australia and Canada indicate that life first appeared on earth about four billion years ago. Hominid fossils as well as stone artifacts found in central East Africa place the emergence of protohumans back over three million years. Yet there are those who, on merely emotional grounds, reject all this compelling evidence, with its obvious inferences and inescapable consequences.

In Victorian England, Philip Henry Gosse (1810–1888) made an unusual attempt to reconcile the glaring discrepancy between the empirical facts of the earth sciences that supported the theory of evolution and the revealed beliefs of traditional religion.[2] As a fundamentalist creationist of the Plymouth Brethren sect, he held to a strict and literal interpretation of the story of Genesis; this Mosaic cosmogony required rigidly accepting both the sudden creation of Adam and the Noachian Deluge. Yet, as an avid naturalist, Gosse was unable to ignore that growing and compelling but contrary evidence in historical geology and comparative paleontology that supported the rational claims of the early evolutionists in the last century. His own scientific interests included ornithology and marine zoology. This well-traveled author and illustrator wrote nearly fifty volumes and even invented the aquarium. Yet, he rejected outright all evolutionary concepts.

Gosse hoped to resolve the contradiction between evolutionist science and fundamentalist religion by offering his own unique account of the planet, which would seemingly, at least for him, do justice to both conflicting conceptual worldviews. He presented a ridiculous explanation of terrestrial creation in two books, *Life* (1857) and *Omphalos: An Attempt to Untie the Geological Knot* (1857), which appeared in print just two years before the publication of Darwin's *On the Origin of Species* (1859).

Gosse's argument clearly gave preference to biblical authority and direct experience rather than to empirical facts and rational inquiry. It is grounded in a crucial distinction between two modes of existence: preternatural, prochronic, or ideal time in the infinite and perfect mind of the personal deity before the act of divine creation and natural, diachronic, or real time of actual objects and events and relationships in our material world of change and development. Furthermore, the course of everything in both inorganic and organic nature is claimed to be a circle.

In his bizarre attempt to reconcile the biblical cosmogony with scientific evolutionism, Gosse firmly argued that the geological column with its sequence of rock strata and record of fossil patterns was divinely created all at once, along with the earth's existing flora and fauna types (including the human animal). Therefore, our planet suddenly came into being as an ongoing world containing a developmental series of fossils in rock layers only to suggest both an extensive physical past and a creative biological evolution before its actual origin in the real space and diachronic time of nature itself.

In short, the earth is at the same time as old as demonstrated by science and as young as claimed in the Sacred Scriptures. How is this possible? It is, if one assumes with Gosse that God created our planet with a built-in past: the stratigraphic sequences of rocks were suddenly formed at the instant of creation along with all plant and animal fossils in their respective places. This appearance of evolution accepts the results of the process while rejecting the process itself.

As such, according to Gosse, the appearance of all geopaleontological evidence merely implies the preexistence of rocks and fossils prior to the abrupt moment of a special creation through the omnipotent fiat of a divine creator only about 6,000 years ago. The formation of the earth occurred at a distinct point in prehistoric time, and the origin of life on the planet, along with our own species, was a violent irruption of organisms into the continuing circles of cosmic reality.

Consequently, following this argument, all the geopaleontological evidence for a succession of countless epochs over millions of years as well as for the progressive development of life on earth is utterly deceptive, since preexistence is simply an illusion. The pristine world of divine creation had instantly intruded into diachronic time as if it had been partaking in an ordered continuity of early history spanning untold eons. Briefly, this unorthodox interpretation allowed Gosse simultaneously to admit and dismiss all the accumulating geological, paleontological, and biological data for the enormous antiquity of this planet and the vast inventory of organic evolution upon it.

Gosse rejected both the eternity of inorganic and organic matter as well as the mutability of species in the biosphere, since the flora and fauna of our earth are fixed and only a few thousand years old. In arguing his position, he even performed a thought experiment: although stressing that there was no human observation of antediluvian history, Gosse boldly imagined himself observing the first plants and animals of earth just after the moment of creation. In keeping within his view of things, these growing organisms would appear to have had a history but in fact such an inference is fallacious according to this interpretation.

As if this interpretation were not a sufficiently strange form of natural theology, Gosse even went so far as to propose that all the fossils of so-called early life found in the alleged ancient rocks of our earth were placed in these geological strata by a personal God in a plausible paleontological sequence deliberately to test the faith of their discoveries or perhaps even cunningly deceive the faithful. For example, he argued that Adam had been created with a navel by this same clever deity in order to simulate a natural birth.

For Gosse, evolutionists are simply studying the divinely imprinted appearance of a seemingly ancient globe, because he claimed that there had been no gradual change of the earth's surface over the supposed vast eons of planetary time and no slow evolution of plant and animal species throughout the equally assumed ages of organic history. Needless to say, this preposterous argument of *Life* and *Omphalos* was not taken seriously and had no major effect on the scientists, philosophers, theologians, or even the general public of the time.

Gosse was heartbroken and could not understand why his bizarre effort at reconciling evolutionary science with fundamentalist religion was rejected with equal force by both naturalists and religionists. Actually, since he believed that divine creation had taken place instantly, his own theistic cosmogony is not in step with the traditional six-day story of Genesis.

Gosse's worldview is neither empirically convincing nor logically defensible. Surely, it is scientifically untestable in principle and ignores the law of parsimony. There is simply no reason in the heavens or on earth to appeal to such a fantastically improbable possibility that today flies in the face of all empirical evidence and common sense (let alone its unacceptability to modern process theology). However, a modified form of this position does surface now and then in the fundamentalist creationist literature of today. Clearly, this whole scheme remains as the residue of a desperate attempt to synthesize the old beliefs of an outmoded religious cosmogony with the established facts of the advancing evolutionary sciences.

The creation/evolution controversy intensified in the middle of the nineteenth century, especially in the private but serious disagreement that emerged between strict biblical literalism as dogmatically defended by a myopic Robert FitzRoy and scientific evolutionism ("descent with modification" or the mutability of species) as convincingly argued by the free-thinking naturalist Charles Darwin. In microcosm, this conceptual dispute actually anticipated the contemporary international debate of the macrocosm.

During the five-year voyage on the *Beagle* (1831–1836), Darwin's own critical observations of life forms and rock formations, particularly his discovery of giant mammal fossils in Argentina, resulted in an awakening for him of a sober doubt concerning the then prevalent teaching of special creation and the alleged immutability of plant and animal species. Soon enough, a personal conflict ensued between the biblical literalist FitzRoy and the would-be clergyman-turned-scientist. As such, one may safely argue that the creation/evolution controversy had an early inception aboard the *Beagle* as it circumnavigated the globe. The captain dogmatically clung to that conventional belief in the eternal fixity of flora and fauna types, while his scientific companion slowly but steadily advanced toward a dynamic interpretation of life as a vast process of incessant transformation and continuous variation as a result of natural causes (evolutionary forces).

During this famous trip, the relationship between Darwin and FitzRoy became seriously strained. The more empirical evidence was amassed by young Darwin and the more he critically reflected on his own experiences within nature (as well as the far-reaching consequences of this data), the more he became convinced that divine creation, a series of special creations, or the vitalist interpretation of life is not a true explanation for the origin and history of living things on this planet. In fact, all the empirical evidence suggested an incredibly different story indeed.

Darwin's scientific attitude was now clearly challenging the age of the earth, the accounts of the origin and development of all life on the planet, and even that assumed special place of humankind within the material universe.

Darwin was curious, open-minded, and always alert to the overwhelming facts and wide-ranging experiences during this unique trip: reason was his teacher and nature was the classroom. However, FitzRoy was generally indifferent to the emerging natural sciences of his day. He remained both closed-minded to the implications of a rigorous study of nature as well as rigidly dogmatic in his own religious commitment to understanding and appreciating humankind in the cosmos. Faith was his only guide and the Holy Bible was the only authority.

Darwin took seriously time, change, and finally evolution as the gradual "descent with modification." The private controversy between Darwin and FitzRoy represented the ongoing and unfortunate struggle between science and religion. Darwin represented the emerging evolutionary interpretation of humankind, life, and earth history grounded in science and reason. On the other hand, FitzRoy clung to the dogmatic traditional Judeo-Christian view of the special place occupied by our own species within this seemingly static cosmos of fixed laws and eternal forms.

After the end of his voyage on the HMS *Beagle*, Darwin was convinced that organic evolution had taken place and would continue to bring about new living forms on the planet. FitzRoy felt responsible for Darwin's evolutionary interpretation of life on earth and developed a deep guilt about having accepted the young naturalist as a member aboard his ship. Darwin lived to see his theory of evolution accepted by most of the scientific community (although modifications are now required as our empirical understanding of and conceptual appreciation for the origin and history of life on earth increase, thanks to continued scientific inquiry and rational deliberation).

FitzRoy eventually committed suicide. One may suspect that his guilt in having aided the scientific founder of organic evolution played a key role in this tragic decision. Symbolically, Darwin's triumph argued in favor of a naturalist overview of process reality while FitzRoy's death reflected that the old, fixed interpretation of things grounded merely in blind faith had been superseded by science and reason.

Darwin was greatly influenced by the Hutton/Lyell geological theory of uniformitarianism (which argued for slow alterations of the crust of the planet), a fortunate five-week visit to the unique Galapagos Islands in the Pacific Ocean, and Thomas Robert Malthus's essay on population. He became convinced of the scientific truth of biological evolution primarily by means of natural selection. It could explain the creative history of all life on the planet far better than any alternative hypothesis or theory, let alone having recourse to a religious myth. After about two decades of hesitation and indecision, Darwin was finally prevailed upon by sincere friends to publish at last his long-maturing major work, On the Origin of Species (November 24, 1859). It remains a comprehensive and intelligible argument for the truth of biological evolution in terms of science and reason, although expectedly both modified and complemented by those discoveries made since Darwin's time. The venerable Englishman, however, never debated or defended (in public or in print) his own theory of evolution, with its awesome theoretical implications and disquieting cosmic consequences.

On 30 June 1860, less than a year after the publication of Darwin's Origin, empirical science and organized religion openly collided in a now-notorious public confrontation under the auspices of the British Association for the Advancement of Science held in Oxford's University Museum Library. The clash between these two institutions occurred in the persons of the eminent vertebrate zoologist and natural philosopher Thomas Henry Huxley[3] and the Anglican Bishop Samuel Wilberforce. (Recall the furious controversy between the ovists and the spermatists in the seventeenth century, as well as the debate between the Vulcanists and the Neptunists in the eighteenth century.)

Wilberforce impudently and unconvincingly railed against the Darwinian framework, which he was simply not competent to understand or qualified to discuss. The theory of evolution, however, was both eloquently and successfully defended by Huxley. In his attempt to refute the emerging theory of biological evolution, Wilberforce, as a Prince of the Church of England, revealed his complete ignorance of the contemporary natural sciences and consequent incompetence to speak on this grave scientific and philosophical issue. Although the bishop's delivery was eloquent, his emotionally charged and not very Christian invective got him into unexpected trouble. This incident can be summarized as follows. After concluding his arguments against Darwinism (suggested by Owen), the pompous, saccharine Wilberforce sarcastically inquired whether the learned scientist Huxley considered himself descended from the monkey through his grandfather's or his grandmother's side. With exemplary dignity and self-control, the great evolutionist purportedly replied: "If the question is put to me, 'Would I rather have a miserable ape for a grandfather, or a man highly endowed by nature and possessed of great means and influence, and yet who employs these faculties and that influence for the mere purpose of introducing ridicule into a grave scientific discussion'—I unhesitatingly affirm my preference for the ape." With this reply, Wilberforce was defeated. (Although Darwin was not present, his theory obviously won its first major public victory.)

Two decades later in Canada, in an exchange reminiscent of the Huxley-Wilberforce clash of ideas and values, the free-thinking intellectual William Dawson LeSueur (1840–1917) debated in print John Travers Lewis, the orthodox Anglican Bishop of Ontario.[4] In this conflict of interests, LeSueur defended the facts for evolution and liberal thought against Lewis's insufficient knowledge and fundamentalist obscurantism. The former championed the splendid advancements of modern science, while the latter clung to the reactionary tenets of a myopic theology. Like so many other evolutionists, LeSueur was also sincerely devoted to that lonely pursuit of the truth wherever it might lead him and whatever the consequences may be.

As a semi-invalid and quasi-recluse at Down House in Kent, the shy and gentle Charles Darwin had rigorously reexamined his earlier traditionalist religious beliefs. As an admitted agnostic, although never an intolerant materialist or militant atheist, his pervasive naturalism and humanism were none the less distressing and unacceptable to his affectionate and loyal wife, Emma Wedgwood Darwin. She deplored the probability that her beloved husband would be wrongly remembered as a blasphemous infidel rather than the kindhearted man and towering genius of science that he actually was. Is it to Darwin's lasting credit that he kept his own personal religious opinions discreetly to himself because he viewed them as irrelevant to the public import of his scientific legacy? How dearly would one love to know Darwin's own final thoughts on both theism and materialism!

Charles Darwin died on 19 April 1882, and is buried in Westminster Abbey with the other greats of Britain's artistic, political, and intellectual history. Unfortunately, as can be seen even today, the biblical fundamentalist/scientific evolutionist controversy did not die with him.

In Germany during the first decade of the nineteenth century, the speculative naturalist Lorenz Oken (1779–1851) wrote that nature is an eternal transformation of God into this world. Therefore, the task of philosophy is to investigate this evolving universe from its primal state of mere potentiality or pure nothingness through the formation of cosmic elements and then celestial bodies to a development of organisms manifesting an increase in both complexity and diversity throughout earth history until the eventual appearance of man as the only rational animal on this planet. In this natural theology of teleological evolution, the highest point of cosmic history in terms of living things is reached in man. As such, the human being is that creature in which the god-time becomes conscious of itself.

Briefly, one has a dynamic cosmogony in which an antecedent absolute as time itself metamorphoses into the actual universe of change and progress or creative evolution. However, it is obvious that Oken's process metaphysics as pantheistic mysticism is diametrically opposed to Darwin's evolution theory grounded in mechanistic materialism. Nevertheless, the former naturalist did anticipate those visions of Joseph Ernest Renan, Don Miguel de Unamuno, Charles Sanders Peirce, Samuel Alexander, Pierre Teilhard de Chardin, and Paolo Soleri (among others).

In the middle of the last century, Joseph Ernest Renan (1823–1892) abandoned the priesthood and subsequently left the Roman Catholic Church. His controversial ideas were greatly influenced by a philological study of the Holy Bible, the advances in the natural sciences (particularly chemistry), and German thought (especially the dynamic worldview of Hegel). He sought to expand scientific rationalism and to reconcile it with idealist philosophy within a process interpretation of human history as endless progress. Like Baruch Spinoza and August Comte, Renan was interested in the historical development of languages and religions. He taught that the further evolution of the human mind is the key to understanding and appreciating the directional development of the concrete universe. He also claimed that the so-called mysteries of the cosmos would inevitably yield before those advancements of knowledge in both the natural and social sciences as well as the use of reason.

In his volume *Philosophical Dialogues and Fragments* (1871), Renan speculated on the higher organisms of the centuries to come.[5] Although his interests ranged from archeology to philosophy, this unorthodox thinker voiced his anxieties about the future of science and humanity.

Renan held that paleontology not only reveals the remote history of our species but also suggests the ongoing direction of organic evolution (characterized by ever-increasing consciousness): the immediate goal of the earth is the development and fulfillment of mind as the realization of God in the fullness of planetary time.

As did Nietzsche, Renan envisioned an immense advancement of human consciousness resulting in the future emergence of an elite group of superior individuals devoted to the irrefutable evidence of science and the power of reason. Through continued progress in biology and selective breeding, he saw an improvement of the human intellect leading to this superior race of beings above and beyond the present capabilities of reality. He imagined that these divine beings would be as far beyond our present species as it is now above the lower animals.

Although Renan anticipated the persecution of scientists, particularly physiologists and chemists, he nevertheless foresaw the inevitable triumph of the mind and the supremacy of science.

Within an organismic view of dynamic nature, Renan suggested that this whole universe as an emerging cosmic zygote might become a single being that enjoys perceptions and sensations analogous to those enjoyed by finite humanity on earth. In fact, everything that has ever existed in this entire world would eventually culminate in the distinct center of ultimate consciousness as the last goal or final term of a God-developing cosmic evolution. As the totality of the universe, this celestial being of trillions of beings will emerge out of living multiplicity into organic unity (just as a terrestrial animal emerges as an organism of billions of cells from the countless atoms making up its genetic information). The French visionary and futurist believed that all existence will participate in this perfect unity of infinite diversity.

In brief, Renan conceived of a future age in which all the matter in the infinity of space that is now present in countless astral systems or galaxies would become organized into a single being: a living universe that thinks and enjoys not as trillions of individuals but rather as a distinct organism pervading reality. Consequently, for Renan, God is that cosmic consciousness yet to arrive in which everything will be reflected and echoed back throughout all eternity.

In Spain at the beginning of this century, the Basque philosopher and poet Don Miguel de Unamuno y Jugo (1864–1936) gave a passionate view of life that is more emotive than intellectual; he thought with his heart. Like Kant, Unamuno never traveled out of his own country (Salamanca was his Königsberg). He was a concrete man of flesh and blood whose eyes penetrated into nature, life, and the existential condition of the human situation. He emphasized the supreme value of an individual and was committed to belief and action. His intense mind embraced the spiritual origin and development of humankind. It also gazed into that vast mystery of the changing universe itself: he was preoccupied with the destiny of the whole human species and obsessed with an endless longing for personal immortality (one is reminded of Nietzsche's ardent wish to live forever).

Unamuno's masterpiece is The Tragic Sense of Life (1912), a volume rich in ideas whose essential theme is the quest for eternal life despite the excruciating conflict between faith and reason.[6] Above life's doubt and agony and despair, there is the religious hope for the personal immortality of an individual soul beyond the evidence of science and the argument of logic. This author longed to overcome human finality through the passionate hope of eternal life. He gravitated to feelings, the will, and the imagination rather than to science, logic, and mathematics. His faith in God and the universe may be summed up in Tertullian's phrase: *credo quia absurdum* (I believe because it is absurd).

Unamuno rejected the evolutionists who, as did Darwin and Haeckel, understood organic history merely in terms of science and reason. His own interpretation of consciousness in the cosmos gives preference to feelings and beliefs. His vision is far closer to the process systems of Bergson, Whitehead, and Teilhard de Chardin than to any of those proponents of a mechanist/materialist explanation for the emergence of life and thought on earth. His view owes more to the ideas in process philosophy (Gottfried Wilhelm von Leibniz, Georg W. F. Hegel, Arthur Schopenhauer, and Friedrich Nietzsche) than to the facts of the evolutionary sciences.

Since everything that exists seems to be doomed to nothingness, Unamuno has universal pity for all finite things in this universe (from lowly amoebas through human beings to distant galaxies). Organic evolution is simply the perpetual struggle of beings to actualize their consciousness through suffering (despite death) in an endless aspiration to preserve their existence while at the same time striving to become a part of something above and beyond their own limits: this applies to the whole range of natural objects from stars and rocks to plants and animals. An objective idealist, the Spanish thinker saw transitory matter as unconsciousness evolving into spirit as consciousness. In this process metaphysics of existence one has a brilliantly intuitive but nevertheless philosophically incomplete conception of humankind within nature.

Unamuno was deeply concerned with the "why" of human origin and the "wherefore" of its destiny. He argued that the Darwinian zoology sees hydrocephalous and erect-standing humans as merely a "diseased" species of orangutan, chimpanzee, or gorilla (a curious and intelligent species with an opposable thumb and articulate speech). The great apes and their kind must look upon humans as feeble and infirm animals, whose strange custom it is to store up and guard their dead. In fact, Unamuno suggested that perhaps feeling rather than reason actually differentiates humans from all the other animals; for the exclusively rational human is merely an aberration and nothing but an aberration.

Must the end of all personal consciousness (the future of all human spirit on this earth) be that absolute nothingness from whence it sprang? Will this entire world be extinguished, with only naked silence and profound stillness remaining? Unamuno maintained that humankind philosophizes to live and yearns to reach all space, all time, and all being: "Either all or nothing!"

Unamuno introduced a mystical vision of cosmic evolution grounded in vitalism, pervasive consciousness, teleological progress, and a profound faith in a personal God as both the beginning and the end of this spiritual universe. From his panpsychic and organismic view of all things as well as through an extraordinary use of analogy, he saw everything, from minute living creatures through cells and organs to organisms, as ever-higher and more complex distinct units of consciously communicating arrangements of elements, fancying that cells may actually express an awareness that they form a part of an advanced organism endowed with a collective personal consciousness. Through extensive extrapolation, Unamuno entertained the possibility that human beings are more or less similar to globules in the blood of a supreme being, whose own personal consciousness is the collective consciousness of this entire universe. Just as all the cells of a human animal actively interrelate to produce the consciousness of a personal being with an immortal soul, then perhaps the immense Milky Way galaxy (a cosmic ring or stellar group of which our solar system is merely an insignificant part) is but a sidereal fragment of the body of God as that ultimate collective of all consciousness throughout reality that is preserved forever in a supreme consciousness. As such, this whole cosmos mirrors the human organism, and love leads one to personalize the all of which everything is a part: it is universal love that sustains and fulfills the spirit of this world. In short, there is a dialectic between a living and personal God and all the immortal souls of humanity. The end of this universe is that supreme person who, at least in its memory, gives everlasting life to all that has ever existed.

Unamuno's evolutionist and mystical cosmology, imagination-bound and human-centered, is an eschatology rooted in the persistence of all consciousness. Like Whitehead and Teilhard de Chardin, he embraced a form of panentheism in which every spiritual thing and God are continuously interrelating in a divine society, a collective unity of imaginations through which God and this universe eventually merge into a single will of the cosmos.

Unamuno suggested that in this vast universe other worlds may be inhabited by living organisms (even consciousness akin to our own) and that there may also be a plurality of universes.

As did Galileo, the unorthodox Miguel de Unamuno spent the last years of his life under house arrest. In his incredible speculations on the nature of things, he offered a mystical worldview that attempts to fathom the evolution and goal of consciousness within this dynamic universe in terms of passion and intuition.

In 1925, the contrived and infamous John Scopes "Monkey Trial" of Dayton, Tennessee, reawakened the old arguments between fundamentalist creationism and scientific evolutionism.[7] The gifted orator and lawyer William Jennings Bryan (1860–1925) defended a literal interpretation of Genesis with his usual fire-and-brimstone speeches, while the brilliant criminal lawyer and atheistically inclined free-thinker Clarence Darrow argued for the freedom and dignity of human thought as well as unimpeded and ongoing scientific progress. Although Mr. Scopes was found to be technically guilty of breaking the antievolutionist Butler Act of Tennessee, with just irony this originally state-confined issue brought worldwide attention to the growing empirical evidence for and the logical implications of the scientific theory of organic evolution.

At present, Carl Sagan[8] and Eric Chaisson[9] give astronomical evidence to support even the nonbiological concepts of cosmic evolution, while Stephen Jay Gould and Ashley Montagu present the synthetic biological theory of terrestrial evolution in modern science chiefly in terms of the accumulating geological, paleontological, and genetic evidence. Yet, astonishing as it may seem, the struggle between fundamentalist creationism and scientific evolutionism continues in some quarters unabated and even essentially unmodified more than a century after the Huxley-Wilberforce confrontation.

Darwin never claimed that humankind descended from one of the now living greater or lesser apes, let alone from monkeys. Rather, he argued that the pongids and our own species share a very distant common ancestry in Africa. Nevertheless, his position continues to be grossly misrepresented, either through malice or ignorance or as a result of both.

The field of primate paleontology and comparative studies of primate behavior combined with other special sciences (such as comparative anatomy, physiology, embryology, immunology, genetics, and biochemistry) clearly demonstrate that humankind is, in fact, closer to its evolutionary cousins the three great apes (the orangutan, chimpanzee, and gorilla) than even Huxley, Haeckel, or Darwin himself could have imagined in the nineteenth century.

In making a necessary distinction between the supernaturalist account of the origin of life in general (and humankind in particular) and the naturalist theory of evolution as the only plausible empirical alternative, one must always keep in mind the essential and crucial difference between the metaphoric imagery of religious myth and poetry on the one hand and the literal truth of scientific literature and reason on the other. Those who fail to make this quintessential distinction between theology and science are insensitive to both the Holy Bible as great literature and to science as our most powerful tool and reliable method of empirical/rational inquiry. Imaginatively interpreted, even the obviously legendary and didactic parts of the Sacred Scriptures (such as the stories of Creation and the Great Flood) do contain some symbolic value and perhaps even useful instruction. Taken literally and imposed as would-be substitutes for scientific fact and critical inquiry, these dogmatic religious traditions may become deadly impediments to free inquiry and consequently the further advancement of empirical knowledge and philosophical wisdom.

What is science? It is an ongoing search for truth based on an empirical method of formulating and testing hypotheses and theories validated wherever possible by evidence, experience, experimentation, critical reasoning, and other rigorous means of verification. As such, science seeks systematically related bodies of factual knowledge and causal relationships that lead to predictions in terms of degrees of statistical probability, if not even empirical laws or principles of the natural world.

Unlike dogmatic religion, advancing science is primarily rational, self-critical, self-correcting, self-revising, and self-improving (it is essentially and irreconcilably antidogmatic, selectively cumulative, and always open-ended). Of course, one must distinguish between science and scientists, just as one notes the difference between religion and its practitioners. Many religionists delight in speaking of evolution as if it were merely a theory, confusing this concept with that of a scientific hypothesis. Far from being a mere hypothesis, the theory of evolution is a tightly reasoned explanation of natural phenomena now sufficiently supported by overwhelming empirical evidence, past and present human experience, and rigorous logic, with no resort or appeal to an alleged supernatural influence however personally believed and theologically defended.

In Western history, other theories and worldviews at first vehemently rejected by religion were later (after a reasonably short period) ultimately accepted by both the scientific community and the general public. One cannot help but recall the cases of Nicholas of Cusa, Nicolaus Copernicus, Giordano Bruno, Galileo Galilei, Johannes Kepler, and Albert Einstein as well as other pioneers of critical thought and modern science. Those profoundly significant facts and concepts of these and other scientists and natural philosophers challenged the emotionally charged and well-entrenched beliefs of their day so strongly that the immediate reaction of the religious and political establishments threatened their very lives. Yet, not too long thereafter, these remarkably insightful new paradigms resulted in scientific discoveries that soon became accepted knowledge, illuminated and confirmed by ever-growing empirical evidence as well as the use of systems of mathematics and technological advances (especially in the reach and precision of recent scientific instruments, from the electron microscope to the radio telescope).

The modern conceptual worldview is grounded in those iconoclastic ideas of Charles Darwin, Karl Marx, Friedrich Nietzsche, Sigmund Freud, Pierre Teilhard de Chardin, and Albert Einstein (among others). It has also been represented by great artists including Salvador Dali in painting, Thomas Mann in literature, and Richard Wagner in the music-drama.

Why, then, is the Darwinian framework still being greeted with intense hostility and unyielding opposition? Surely this resistance is motivated in some religionists by the mere thought of a serious materialist alternative to what they have dogmatically believed in as that eternal truth concerning the divine creation of the finite world in terms of spirit.

It was one thing for Cusa, Copernicus, Bruno, Galileo, Kepler, Newton, and Einstein to argue against geocentrism (thus removing the earth from its wrongly alleged central position in this universe), but quite another for Darwin and his followers to demonstrate in terms of science and reason that the anthropocentric position is also a conceited illusion. The glaring truth remains that our human species is a long-emerging product of primate evolution in particular and rooted in the history of life in general: it is a recent product of and totally within this material world. In short, our planet does not occupy a central position in the cosmos, and our own species is not a privileged animal in the universe.

Concerning the age of the planet, most recent astrophysical and astrochemical evidence clearly demonstrates that the universe (including the earth) is billions of years old. Obviously, these facts cannot be harmonized with a strict and literal interpretation of Genesis.

It is bewildering that any enlightened thinker should take seriously this ongoing conflict between fundamentalist creationism and scientific evolutionism. Is not the process of organic evolution richly creative? Is not creation a process of pervasive change? It follows that creative evolution and evolutive creation (properly interpreted) are only different conceptions for the complementary scientific understanding of and philosophical appreciation for the same cosmic process and planetary history.

As for God, some scientific evolutionists even continue to believe in his existence. For example, God may be held to be ontologically and logically identified with nature itself as the all-encompassing totality and essential unity of infinite being as eternal becoming.

Or, as in the recent example of Carl Sagan, one may simply state that evolution is a fact of material reality, irrespective of one's being a self-professed atheist. Finally, as in the case of Pierre Teilhard de Chardin, one may make a heroic but unsuccessful attempt at reconciling an acceptance of the scientific facts of evolution with an adherence to personal religious beliefs.

Surely, the numerous interpretations of the theory of evolution in the world literature do differ in emphases and details as well as mechanisms and implications. However, this does not invalidate either the basic fact of or the empirical evidence for organic evolution. To turn away from modern science is to doom oneself to needless self-isolation and eventual self-defeat. Regardless of all our wishful thinking, material reality is continuously reasserting itself in utter indifference to humankind with its needs and wishes and illusions. The theory of evolution remains an amply demonstrated fact that human biases, prejudgments, and emotional insecurities can never disprove: just as this earth moves, so do species change.

Once again the scientific theory of biological evolution has come under scathing criticism by special creationists, particularly the fundamentalist movement.[10] Even after the Darwin-FitzRoy conflict (1831–1836), the Huxley–Wilberforce confrontation (1860), and the John Scopes "Monkey Trial" (1925), the evolution framework continues to be challenged or rejected by orthodox religionists and static philosophers. Given the strong differences today in commitments to dogmatic faith or open science, the ongoing controversy between the Holy Bible and evolution is perhaps unavoidable. A critical examination of the arguments given to refute the claims of an evolutionary viewpoint is required. Such a serious investigation shows the fundamentalist creationist position to be clearly untenable as empirical science or rational philosophy or even natural theology.

Proponents claim that the fundamentalist creationist worldview is scientific, but they actually disregard all facts that contradict their own religious viewpoint. More alarming, they are even insisting that special creationism be taught on an equal basis with the theory of evolution in science courses (especially in biology classes) in public schools. Some fundamentalist creationists have introduced bills in several states to force school board committees to choose only those textbooks that include the biblical story of creation along with the scientific theory of organic evolution. This is obviously a glaring violation of the separation of church and state as well as an attack on the freedom of thought. This movement is primarily the result of the efforts of the Institution of Creation Research in San Diego, California. It is made up of chemists and engineers, but very few biologists. Several staff scientists do have doctoral degrees, but they spend most of their time promoting only fundamentalist creationism on college and university campuses and writing books advocating their own position (these texts are published by the Creation-Life Publishers, also in San Diego), rather than dedicating their efforts to scientific research.

In sharp contrast to this religious fundamentalism, what is the case for scientific evolutionism?

Influenced by Lyell and Malthus, both Darwin and Wallace argued for biological evolution primarily by means of natural selection or the survival of the fittest. In both On the Origin of Species (1859) and The Descent of Man (1871), Darwin presented a mechanist and materialist explanation for organic evolution, which still remains the essential foundation of modern biology. As a theory, evolution is a complex system of facts and ideas that explains both the similarities and differences among organisms within geographical space and throughout planetary time.

The evolutionist holds that new species of plants and animals emerge from earlier, different forms over vast periods of time and change while other species become extinct. As a result of selective forces (especially the major explanatory mechanism of natural selection) and the chance appearance and accumulation of beneficial mutations, whether slight variations or major modifications, some individuals in a population have an adaptive and survival advantage and therefore a reproductive advantage over other individuals. These mutated individuals are favored over others in the ongoing struggle to exist in a changing environment.

Fundamentalist creationists now refer to their religious view as "scientific creationism" while at the same time rejecting those empirical facts that do support organic evolution. Recently, some fundamentalist creationists have admitted that new varieties may appear within a species (microevolution) and that our planet is somewhat older than 6,000 years. Yet, they still will not accept macroevolution and its implications or the earth's enormous age within cosmic history.

Nevertheless, the empirical facts to support biological evolution are sufficient to convince any open-minded intelligent thinker. Although admittedly incomplete, the fossil record is the single most important body of evidence to support the truth of organic evolution. Continuous discoveries in geopaleontology are filling up the major gaps in the rock and fossil records. They support the evolution model rather than the special creations and worldwide Noachian Deluge myths as presented in Genesis (likewise, although rare, transitional fossil forms do exist in the known paleontological evidence).

Both modern taxonomy and comparative studies, from biochemistry and embryology to anatomy and physiology, support the implications of the scientific theory of biological evolution. Today, genetic research demonstrates the historical continuity and essential unity of all life on earth (especially in terms of amino acids and the DNA molecule). Like does not beget like, and kind does not beget kind. All populations are variable and subject to the forces of evolution. In fact, no two plants or animals even in the same population are ever absolutely identical. The biological evolutionist argues that over long periods of time the accumulation of genetic changes in a population interacting with a changing environment may result in a variety becoming a new species in its own right. More time and change may result in the emergence of higher taxonomic categories.

Biogeography demonstrates the explanatory mechanism of natural selection. The distribution of organisms and their physical/behavioral adaptations to different environments clearly support the survival of the fittest. There is a direct and ongoing relationship between living beings and their natural habitats. This is especially illustrated among the unique plant and animal forms throughout the Galapagos Islands (e.g., the differing species of finches, iguanas, and tortoises inhabiting their own specific environments within this oceanic archipelago).

Finally, the fact of evolution does not violate the second law of thermodynamics (entropy). The earth is not a closed system, for energy from the sun is always available for the creative process of organic evolution to occur over billions of years.

Taken together, all of this evidence provides sufficient ground for accepting the scientific fact of organic evolution. However, even today, biblical fundamentalists believe in a strict and literal interpretation of the scriptural account of divine creation. They still hold to a sudden appearance of all living forms within a fixed and ordered view of nature. Yet, the Holy Bible is not a scientific document. Special creationism is theology and not an empirico-logical explanation for the origin and history of life on this planet. As a religious view, it appeals to a supernatural God and the authority of Genesis (not to mention that it is biased in excluding all other creation stories except the Judeo-Christian account).

Fundamentalist creationism is neither falsifiable nor verifiable in principle. It ignores the established facts of the evolutionary sciences (e.g., the evidence from historical geology, comparative paleontology, and prehistoric archeology as well as the recent advances in biochemistry, genetics, and use of radiometric dating techniques). Fundamentalist creationism consistently misrepresents all this evidence for evolution and ultimately breaks down under rigorous scientific and logical scrutiny. It is irrational in principle and, therefore, not admissible as a scientific doctrine. To suggest that a scientific investigation of and rational explanation for the rock and fossil and artifact records support a strict and literal interpretation of Genesis is clearly ludicrous: such a view ignores both facts and logic in favor of religious assumptions and biblical authority.

In sharp contrast, the evolution model has not been refuted by either empirical tests or the principle of falsifiability. Evolution remains a powerful fact and meaningful theory in its explanation of evidence and prediction of events in modern biology, including physical anthropology. Actually, most Western religions do accept the fact of evolution as that creative process throughout natural history that has produced all life on earth. God remains as the first cause of this dynamic universe, if not also the creator of only the common source or first forms of all life on our planet.

Of course, there is a crucial distinction between the scientific fact of evolution and the various interpretations and perspectives of this natural process in the world literature. Bold attempts to reconcile scientific evolution with religious beliefs have failed, being poor compromises grounded in obscurity. The natural and supernatural are not compatible in terms of facts and logic. Such undertakings by Wallace, Bergson, and Teilhard de Chardin resulted in giving preference to process spiritualism (*The World of Life*, 1910),[11] intuitive metaphysics (*Creative Evolution*, 1907),[12] and panentheistic mysticism (*The Phenomenon of Man*, 1940),[13] respectively.

Does a personal God design harmful mutations, allow the spread of fatal diseases, destroy countless plant and animal species from time to time through mass extinctions (only to replenish the earth with new forms of life), and deliberately confuse believers with the age and sequence of rocks and fossils and artifacts? If one believes in a supreme being, then could both Darwin and Teilhard have been divinely inspired to understand organic evolution in order to appreciate the creative power of a personal God?

Darwin's importance in intellectual history is enormous, and his theory of evolution with appropriate modifications remains as powerful as ever. No alternative to organic evolution gives an equally comprehensive and intelligible explanation of those facts and events and relationships in astronomy, biochemistry, genetics, geology, paleontology, archeology, anthropology, sociology, and even psychology. The modern synthetic theory of biological evolution is grounded primarily in the explanatory mechanism of natural selection (Darwin and Wallace) and genetic variability (Mendel and DeVries).[14] Organic evolution is ongoing genetic changes in populations throughout the history of life on earth resulting at times in the appearance of new species and eventually higher taxonomic categories. It does not necessarily result in progress, but such progress is not always excluded. Furthermore, evolution neither supports the necessity of increasing complexity, consciousness, or perfection nor demonstrates a preestablished meaning, purpose, direction, end, or goal (teleology) within this cosmos or throughout earth history. In biology, the guiding principle is simply survival by adaptation to a changing environment and the subsequent reproduction of life.

Although Darwinian natural selection still remains an important aspect of modern biology, science has not as yet exhausted an understanding of all those mechanisms or forces of organic evolution (especially in the areas of genetics and group behavior). There is even a remote possibility that some Lamarckian element plays a minor role in biological inheritance, perhaps in earlier or lower forms of life. However, present uncertainty or incompleteness in evolutionary science does not warrant turning to the viewpoint of fundamentalist creationism. There is often a psychological need to believe in the supernatural, and this human desire itself is subject to scientific investigation within a sociohistorical framework.

The biblical creation story may be taught in a literature class, but to compel its teaching in a biology course is clearly not appropriate (just as astrology and faith healing are no substitutes for astronomy and medicine, respectively). As dogmatic religion, fundamentalist creationism is an ongoing danger to freedom and science and education, especially if it discourages serious empirical inquiry and rigorous philosophical reflection.

Fundamentalist creationism blindly ignores the overwhelming scientific evidence that now supports the fact of organic evolution. Therefore, it actually threatens the advancement of free thought and the special sciences. In principle, modern science does not acknowledge the supernatural but does require open inquiry into the universe and earth history as well as human existence. Scientific investigation is a liberation from blind faith, vacuous myths, human emotions, and religious dogmatism. In fact, all empirical explanations as such are essentially naturalistic and rational in principle.

In *Eupraxophy* (1989), the third volume of a trilogy on modern secular humanism, Paul Kurtz presents his thesis that an authentic moral life is indeed possible without religion.[15] He argues that a human being can lead a productive and rewarding life totally free from all religious beliefs and practices; in fact, millions of people do. His bold position is based on both a historical approach and a cross-cultural orientation. Grounded in science and reason, this naturalist philosophy acknowledges the fact of evolution and a cosmic perspective.

Kurtz has coined the term "eupraxophy" to mean good conduct resulting from scientific knowledge and philosophical wisdom. The conceptual roots of eupraxophy are found in the pioneering thoughts of Darwin, Marx, Nietzsche, Freud, and Dewey (among others). Use is made of the empirical evidence from both the natural and social sciences, especially anthropology and psychology; one recalls the ideas of Ludwig Feuerbach (1804–1872). For example, the obvious materialist implications of an evolutionary framework have dealt a severe blow to traditional religion, particularly the biblical fundamentalism of the creationist movement in this century.

The life stance of a secular humanist as eupraxopher is grounded in this material universe as well as human nature and social action. It is committed to science (including technology and industry), reason, free inquiry, and ethical wisdom. Of primary value is the critical intelligence of a responsible person.

Kurtz gives a brief but penetrating introduction to comparative religion. His evaluations of theistic religions as well as Buddhism, Hinduism, and Confucianism are very revealing; one could do the same for Taoism. In general, he shows that religion has been a serious impediment to the growth of rational thought and scientific understanding. In agreement with Corliss Lamont and Sidney Hook, Kurtz rejects any religious definition of humanism. Consequently, his own secular humanism is a far more honest and rigorous position than even the naturalist philosophy held by John Dewey (1859–1952)!

Of crucial importance is Kurtz's use of a cosmic outlook to properly assess the true place of humankind within a sidereal view of all things. This awesome perspective has not been established in modern thought without an ongoing confrontation between religious dogmatism and scientific inquiry. One need only recall the personal struggles of Kepler, Galileo, Corpernicus, and especially Bruno in their courageous attempts to overcome those errors in the Thomistic/Aristotelian worldview. In sharp contrast to idealist philosophy and theistic religion, modern secular humanism places our own species completely within this dynamic material universe, which is indifferent to both human existence in general and the brute reality of each individual's finitude in particular.

As a direct outgrowth of his orientation, Paul Kurtz offers a secular humanist mandate. In essence, the eupraxopher ought to be both knowner and doer with convictions and commitments based on the resolute courage to create a better tomorrow for our global species and its future universal community. Furthermore, Kurtz proposes an ambitious program to enhance the growth of eupraxophy: it involves the establishment of institutions dedicated to teaching a comprehensive view of humankind and encouraging human fulfillment in terms of social action, the special sciences, and holistic philosophy.

There is an urgent need for such an enlightened and relevant worldview free from intellectual apathy and religious obfuscation. Of course, one cannot predict the future of the human animal hundreds or thousands of years from now. Nevertheless, devoted to the unending quest for truth and wisdom, modern secular humanism does provide a realistic conceptual framework for the ongoing ethical and social development of our species on this planet and elsewhere.

Evolution is a natural process and scientific framework supported by facts, experience, and reason. As an explanation, evolution is open to modifications in terms of new empirical evidence and logical reflection. Unlike the fundamentalist creationist, an evolutionist as materialist (naturalist and humanist) accepts the implications and consequences of the evolutionary sciences: this planet is not the center of reality, life may exist elsewhere in the cosmos, and our own recent species does not occupy a privileged position within the dynamic history of this material universe.

Although many questions remain to be answered at this time, modern evolutionists continue to make great progress through the ongoing abductive method of scientific investigation. What is clearly needed is more science, i.e., that continued free and rational inquiry into the natural origin and historical development of all life on earth and its probable existence elsewhere in the evolving universe.

In accordance with facts and logic, one goal for the evolutionist as materialist is to establish a sane and humane cosmology for the further development of our own species.

ENDNOTES AND SELECTED REFERENCES

1. Cf. Ronald Lane Reese, Steven M. Everett, and Edwin D. Craun, "The Chronology of Archbishop James Ussher," *Sky and Telescope*, 62(5):404–5.

2. Cf. Philip Henry Gosse, *Creation (Omphalos): An Attempt to Untie the Geological Knot* (London: John Van Voorst, 1857).

3. Cf. William Irvine, *Apes, Angels, and Victorians: The Story of Darwin, Huxley, and Evolution* (New York: McGraw-Hill, 1955).

4. Cf. William Dawson LeSueur, *Evolution and the Positive Aspects of Modern Thought* (Ottawa: A. S. Woodburn, 1884).

5. Cf. Joseph Ernest Renan, "The Higher Organisms of the Centuries to Come," in *French Utopias: An Anthology of Ideal Societies*, ed. Frank E. Manuel and Fritzie P. Manuel (New York: The Free Press, 1966), pp. 381–91.

6. Cf. Miguel de Unamuno, *The Tragic Sense of Life* (New York: Dover, 1954).

7. Cf. Ray Ginger, *Six Days or Forever? Tennessee v. John Thomas Scopes* (New York: Signet Books, 1960). Also refer to Irving Stone, *Clarence Darrow for the Defense: A Biography* (Garden City, N.Y.: Garden City Publishing, 1943), pp. 422–65.

8. Cf. Carl Sagan, *Cosmos* (New York: Random House, 1980).

9. Cf. Eric Chaisson, *Universe: An Evolutionary Approach to Astronomy* (Englewood Cliffs, N.J.: Prentice Hall, 1988).

10. Cf. Ashley Montagu, ed., *Science and Creationism* (New York: Oxford University Press, 1984).

11. Cf. Alfred Russel Wallace, *The World of Life: A Manifestation of Creative Power, Directed Mind and Ultimate Purpose* (New York: Moffat/Yard, 1916).

12. Cf. Henri Bergson, *Creative Evolution* (New York: Modern Library, 1944).

13. Cf. Pierre Teilhard de Chardin, *The Phenomenon of Man*, 2d. ed. (New York: Harper Torchbooks, 1965).

14. Cf. Julian Huxley, *Evolution: The Modern Synthesis* (New York: John Wiley & Sons, 1964).

15. Cf. Paul Kurtz, *Eupraxophy: Living Without Religion* (Buffalo, N.Y.: Prometheus Books, 1989). Also see Kurtz's *Forbidden Fruit: The Ethics of Humanism* (Buffalo, N.Y.: Prometheus Books, 1988) and *The Transcendental Temptation: A Critique of Religion and the Paranormal* (Buffalo, N.Y.: Prometheus Books, 1986).

FURTHER READINGS

Chaisson, Eric. *Cosmic Dawn: The Origins of Matter and Life.* Boston: Little/ Brown, 1981.

———. *The Life Era: Cosmic Selection and Conscious Evolution.* New York: Atlantic Monthly Press, 1987.

———. *Universe: An Evolutionary Approach to Astronomy.* Englewood Cliffs, N.J.: Prentice-Hall, 1988.

Clements, Tad S. *Science Versus Religion.* Buffalo, N.Y.: Prometheus Books, 1990.

Ecker, Ronald L. *Dictionary of Science and Creationism.* Buffalo, N.Y.: Prometheus Books, 1990.

Eldredge, Niles. *The Monkey Business: A Scientist Looks at Creationism.* New York: Pocket Books, 1982.

Flew, Antony. "The Erosion of Evolution: A Treason of the Intellectuals." *Free Inquiry* 2, no. 2 (Spring 1982):19–23.

Godfrey, Laurie R., ed. *Scientists Confront Creationism.* New York: W. W. Norton, 1983.

Gosse, Edmund. *Father and Son: A Study of Two Temperaments* (1907). New York: W. W. Norton, 1963.

Gosse, Philip Henry. *Creation (Omphalos): An Attempt to Untie the Geological Knot.* London: John Van Voorst, 1857.

Hawking, Stephen W. *A Brief History of Time: From the Big Bang to Black Holes.* New York: Bantam Books, 1988.

Holland, Clifford G. "William Dawson LeSueur." Dictionary of Literary Biography 92:198-200. Detroit, Mich.: Gale Research Inc., Bruccoli Clark Layman, 1990.

Kitcher, Philip. *Abusing Science: The Case Against Creationism.* Cambridge, Mass.: MIT Press, 1982.

Kurtz, Paul. "Eupraxophy: The Need to Build Secular Humanist Centers." *Free Inquiry* 10, no. 1 (Winter 1989/1990):24–31.

———. "Humanist Ethics: Eating the Forbidden Fruit." *Free Inquiry* 9, no. 2 (Spring 1989):25–29.

———. *Philosophical Essays in Pragmatic Naturalism.* Buffalo, N.Y.: Prometheus Books, 1990.

Lamont, Corliss. *The Philosophy of Humanism.* 6th ed. New York: Frederick Ungar, 1982.

McCollister, Betty. "Our Traditional Species." *Free Inquiry* 8, no. 1 (Winter 1987/1988):55.

McGowan, Chris. *In the Beginning . . . : A Scientist Shows Why the Creationists Are Wrong.* Buffalo, N.Y.: Prometheus Books, 1984.

McKown, Delos. " 'Scientific' Creationism: Axioms and Exegesis." *Free Inquiry* 1, no. 2 (Summer 1981):23–28.

Montagu, Ashley, ed. *Science and Creationism*. Oxford, England: Oxford University Press, 1984.

Nelkin, Dorothy. *The Creation Controversy: Science or Scripture in the Schools*. New York: W. W. Norton, 1982.

Newell, Norman D. *Creation and Evolution: Myth or Reality?* New York: Columbia University Press, 1982.

Nielsen, Kai. *Ethics Without God*, rev. ed. Buffalo, N.Y.: Prometheus Books, 1989.

Rohr, Janelle, ed. *Science & Religion: Opposing Viewpoints*. St. Paul, Minn.: Greenhaven Press, 1988, esp. pp. 29–52, 145–75.

Ruse, Michael. *Darwinism Defended: A Guide to the Evolution Controversies*. Reading, Mass.: Addison-Wesley, 1982.

———., ed. *But Is It Science?: The Philosophical Question in the Creation/Evolution Controversy*. Buffalo, N.Y.: Prometheus Books, 1988.

Sagan, Carl. *Cosmos*. New York: Random House, 1980.

Scopes, John Thomas. *Center of the Storm*. New York: Holt, Rinehart, and Winston, 1967.

Shaeffer, Robert. "Nietzsche's *Der Antichrist*: Looking Back from the Year 100." *Free Inquiry* 9, no. 1 (Winter 1988/1989):52–56.

Soleri, Paolo. *The Omega Seed: An Eschatological Hypothesis*. Garden City, N.Y.: Anchor Press, 1981.

Sprague de Camp, L. "The Continuing Monkey War." *Free Inquiry* 2, no. 2 (Spring 1982):12–17.

Stone, Irving. *Clarence Darrow for the Defense: A Biography*. Garden City, N. Y.: Garden City, 1943, esp. pp. 422–65.

Strahler, Arthur N. *Science and Earth History: The Evolution/Creation Controversy*. Buffalo, N.Y.: Prometheus Books, 1987.

Wilson, Edward O. "Biology's Spiritual Products." *Free Inquiry* 7, no. 2 (Spring 1987):13–15.

Zetterberg, J. Peter, ed. *Evolution Versus Creationism: The Public Education Controversy*. Phoenix, Ariz.: Oryx Press, 1983.

8

Conclusion

On Life within this Universe

Following the publication of Darwin's *Origin* (1859) and *Descent* (1871), the scientific theory of biological evolution slowly but steadily became the major explanatory instrument of modern astronomy, cosmology, and the biosocial sciences. As a result, today the theory and the fact of evolution represent the principal conceptual thread that runs through the whole of existence, from sidereal galaxies to human ethics.

The survival and fulfillment of humankind seem to require its eventual venturing to the stars as a continuing outgrowth of the as yet unfinished terrestrial evolution. Through modern science and technology, our own species should soon be able to adapt, survive, and thrive elsewhere on other worlds. However, there are many serious obstacles that must be overcome. A favorable convergence of the space sciences with genetic engineering should remove many of the physical and pyschosocial barriers challenging the fragile and vulnerable human animal as it responds to the cosmic call and vital imperative: explore and colonize this material universe or perish prematurely on planet earth.

This dynamic cosmos is about twenty billion years old, and earth is but a tiny fragment within seemingly eternal time, infinite space, and endless change. With the relatively recent emergence of our own species, this evolving universe has at least once become aware of itself through many ephemeral personal centers of partial and imperfect consciousness. Yet, human curiosity and research need not remain forever earth-bound or confined to the solar system or even limited to this galaxy.

From Thales to Stephen W. Hawking,[1] cosmologists have wondered about the origin, scope, and destiny of the universe. Other stars, planets, and celestial phenomena are subject to critical scientific inquiry and ongoing rational speculation. Therefore, the ongoing challenge of both understanding and appreciating the emergence, nature, and evolution of life within the dynamic cosmos may be endless in the realm of human thought and research.

In the last century, naturalists took time, change, and development seriously. As the pioneering father of the scientific theory of biological evolution, Darwin explored the historical continuity and essential unity of all life on earth. He was aware of the creative and destructive interactions between organisms and their environments. Nevertheless, he was unfortunately unaware of Mendel's pioneering discovery of the basic principles and algebraic patterns of heredity. In all cells, chromosomes contain the hereditary units of life. For the most part, behavioral traits and physical features are potentially determined by these particles called genes or alleles. These units, both singly and in intricately combined association and interaction, transfer biological information from generation to generation and, through chemical mutations (slight variations or major modifications), may in time bring about new varieties and eventually perhaps even new species as well as higher taxa.

Chemical mutations in the genetic makeup or genotype are likely to be externally expressed as physical and/or behavioral alterations in an organism's phenotype: slight variations or major modifications in the organism's appearance may (at any one time) be of temporarily positive, neutral, or negative value. Until the advent of human culture, life was subject to the two major forces of random biological evolution: chance genetic variability and necessary natural selection.

Increasing human interference with hitherto predominantly natural selection through culture (especially science and technology) has already resulted in new varieties or subspecies of cultivated plants, domesticated animals, and the ongoing civilizing of our own zoological group. The early experiments with garden peas, evening primroses, and fruit flies may now be extended to alter higher or later forms of life, including our own species in accordance with ethically, aesthetically, and rationally guided human planning and human designing for the future. Such directed and controlled evolution potentially holds both great promise and cataclysmic peril. Certainly, the sobering complexity of the enormous risks and vast opportunities involved cause one to take pause and seriously reflect upon the possibly ominous intellectual implications and staggering moral responsibilities of all human programming of life. Even minor errors could have major consequences, even in the best of naturalist and humanist intentions.

In the middle of this century, James D. Watson and Francis H. C. Crick discovered the chemical, structural, and functional characteristics of the DNA molecule.[2] This material is the essential code of all life on earth. Altering the DNA molecule, either by accident or deliberately, brings about changes in living forms that may enhance or endanger their survival and reproductive potential. The human organism is also determined by its genetic makeup in interaction with its natural environment as well as its psychosocial and cultural setting (i.e., those physical and symbolic aspects of material reality).

Humankind with its societies and cultures is completely dependent upon nature. The existence of our own species is incontrovertibly dependent upon and historically tied to both the mutable genes within and the elusive stars beyond. The human being is a recent product of and totally within cosmic evolution.

For about four million years, the emerging human animal has been more or less successfully adapting to various habitats on this planet. Yet, cradle earth is only the birthplace of our zoological group. Humankind already has the science, medicine, technology, and mathematical skill to leave the physical confines and material limitations of this exquisite world. No longer earth-bound, humankind is becoming a citizen of the dynamic universe. It may now daringly soar starward, exploring and enjoying the cosmic spectacle. However, to reach distant suns and prosper on other planets, our form of life will have to adapt to deep space and to the strange environments of unfamiliar worlds.

Human curiosity and imagination are insatiable; even before Carl Sagan and Eric Chaisson, scientists envisioned engineering other worlds for the sake of peopling the galaxies. The exploration, colonization, and utilization of the universe may require even a partial reengineering of the human animal to overcome some of the major obstacles to life in outer space.

Beyond certain arbitrary legal restrictions on the planet, genetic research in deep space could result in knowledge that will help solve the problems that humans will face as they journey among the stars. In short, this universe would become a celestial school and cosmic laboratory for accelerated intellectual and behavioral evolution.

Humankind is only beginning to understand and appreciate all the wondrous potential within a living cell and the rich resources throughout the sidereal depths of the universe. Through genetic engineering, the blind necessities of nature become the deliberate choices of humanity. To succeed in this cosmic quest, the human organism may need to be partially redesigned to overcome the physical challenges of differing gravities, energies, and as yet unimaginable other forces and impediments (including the relativity of time, space, and change).

The current military and political issues of space exploration and genetic engineering should not lastingly blind us to all the clear advantages of ongoing research in these two areas so vital to the qualitative continuation and possible improvement of our own species and other forms of life.

From worm through ape to humans, organic evolution on earth has had to struggle against the forces of death and entropy. Time and mortality are the ravages of life as we know it. However, through genetic engineering, the biological sciences and medical technologies of the future may give our own species such a dramatic extension of average life expectancy as would amount to what essentially could be considered practical immortality (approximating in the physical sense our deep wish for eternity). Along with retarding and eventually halting the aging process, the human senses may be intensified and the mind expanded to overcome the disturbing periods of social isolation and the possibility of cosmic boredom. Also, future science should be able to deal successfully with those as yet unknown illnesses and diseases of outer space.

Gene splicing (recombinant DNA/RNA) and cloning should allow humanity to seed this universe with new forms of plants and animals as well as our own kind.[3] Our species would thereby spread the sparks of life and consciousness throughout the dark abysses of endless space and eternal time. Human hope with visionary science may elevate and transform us not only physically and intellectually but also, more importantly, perhaps even in the ethical and moral realms. There could be an eventual holistic transformation of humanity as the cosmic primate into a higher being worthier of the challenges and rewards of its sidereal destiny.

The advancement of both scientific knowledge and human wisdom has always required courageous and bold individuals working in an open and pluralistic society that encourages critical inquiry within the rational guidelines of ethically and legally responsible freedom of choice and behavior.

In this obvious and drastic tampering with the stuff of life, one should not (on any a priori grounds) fear the unusual and the unknown. On the contrary, one should approach them with tempered courage and proper respect. Of course, the awesome and disturbing prospects of a scientific convergence of modern technology with genetic engineering demand considerable caution and wise enlightenment. The cosmic quest should actually be enhanced through the possibilities of genetic engineering. In fact, as a result, our embryonic species may yet enjoy its finest hours amid the intergalactic lights of this and perhaps other universes.

Still, with all its frightening aspects and magnificent splendors, this material cosmos is totally blind to human evolution and its destiny. Consequently, the more we realize that this universe is indifferent to us, the more concerned we should become with one another's needs and desires and vulnerabilities. Scattered throughout the vastness of space-time, we must continue to respect the freedom and dignity of human thought and feeling as well as value the adaptive plasticity and emerging creativity of life everywhere. In a word, let us humanize the cosmic quest!

Although full of ice and snow, not all of reality is frozen: nature is flux, change, and evolution. Yet, pervasive nothingness or a frigid eternity may well be the inevitable outcome of cosmic history for our own species. Nevertheless, humanity with its delicate genes and mighty space vehicles will probably have made at least a valiant attempt to fathom that magnificent tragicomedy and perhaps the ultimate absurdity of it all. One may hope that this ennobling effort will, however temporarily and fleetingly, increase the sum of human happiness and wisdom. For what benefit would it be if humankind were to gain this entire universe at the price of losing, in the process, its humanity? The future course our own species takes is up to all of us, and the responsibility for its consequences is all of ours.

The earth is nearly five billion years old, and life as we know it emerged approximately four billion years ago. That physicochemical transition from geogenesis to biogenesis was a major advance in the ongoing flux of material reality. With the recent origin of our own species, the evolution process becomes aware of itself at least once in organic history. However, is there life or intelligence or civilization on other clement earthlike planets elsewhere in this dynamic universe?

No doubt prehistoric people gazed at the night skies, curious about the distant shining objects strewn across the ominous heavens. In their lack of scientific understanding and fear of the unknown, they resorted to magicoreligous thoughts and activities as well as legends and myths to explain a world seemingly inhabited by both good and evil spirits.

In antiquity, some natural philosophers as cosmologists not only advocated the empirical/rational investigation of nature but also speculated on living things elsewhere in this universe, e.g., Xenophanes, Anaxagoras, Epicurus, and Lucretius.[4] The following church-dominated Dark Ages and medieval period suppressed scientific inquiry. Turning from nature to the supernatural, Augustine wrote that he could no longer dream about the stars. Later, however, Nicholas of Cusa (1401–1464) did envision inhabitants in every region of the cosmos.[5]

During the subsequent Italian Renaissance, Giordano Bruno (1548–1600) boldly initiated the modern cosmology,[6] an orientation revitalized by astronomers in this century.

Bruno went far beyond the Aristarchus-Copernicus heliocentric theory by claiming the universe to be eternal in time, infinite in space, and endlessly changing. His own sweeping worldview is grounded in the metaphysical principles of plurality, uniformity, and cosmic unity. He rejected the peripatetic terrestrial/celestial dichotomy in favor of an infinite number of inhabited worlds more or less analogous to our earth. For him, life and intelligence necessarily exist throughout this eternal, infinite, and dynamic universe.

Bruno overcame all geocentric, zoocentric, and anthropocentric models of reality. His theme of extraplanetary life as well as extraterrestrial intelligence was later supported by Johannes Kepler, John Wilkins, Christiaan Huygens, Bernard le Bovier de Fontenelle, Immanuel Kant, Ernst Haeckel, Percival Lowell, Camille Flammarion, and Fred Hoyle as well as Carl Sagan and Eric Chaisson (among others). In imagining the future of humankind, even paleoanthropologist and wildlife conservationist Richard E. F. Leakey considers the possibility of contact with interstellar planetary travelers.

In the last century, the scientific search for fossils led to a discovery of the unusual life forms of the remote past. The evolutionary framework held our human species to be a recent product of organic history. Yet, how accidental or inevitable is biological evolution? Although living forms did emerge at least once in this noisy and violent universe, exobiologists do speculate that they will be found elsewhere (perhaps based on different chemistries other than carbon, e.g., silicon or ammonia). Thus, humankind's intellectual penetration into the infinity of cosmic space was followed by an equally important intellectual penetration into the depths of evolutionary time.

Logic does not dictate that humanity or the earth must be unique. The birth of stars may inevitably entail the formation of planets engaged in prebiotic chemistry producing organic molecules like nucleotides and/or amino acids (hydrogen, liquid water, and energy are abundant throughout this cosmos). If the basic laws of biology and physics are applicable everywhere in the universe, then the emergence of life from inorganic matter seems inevitable given the environmental prerequisites.

To date, the life sciences have restricted their concerns to this earth. However, humankind is no longer confined either physically or intellectually to the finite sphericity of this globe. The new challenge is to overcome that myopic planetary view of things by adopting a properly contextual cosmic perspective of reality. Although there is as yet no irrefutable empirical evidence for astrobiology or sentient life beyond the planet, this alone should not deter us from the cosmic quest; it is too soon to abandon rational speculation about and scientific inquiry into exobiology. No dogmatic closure should be placed on human wonder or free inquiry in an intellectually open and ethically concerned society.

Fortunately, there are those who are still eager to engage in a serious search for organisms and civilizations elsewhere among innumerable galaxies. Unmanned interplanetary and interstellar vehicle probes as well as radio telescopes penetrate the abysses of outer space in the hope of detecting empirical evidence of extraterrestrial life, including intelligent forms of nonhuman life. Further advances in space technology, incorporating minicomputers and cosmic telescopes, will greatly enhance the human search for organic existence and intelligence on other celestial bodies throughout this material universe.

Apparently the moons and other planets of our own solar system do not support indigenous living things. However, astrobiology maintains that the physical universe may be a zoocosmos teaming with life. Beyond our Milky Way are billions of other galaxies, each with billions of stars, if not also countless planetary systems. A myriad of other worlds may be swarming with awesome biological forms, disquieting alien beings, and perhaps even organic phenomena now beyond our imagination and cognition. Thus, although as a form of highly organized matter life may be a relatively rare occurrence in each galaxy, it is statistically speaking still bound to be very abundant in the cosmos as a whole in view of the staggering number of existing galaxies.

Special stony carbonaceous chondrite meteorites contain water, minerals, and complex organic molecules such as hydrocarbons or amino acids (the precursors of life on earth). Comets, asteroids, and cosmic gas/dust clouds do harbor the molecules of prebiotic chemistry that anticipate organic phenomena. Physical uniformity or the seemingly pervasive and inexorable laws of nature, statistical probability, empirical inferences, and a temporal framework of billions of years all support both the logical possibility and empirical probability of life-yielding star/planet systems elsewhere in this evolving universe.

Other universes with life forms may have existed before, are coexisting with, or may even exist after this particular cosmos. If so, then our own species will always remain ignorant of the total manifestation of organic beings throughout reality.

Of course, the existence of intelligent beings elsewhere in the material cosmos would require an awesome convergence of possibilities. Even if this event is centuries from now, the anticipated conclusive empirical evidence for extraplanetary life or human contact with extraterrestrial intelligence will be the most profound discovery in the history of scientific inquiry and a singularly awesome turning point in the life of our own species. This event will stimulate a retrospective reevaluation of the nature of existence and at the same time revolutionize our conception of the meaning, purpose, and destiny of humankind itself.

Human beings have never been the ultimate measure of all things. A new and challenging paradigm is emerging; this universe is our ultimate frontier. In the search for life/intelligence beyond our earth, sidereal biophysics is emerging as an intriguing and legitimate area of scientific inquiry. In its continuing exploration of the universe, let us hope that humankind will take (with the rest of its nature) love, compassion, and ethical wisdom to the stars and even beyond. It is this cosmic quest that ennobles and dignifies our human species, and the saga is just beginning.

In his final chapter of *The Voyage of the Beagle* (1839), Charles Darwin, while walking through the dense, wild, and luxuriant splendor of a Brazilian tropical rain forest in early August, wrote the following words: "How great would be the desire in every admirer of nature to behold, if such were possible, the scenery of another planet!"[7] There will probably be future naturalists who will one day study life forms, if not even intelligent beings, on other celestial bodies elsewhere in this physical universe. A continuation of exobiology, exoevolution will be the science that studies the origin and history of life on a different world. One exciting result of such activity would be the science of comparative exoevolution.

From Thales to Hawking, philosophical speculation about and scientific inquiry into the origin and nature of this universe and life within it have preoccupied the best thoughts of the greatest thinkers. Darwin is clearly one of these. As can be seen from the above quotation, his vision caught a glimpse of what may yet be the future of both the theory and fact of cosmic evolution. From this sidereal perspective, Darwin's intellectual legacy will remain as conceptually relevant as it is rich, sound, and indispensable.

The vision of Pierre Teilhard de Chardin, earth-bound and human-centered as it is, needs to be surpassed by a bold new worldview that acknowledges both the supremacy of science and a cosmic perspective.[8] In the final analysis, neither Darwin nor Teilhard had the last word on interpreting evolution from any perspective or within any framework.

ENDNOTES AND SELECTED REFERENCES

1. Cf. Stephen W. Hawking, *A Brief History of Time: From the Big Bang to Black Holes* (New York: Bantam Books, 1988).

2. Cf. James D. Watson, *The Double Helix: A Personal Account of the Discovery of the Structure of DNA* (New York: Mentor Books, 1969).

3. Cf. Alan E. H. Emery, *An Introduction to Recombinant DNA* (New York: John Wiley & Sons, 1984).

4. Cf. Benjamin Farrington, *Greek Science: Its Meaning For Us* (Baltimore, Md.: Penguin Books, 1966); and Giorgio de Santillanan, *The Origins of Scientific*

Thought: From Anaximander to Proclus, 600 B.C. *to* 500 A.D. (New York: Mentor Books, 1961).

5. Cf. John Patrick Dolan, ed., *Unity and Reform: Selected Writings of Nicholas de Cusa* (Notre Dame, Ind.: University of Notre Dame Press, 1962).

6. Cf. Giordano Bruno, *Cause, Principle and Unity: Five Dialogues* (Westport, Conn.: Greenwood Press, 1976).

7. Cf. Charles Darwin, *The Voyage of the Beagle* (1839) (Garden City, N.Y.: Anchor Books, 1962), p. 494.

8. Cf. Pierre Teilhard de Chardin, *The Phenomenon of Man*, 2d. ed. (New York: Harper Torchbooks, 1965).

FURTHER READINGS

Arrhenius, Svante. *Worlds in the Making: The Evolution of the Universe.* New York: Harper, 1908.

Asimov, Isaac. "Where Is Everybody?" *American Way* (December 15, 1987) 20(24):20, 22.

Bova, Ben, and Byran Preiss, eds. *First Contact.* New York: NAL Books, 1990.

Bylinsky, Gene. *Life in Darwin's Universe: Evolution and the Cosmos.* Garden City, N.Y.: Doubleday, 1981.

Chaisson, Eric. *Cosmic Dawn: The Origins of Matter and Life.* Boston: Little/ Brown, 1981.

——. *The Life Era: Cosmic Selection and Conscious Evolution.* New York: Atlantic Monthly Press, 1987.

Dyson, Freeman J. "Time without End: Physics and Biology in an Open Universe." *Reviews of Modern Physics* 51(3):447–60.

Easterbrook, Gregg. "Are We Alone?" *The Atlantic Monthly* (August 1988) 262(2):25–38.

Feinberg, Gerald, and Robert Shapiro. *Life Beyond Earth: The Intelligent Earthling's Guide to Life in the Universe.* New York: William Morrow, 1980.

Foglino, Annette. "Is Anyone Out There?" *Life* (July 1989) 12(8):45–53, 57.

Gribbin, John. *Genesis: The Origins of Man and the Universe.* New York: Delacorte Press, 1981.

Hawking, Stephen W.: *A Brief History of Time: From the Big Bang to Black Holes.* New York: Bantam Books, 1988.

Heppenheimer, T. A. "After the Sun Dies." *Omni* 8(11):36–38, 40.

Kamshilov, M. M. *Evolution of the Biosphere.* Moscow: Mir, 1976.

Krishtalka, Leonard. *Dinosaur Plots, and Other Intrigues in Natural History.* New York: Avon Books, 1990, esp. pp. 103–113.

Metchnikoff, Élie. *The Nature of Man: Studies in Optimistic Philosophy.* New York: G. P. Putnam's Sons, 1903.

Rothman, Tony. "This Is the Way the World Ends." *Discover* 8(7):82–93.

Sagan, Carl. *Cosmos.* New York: Random House, 1980.

Sagan, Carl, and Frank Drake. "The Search for Extraterrestrial Intelligence." *Scientific American* 232(5):80–89, 122.

Sagan, Carl, and I. S. Shklovskii. *Intelligent Life in the Universe.* San Francisco: Holden-Day, 1966.

Shapley, Harlow. *The View from a Distant Star: Man's Future in the Universe.* New York: Delta, 1964.

Shostak, Seth. "Where are the Extraterrestrials Hiding?" *The Saturday Evening Post* 259(6):62–65.

Weinberg, Steven. *The First Three Minutes: A Modern View of the Origin of the Universe.* New York: Bantam, 1977.

Epilogue

A Cosmic Perspective

No human being experienced the origin and development of this universe. Yet, it is intriguing to imagine the existence of an ageless and indifferent observer watching the awesome sweep of cosmic evolution. What incredible objects, events, and relationships would this galactic spectator view during the twenty billion years of sidereal history?

This witness would be present at the instant emergence of space, time, energy, and finally matter itself, as well as the dimensions and forces of dynamic reality. However, what unique manifestations of energy and matter now exist beyond our present glimpse of the cosmos? There are probably numerous other disparate universes, each locked within its own configurations and, as such, perhaps isolated forever from human inquiry.

Throughout eons, swirling clouds of gas and dust have condensed to form billions and billions of galaxies. One of these clusters is our spiraling Milky Way, which includes our solar system and countless other stars. The sun appeared about five billion years ago, and its planets emerged shortly thereafter. Within a cosmic framework, the earth is merely a recent speck in this noisy and violent universe. At the midpoint of planetary history, our own species emerged as the only animal aware of the very process that has brought it into existence: at least once, cosmic evolution is aware of itself. Literally, humankind owes its birth and development and perhaps death as well to the stars.

There is in this grand view of the totality of things a profound awareness of the creative continuity and essential unity of material reality. The seemingly fixed and rigid earth is actually a dynamic world of pervasive change. Across the enormous eras of planetary time, continents are altered by internal and external causes while the complex pyramidal web of flora and fauna evolves in response to both terrestrial and celestial forces. Biogenesis is superimposed upon but bound to geogenesis.

During the Italian Renaissance, the creative and bold philosopher Giordano Filippo Bruno (1548–1600) challenged the entrenched beliefs of the Roman Catholic faith as well as the metaphysical biases of his contemporary astronomers and physicists. He even rejected that unrelenting authority given to the closed Aristotelian worldview. No doubt, the iconoclast saw himself as a citizen of this dynamic universe.

Bruno's daring cosmology rigorously rejected the geostatic, geocentric, anthropocentric, and finite-because-spherical model of this universe found in the Aristotelian writings of antiquity that were later supported by the Church and state. It held reality to be eternal in time, infinite in space, and endlessly changing. The inspiring Brunian worldview is a remarkable interpretation of the universe that far surpasses the conceptual contributions made in astronomy by Copernicus, Brahe, Kepler, Galileo, and Newton. In fact, Bruno stood alone in foreshadowing a modern understanding of and appreciation for space, time, matter, and life itself (particularly the place of humankind within a cosmic scheme of things).

It was Bruno who repudiated the peripatetic terrestrial/celestial dichotomy and, instead, maintained that the same physical laws and material elements of this earth also exist throughout an eternal and infinite universe. In doing so, he was able both to correct and to surpass that merely planetary perspective expounded by Aristotle, Ptolemy, and Aquinas (the Brunian vision even advancing beyond that sun-centered astronomy of his contemporaries).

Bruno realized that dogmatic religious beliefs are doomed in light of the scientific method and ongoing empirical discovery. He taught that there are an infinite number of perspectives, with no privileged or fixed frame of reference. Bruno was thrilled by the idea of infinity. He was not willing to set limits to the possibilities and probabilities inherent in this universe. His imagination thrived on the plausibility of extending the concept of infinity to embrace all aspects of material reality. For Bruno, this creative and endless cosmos is infinite in both potentiality and actuality. As such, no fixed ceiling of a finite number of stars sets a spherical boundary to this physical universe and, likewise, no dogmatic system of thoughts and values should imprison that free inquiry so necessary for human progress.

Bruno claimed that in this cosmos there are an infinite number of stars and planets (as well as comets and moons) more or less analogous to the sun and the earth, respectively. He envisioned an infinite number of solar systems, sidereal galaxies, and island universes strewn throughout the boundless cosmos. Likewise, his vision taught that there is life (including intelligent beings) on countless other worlds. Consequently, Bruno anticipated the emerging science of exobiology. For this cosmologist, neither our species, this planet, nor the sun is unique within the incomprehensible vastness of material reality.

Bruno professed a pantheistic view of reality, espousing the idea that a supreme single necessary substance is God or nature; it encompasses every particular object and event and relationship that exists potentially or actually in this universe. As a mystic, Bruno grasped God as the essential unity of reality itself.

In the history of philosophy, these unorthodox ideas and the cosmic perspective of Giordano Bruno remain a symbol of creative thought and free inquiry. In essence, he alone ushered in the modern cosmology.

In his numerous writings, astronomer and evolutionist Carl Sagan has emphasized a cosmic perspective and the place our own species occupies within both the material universe and organic history on earth. He is especially dedicated to the emerging science of exobiology (that search for life, intelligence, and civilization elsewhere on other clement planets among those billions and billions of galaxies throughout physical reality).

Stretching the human mind from its earth-bound orientation to a cosmos-expanding framework, Sagan focuses on space exploration and even the sidereal objects beyond our Milky Way (including envisioning such awesome projects as astroengineering and the intellectual unification of this evolving universe). His critical investigation encompasses both the marvels of the human mind and the wonders of outer space. His unabashed praise of science and technology within the open spectrum of naturalism and humanism is honest, refreshing, and optimistic.

Sagan traces the organic evolution of the brain from early forms of life to the extraordinary potentialities of the human animal. Our own species is becoming increasingly aware of its lowly origin as well as its cosmic destiny. Sagan argues that, in general, the entire fossil record of this planet illustrates a progressive tendency toward intelligence. Furthermore, he envisions the future convergence of artificial intelligence with human intelligence (this will greatly enhance that ongoing quest to discover life, intelligence, and civilization among the countless stars of expanding reality).

In going beyond the limitations of planetary chauvinism, Carl Sagan has contributed to our scientific understanding of and rational appreciation for this cosmos. He has brought the starry heavens above and earth history below into the mental reach of both the modern specialist and the general reader. His own inquiry is free from pessimism, superstition, metaphysical illusions, and theological obfuscations. In fact, one may argue that the continued investigation of the material universe is the most worthwhile endeavor of our entire species. In the distant future, no longer confined to this planet or the solar system, our species will no doubt explore the endless abysses of deep space and perhaps enjoy an indeterminate life span. Even human contact with beings elsewhere in cosmic reality is very probable, given the size and nature of this material universe.

For its survival and fulfillment, it is imperative that the human animal keep the freedom of choice alive. Likewise, the quest for truth demands a rigorous adherence to science and reason. Actually, earth's ecosystem as we now know it may not always be essential for the continuation of our own species. However, should humankind not take serious measures to protect some life forms on the globe (including our own species), the rich genetic reservoir of the planet will be so depleted that, given enough time, the diversified and adaptable insects along with the plants they need may yet inherit the earth.

Evolution has had an enormous impact on ways of thinking and believing: it shattered old views of our universe, earth history, and life itself. In truth, the human animal is related not to the angels and devils of myth but to the apes and monkeys within nature.

This planet once belonged to the invertebrates, then to the fishes, and later to the reptiles. The earth now belongs to us. Our own species is a newcomer to this material universe: it has only recently left the open woodlands and grassy savannahs of blissful ignorance for the methods of and advancements in science, medicine, technology, and reason (logic and mathematics). Humankind is entirely within the purview of empirical inquiry and critical evaluation. Whatever the destiny of our own species, physical reality is utterly blind to all the fleeting thoughts and actions (dreams and goals) of human existence. Should our species suddenly disappear from the surface of the planet, the cosmic objects of this galaxy and all other star clusters would still endure.

A sound assessment of the proper place of both the human animal in general and its mental activity in particular within this evolving universe is a necessary precondition for any intellectually defensible interpretation of the overall sidereal scheme of this concrete world. There is neither right nor wrong (good nor bad) among the stars, and the findings of rigorous cross-cultural research in comparative religion and biblical criticism are sobering indeed. Obviously, both the cosmic perspective and an evolutionary framework are a severe blow to the deeply entrenched and rigidly set Judeo-Christian beliefs and values that one is not likely to abandon easily.

Even if this dynamic universe is doomed to cosmic silence and black nothingness, the erect-walking and large-brained human animal with an opposable thumb and symbolic language will have made a noble attempt to fathom at least in part the nature of things.

Again, the human animal is a newcomer to both earth history and the material cosmos. Its existence goes back to the earliest hominids of equatorial Africa, the brave creatures that ventured out of their tropical forests and onto the open savannahs. Through the complex interplay of genetic variations and selective pressures, our remote ancestors slowly became clever bipeds. They eventually walked erect, then made stone implements, and finally wondered about the stars above and the forces below.

Through symbolic language and growing awareness, humankind developed art and religion and then science and philosophy. Personal beliefs and mere opinions yielded to the power of evidence and logic. It may even be said that our own species is now this evolving universe conscious of itself.

Yet, despite this superiority of the human animal, our own species is forever inextricably bound to the apes and monkeys and prosimians. Although the cosmic primate will probably take its genes and dreams throughout galactic systems, it is always linked to life on earth.

The fossil hominid footprints found at Laetoli are about 3.75 million years old; they clearly attest to the long and successful evolution of our zoological group. Now, human beings have walked on the surface of the moon. This is an incredible accomplishment indeed. In the ages to come, our present probes into deep space will have given cosmic artifacts for future civilizations to ponder.

Perhaps life and thought did not emerge just once in this dynamic universe. If so, our descendants may join with other intelligent beings in a cosmic voyage of ongoing discovery (or we may be utterly alone). In either event, human inquiry will continue. Even after mastering the dimensions of space and the arrows of time, there will still be other worlds for our species to conquer. How radically different the human worldview will be in hundreds or thousands or millions of years from now.

The scientific theory of organic evolution has shed considerable light on both the history and nature of the human animal. Fossil hominid evidence places the emergence of our zoological group back at least four million years; conflicting interpretations of this incomplete record do not discredit the indisputable fact of human evolution. Likewise, Homo sapiens sapiens belongs to the same taxonomic order that includes the prosimians, monkeys, hylobates, and pongids. Comparative evidence, from biochemistry and genetics to morphology and behavior, clearly links humankind to these other primates (especially the three great apes).

The past success of our early ancestors was due to bipedality, toolmaking, and a superior brain. During the Miocene epoch, some hominoids had ventured from the safety of jungles to explore the dangerous savannahs. An open environment put selective pressure on these protohominids. The result was bipedality with hands now freed from locomotion for tool use and, much later, toolmaking. No doubt, there was also an adaptive and survival advantage to living in social units and using forms of communication: vocal, facial, and gesticulative. The onset of articulate speech led to the acceleration of culture (in fact, the emergence of a protolanguage and the use of fire may have occurred as early as two million years ago).

Since the coming of the agricultural revolution and the following industrial/information explosion, the human animal has faced a multitude of new problems. It has yet to come to terms with crowded cities and the use of atomic energy. Also, entrenched beliefs and myopic values are seriously challenged by modern science and ruthless logic. Unfortunately, there is often a blind resistance to naturalism and humanism.

The survival of humankind requires facing reality with courage and wisdom. Clearly, the awesome cosmic perspective sweeps away all arguments for any earth-bound and human-centered view of the material universe. Furthermore, the human animal is not nailed to this planet. Should our hominid descendants endure for another four million years, they will certainly enjoy their finest hours among the myriad stars of distant galaxies.

As the scientific study of humankind, anthropology is dedicated to understanding and appreciating the origin, history, and future direction of our own species. The human animal is seen as a recent product of and totally within material nature (it is linked to the emergence of all primates).

There are two laws of life: change and death. Organic evolution has created endless novelty throughout the living world. Yet, most of the plant and animal forms that have existed on earth are now extinct—a very sobering fact indeed. Actually, humankind must survive for over 140 million years if it is merely to match the remarkable success of all the dinosaurs; only their complete demise made possible the age of mammals.

According to modern astronomy, the sun will still burn for another five billion years. Then, after exploding, it becomes a cinder. As such, the earth will end in both fire and ice. Once again, this solar system will be lifeless. However, long before this event, our descendants will very likely have migrated to other stars and perhaps even beyond the galaxy. They may encounter superior intelligent beings among the sidereal depths of this dynamic universe.

Organic evolution is not restricted to our planet. Today, one is awestruck by the fathomless immensity of the dynamic universe and the possibility of godlike beings existing among the billions of galaxies throughout material reality. The emerging science of exobiology is a serious inquiry into the probability that life and intelligence do exist elsewhere within the cosmos. In the future, there may even be the science of comparative exoevolution.

The idea of a complete annihilation of everything in the universe, including our own species, is clearly unacceptable to most human beings. Some religious faiths do uphold a belief in the personal immortality of an individual elsewhere beyond material death on this planet (an assumption usually involving a continued afterlife in some spiritual realm). However, this belief is simply a religious answer to the real challenges of fear, evil, suffering, and death itself in this imperfect world. In truth, there is no credible empirical evidence that human personality survives the death of the physical brain. Wishful thinking from the heart and imagination is no substitute for facts and logic. A major fallacy in the history of philosophy is the idea that the mind and body of a human being are two distinct and separable things. Actually, they are merely two aspects of the human being that will perish; an ontological dualism is unnecessary in terms of science and reason. In essence, the claim to a spiritual existence that outlives even the end of this universe is grounded only in the supreme egoism of our own species.

Humankind has yet to come to grips with its own cosmic arrogance and an amoral material universe as well as the truth of evolution. The fact remains that humankind is evolving in a cosmos that is totally indifferent to the needs and desires of our own species. Furthermore, a complete and final overview of the cosmos remains an ideal. Yet, science is no illusion: both philosophy and theology must face the undermining discoveries of empirical inquiry in the special sciences, from astronomy and biology to anthropology and psychology. Surely, the ongoing advancement of science in every area of serious investigation is the best hope for the survival and fulfillment of our own species within precarious nature. Likewise, there is the never-ending human quest for knowledge and wisdom as well as increased freedom, happiness, and longevity.

According to modern science, the sun is now halfway through its stellar evolution. Therefore, our planet should endure for about five billion more years. It is an interesting coincidence that there is at this present time a living human being for each year that the earth has existed. Furthermore, this epoch represents a turning point for our own species as well as the planet itself. If humankind is not to drift aimlessly on the globe or elsewhere and if it is not to destroy the natural environment, then all peoples must work together to plan for the ongoing survival and future fulfillment of our own species in the material universe. Without such a design for success, both the cosmic perspective and an evolutionary framework tell us that humankind may one day join that growing and alarming list of extinct animals and plants that contains countless millions of species (including the trilobites and dinosaurs).

Fearful of ongoing evolution, one may suggest that our species use all of its scientific knowledge and collective effort to actually halt future changes in the human world. Then, it is argued, our species could enjoy an endless period of biocultural stasis on earth. Whatever dubious merit this incredible idea might have, it does not seem plausible in light of long-range astronomical events and given the short-range terrestrial ramifications. Furthermore, a mystical unification of humankind among the galaxies and free from all matter, energy, and change seems equally implausible even to the most liberal materialist. In short, both our immediate and distant descendants must adapt to, and survive within, pervasive change. However, one intriguing possibility is that, should humankind migrate throughout this universe, there may be an adaptive radiation of our own species on a cosmic scale. If so, then planet earth itself would have been the cradle of many different intelligent forms of alien life!

The ultimate goal for humankind is the overcoming of the ravages of time, especially the ugly specter and tragic scourge of death itself. If an indeterminate life span (practical immortality) is ever achieved by our species, then all that future human beings would wish for is that those who had come before them could have also enjoyed our glorious achievement.

Finally, the ultimate sweep of cosmic evolution may be from the big-bang singularity to a big-crunch singularity: this universe beginning from and returning to a black hole. Even so, billions of years from now, if the material universe does in fact begin to collapse upon itself, then one wonders if our descendants (or any other form of intelligent life) will be able to prevent this fatal cosmic event.

FURTHER READINGS

Adair, Robert K. *The Great Design: Particles, Fields, and Creation.* New York: Oxford University Press, 1987.

Barrow, John D., and Joseph Silk. *The Left Hand of Creation: The Origin and Evolution of the Expanding Universe.* New York: Basic Books, 1983.

Bartusiak, Marcia. *Thursday's Universe.* New York: Times Books, 1986.

Chaisson, Eric. *Cosmic Dawn: The Origins of Matter and Life.* Boston: Little/Brown, 1981.

————. *The Life Era: Cosmic Selection and Conscious Evolution.* New York: Atlantic Monthly Press, 1987.

Cook, William J., and Joannie Schrof. "Journey to the Beginning of Time." *U.S. News & World Report* (March 26, 1990) 108(12):52–58, 61.

Crease, Robert P., and Charles C. Mann. *The Second Creation: Makers of the Revolution in Twentieth-Century Physics.* New York: Macmillan, 1986.

Davies, Paul. *God and the New Physics*. New York: Touchtone/Simon and Schuster, 1984.

———. *Superforce: The Search for a Grand Unified Theory of Nature*. New York: Touchstone Book, 1985.

Dick, Steven J. *Plurality of Worlds: The Origins of the Extraterrestrial Life Debate from Democritus to Kant*. Cambridge, England: Cambridge University Press, 1982.

Disney, Michael. *The Hidden Universe*. New York: Macmillan, 1984.

Dyson, Freeman J. *Infinite in All Directions*. New York: Harper & Row, 1988.

Einstein, Albert. *Relativity: The Special and the General Theory*. New York: Crown, 1961.

Field, George B., and Eric J. Chaisson. *The Invisible Universe: Probing the Frontiers of Astrophysics*. New York: Vintage Books, 1987.

Freedman, David H. "Maker of Worlds." *Discover* (July 1990) 11(7):46–52.

Greenstein, George. *The Symbiotic Universe: Life and Mind in the Cosmos*. New York: William Morrow, 1988.

Grünbaum, Adolf. "The Pseudo-Problem of Creation in Physical Cosmology." *Free Inquiry* 9, no. 4 (Fall 1989):48–57.

Hawking, Stephen W. *A Brief History of Time: From the Big Bang to Black Holes*. New York: Bantam Books, 1988.

Horgan, John. "Universal Truths." *Scientific American* (October 1990) 263(4):108–117.

Horowitz, Norman H. *To Utopia and Back: The Search for Life in the Solar System*. New York: W. H. Freeman, 1986.

Lightman, Alan, and Roberta Brawer. *Origins: The Lives and Worlds of Modern Cosomlogists*. Cambridge, Mass.: Harvard University Press, 1990.

Morris, Richard. *Time's Arrows: Scientific Attitudes toward Time*. New York: Touchstone Book, 1985.

Munitz, Milton K. *Cosmic Understanding: Philosophy and Science of the Universe*. Princeton, N.J.: Princeton University Press, 1986.

Pagels, Heinz R. *The Cosmic Code: Quantum Physics and the Language of Nature*. New York: Bantam Books, 1983.

Parker, Barry. *Creation: The Story of the Origin and Evolution of the Universe*. New York: Plenum Press, 1988.

———. *Einstein's Dream: The Search for a Unified Theory of the Universe*. New York: Plenum Press, 1986.

———. *Search for a Supertheory: From Atoms to Superstrings*. New York: Plenum Press, 1987.

Peters, Ted, ed. *Cosmos as Creation: Theology and Science in Consonance*. Nashville, Tenn.: Abingdon Press, 1989.

Preiss, Byron, ed. *The Universe.* New York: Bantam Books, 1987.

Preston, Richard. *First Light: The Search for the Edge of the Universe.* New York: Atlantic Monthly Press, 1987.

Sagan, Carl. *Cosmos.* New York: Random House, 1980.

Stenger, Victor J. "God and Stephen Hawking." *Free Inquiry* 9, no. 1 (Winter 1988/1989):59–61.

———. *Not by Design: The Origin of the Universe.* Buffalo, N.Y.: Prometheus Books, 1988.

———. "Was the Universe Created?" *Free Inquiry* 7, no. 3 (Summer 1987):26–30.

Tucker, Wallace, and Karen Tucker. *The Dark Matter: Contemporary Science's Quest for the Mass Hidden in Our Universe.* New York: William Morrow, 1988.

Weinberg, Steven. *The First Three Minutes: A Modern View of the Origin of the Universe.* New York: Bantam Books, 1977.

Wolf, Fred Alan. *Parallel Universes.* New York: Touchtone/Simon and Schuster, 1990.

Bibliography

Abraham, Francis, and John H. Morgan. *Sociological Thought: From Comte to Sorokin.* Bristol, Ind.: Wyndham Hall Press, 1989.

Albritton, Claude C., Jr. *The Abyss of Time: Changing Conceptions of the Earth's Antiquity After the Sixteenth Century.* San Francisco: Freeman/Cooper, 1980.

———. *The Abyss of Time: Unraveling the Mystery of the Earth's Age.* Los Angeles: J. P. Tarcher, 1986.

———. "Geologic Time." *Journal of Geological Education* 32(1):29–37.

Alexander, Samuel. *Philosophical and Literary Pieces.* London: Macmillan, 1939, esp. pp. 316–31, 256–78.

———. *Space, Time, and Deity* (1920). 2 vols. New York: Dover, 1966.

Alland, Alexander, Jr. *Human Nature: Darwin's View.* New York: Columbia University Press, 1985.

Allman, William F. "Elephant Man: Richard Leakey." *U.S. News & World Report* (2 October 1989):58, 61.

Allman, William F., and Joannie M. Schrof. "Endangered Species: Can They Be Saved?" *U.S. News & World Report* (2 October 1989):52–58.

Altmann, Stuart A. "The Monkey and the Fig." *American Scientist* 77(3): 256–63.

Alvarez, Walter, and Frank Asaro. "An Extraterrestrial Impact." *Scientific American* (October 1990) 263(4):78–84.

Andreyev, I. L. *Engels's "The Part Played by Labour in the Transition from Ape to Man"* (1876). Moscow: Progress Publishers, 1985.

Appelbaum, Richard P. *Karl Marx*. Beverly Hills, Calif.: Sage, 1988.

Appleman, Philip, ed. *Darwin*. 3d ed. New York: W. W. Norton, 1990.

Armour, Leslie, and Elizabeth Trott. *The Faces of Reason: An Essay on Philosophy and Culture in English Canada 1850–1950*. Waterloo, Ontario: Wilfred Laurier University Press, 1981.

Auer, J. A. C. Fagginger, and Julian Hartt. *Humanism Versus Theism*. Ames: Iowa State University Press, 1981.

Aulie, Richard P. "The Doctrine of Special Creation." *The American Biology Teacher* 34(4):191–200, and 34(5):261–68, 281.

Avers, Charlotte J. *Evolution*. New York: Harper & Row, 1974.

Badash, Lawrence. "The Age-of-the-Earth Debate." *Scientific American* 261(2):90–94, 96.

Bahm, Archie J. "Evolutionary Naturalism." *Philosophy and Phenomenological Research* 15(1):1–12.

Bailey, Cyril. *The Greek Atomists and Epicurus*. Oxford: Clarendon Press, 1928.

Barash, David P. *The Hare and the Tortoise: Culture, Biology, and Human Nature*. New York: Viking, 1986.

Barbour, George B. *In the Field with Teilhard de Chardin*. New York: Herder and Herder, 1965.

Barlow, Nora, ed. *The Autobiography of Charles Darwin, 1809–1882* (1876, 1887, 1929, 1950, 1958). New York: W. W. Norton, 1969.

Barnett, S. A., ed. *A Century of Darwin*. Cambridge, Mass.: Harvard University Press, 1958.

Barrett, Paul H., ed. *The Collected Papers of Charles Darwin*. Chicago: University of Chicago Press, 1977.

———. *Metaphysics, Materialism, and the Evolution of Mind: Early Writings of Charles Darwin*. Chicago: University of Chicago Press, 1980.

——— et al., eds. *Charles Darwin's Theoretical Notebooks (1836–1844)*. Cambridge, England: Cambridge University Press, 1987.

Barrow, John D., and Joseph Silk. *The Left Hand of Creation: The Origin and Evolution of the Expanding Universe*. New York: Basic Books, 1983.

Baumel, Howard B. "Alfred Wallace: Man in a Shadow." *The Science Teacher* 43(4):29–30.

Beddall, Barbara G., ed. *Wallace and Bates in the Tropics: An Introduction to the Theory of Natural Selection*. Toronto: Macmillan, 1969.

Berger, Carl. *Science, God and Nature in Victorian Canada (the 1982 Joanne Goodman lectures)*. Toronto: University of Toronto Press, 1983.

Bergson, Henri. *Creative Evolution* (1907). New York: Modern Library, 1944.

———. *An Introduction to Metaphysics* (1903). New York: Bobbs-Merrill, 1955.

Berlin, Isaiah. *Karl Marx: His Life and Environment.* 4th ed. Oxford, England: Oxford University Press, 1978.

Bernal, J. D. *Origin of Life.* New York: World Publishing, 1967.

Berry, William B. N. *Growth of a Prehistoric Time Scale Based on Organic Evolution,* rev. ed. Palo Alto, Calif.: Blackwell Scientific Publications, 1987.

Betzig, L. L. *Depositism and Differential Reproduction: A Darwinian View of History.* New York: Aldine, 1986.

Birx, H. James. "The Creation/Evolution Controversy." *Free Inquiry* 1(1): 24–26.

———. "Darwin & Teilhard: Some Final Thoughts." *Proteus: A Journal of Ideas* 6(2):38–46.

———. "Evolution and Unbelief." In *Encyclopedia of Unbelief,* edited by Gordon Stein, pp. 195–211. Buffalo, N.Y.: Prometheus Books, 1985.

———. "Giordano Bruno." In *Encyclopedia of Unbelief,* edited by Gordon Stein, pp. 68–72. Buffalo, N.Y.: Prometheus Books, 1985.

———. *Human Evolution.* Springfield, Ill.: Charles C. Thomas: 1988.

———. "Origin of Life and Unbelief." In *Encyclopedia of Unbelief,* edited by Gordon Stein, pp. 414–20. Buffalo, N.Y.: Prometheus Books, 1985.

———. *Pierre Teilhard de Chardin's Philosophy of Evolution.* Springfield, Ill.: Charles C. Thomas, 1972.

———. "Teilhard and Evolution: Critical Reflections." *Humboldt Journal of Social Relations* 9(1):151–67.

———. *Theories of Evolution.* Springfield, Ill.: Charles C. Thomas, 1984.

Bleibtreu, H. K. *Evolutionary Anthropology.* Boston: Allyn and Bacon, 1969.

Blinderman, Charles. *The Piltdown Inquest.* Buffalo, N.Y.: Prometheus Books, 1986.

Blum, Harold F. *Time's Arrow and Evolution.* Princeton, N.J.: Princeton University Press, 1968.

Boller, Paul F., Jr. *American Thought in Transition: The Impact of Evolutionary Naturalism 1865–1900.* Chicago: Rand McNally, 1971.

Boorstin, Daniel J. *The Discoverers,* esp. pp. 419–76. New York: Random House, 1983.

Boslough, John. *Stephen Hawking's Universe: An Introduction to the Most Remarkable Scientist of Our Time.* New York: Quill/William Morrow, 1985.

Bowler, Peter J. *The Eclipse of Darwinism: Anti-Darwinian Evolution Theories in the Decades around 1900.* Baltimore, Md.: Johns Hopkins University Press, 1983.

———. *Evolution: The History of an Idea,* rev. ed. Berkeley: University of California Press, 1989.

———. "Herbert Spencer and 'Evolution.' " *Journal of the History of Ideas* 36(2):367.

Brackman, Arnold C. *A Delicate Arrangement: The Strange Case of Charles Darwin and Alfred Russel Wallace.* New York: Times Books, 1980.

Braidwood, Robert J. "The Agricultural Revolution." *Scientific American* 203 (3): 130–34, 136, 138, 143–44, 146, 148.

Brandon, Robert N., and Richard M. Burian, eds. *Games, Organisms, Populations: Controversies over the Units of Selection.* Cambridge, Mass.: MIT Press, 1984.

Brooks, Daniel R., and E. O. Wiley. *Evolution as Entropy: Toward a Unified Theory of Biology.* 2d ed. Chicago: University of Chicago Press, 1988.

Brooks, John Langdon. *Just Before the Origin: Alfred Russel Wallace's Theory of Evolution.* New York: Columbia University Press, 1984.

Brown, Michael H. *The Search for Eve.* New York: Harper & Row, 1990.

Buchler, Justus, ed. *Philosophical Writings of Peirce,* esp. pp. 361–74. New York: Dover, 1955.

Bull, James J., and Paul H. Harvey. "Evolutionary Biology: A New Reason for Having Sex." *Nature* 339 (6222):260–61.

Burkhardt, Frederick, and Sydney Smith, eds. *The Correspondence of Charles Darwin,* 5 vols. 1821–1855. Cambridge, England: Cambridge University Press, 1985–1989.

Buvet, R., and C. Ponnamperuma, eds. *Chemical Evolution and the Origin of Life.* New York: Elsevier, 1971.

Bylinsky, Gene. *Life in Darwin's Universe: Evolution and the Cosmos.* Garden City, N.Y.: Doubleday, 1981.

Capouya, Emile, and Keitha Tompkins, eds. *The Essential Kropotkin.* New York: Liveright, 1975.

Carson, Rachel L. *Silent Spring* (1962). Boston: Houghton Mifflin, 1987.

Carter, G. S.: *A Hundred Years of Evolution.* New York: Macmillan, 1957.

Cassirer, Ernst. *The Philosophy of the Enlightenment* (1932). Boston: Beacon Press, 1965.

Chaisson, Eric. *Cosmic Dawn: The Origins of Matter and Life.* Boston: Little/Brown, 1981.

———. *The Life Era: Cosmic Selection and Conscious Evolution.* New York: Atlantic Monthly Press, 1987.

———. *Universe: An Evolutionary Approach to Astronomy.* Englewood Cliffs, N.J.: Prentice-Hall, 1988.

Chambers, Robert. *Vestiges of the Natural History of Creation* (1844). New York: Humanities Press, 1969.

Chancellor, John. *Charles Darwin.* New York: Taplinger, 1976.

Charon, Jean. *Cosmology: Theories of the Universe.* New York: McGraw-Hill, 1970.

Childe, V. Gordon. *Social Evolution* (1951). New York: Meridian Books, 1963.

Cho, Kah Kyung, and Lynn E. Rose. "Marvin Farber (1901–1980)." *Philosophy and Phenomenological Research* 42(1):1–4.

Clark, Robert E. D. *Darwin: Before and After.* Chicago: Moody Press, 1967.

Clark, Ronald W. *The Survival of Charles Darwin: A Biography of a Man and an Idea.* New York: Random House, 1984.

Clarke, Arthur C. *The Lost Worlds of 2001: The Ultimate Log of the Ultimate Trip.* New York: Signet Books, 1972.

———. *2001: A Space Odyssey.* New York: Signet Books, 1972.

Clements, Tad S. *Science and Man: The Philosophy of Scientific Humanism.* Springfield, Ill.: Charles C. Thomas, 1968.

———. *Science vs. Religion.* Buffalo, N.Y.: Prometheus Books, 1990.

Cloud, Preston. " 'Scientific Creationism': A New Inquisition Brewing." *The Humanist* 37(1):6–15.

Cohen, I. Bernard. *Revolution in Science,* esp. pp. 283–300. Cambridge, Mass.: Belknap Press of Harvard University Press, 1985.

Collier, John. "Entropy in Evolution." *Biology & Philosophy* 1(1):5–24.

Collin, R. *Evolution.* New York: Hawthorne Books, 1962.

Collins, Randall, and Michael Makowsky. *The Discovery of Society.* 4th ed., esp. pp. 32–54, 68–82, 87–90, 95–97, 141–66. New York: Random House, 1989.

Colp, Ralph, Jr. "The Contacts Between Karl Marx and Charles Darwin." *Journal of the History of Ideas* 35(2):329–38.

Colver, A. Wayne, and Robert D. Stevick. *Composition and Research: Problems in Evolutionary Theory.* New York: Bobbs-Merrill, 1963.

Condorcet, Marquis de. *Sketch for a Historical Picture of the Progress of the Human Mind* (1795). New York: Noonday Press, 1955.

Coonen, Lester P. "Aristotle's Biology." *BioScience* 27(11):733–38.

Cornford, F. M. *Principium Sapientiae: The Origins of Greek Philosophical Thought.* Edited by W. K. C. Guthrie. New York: Harper Torchbooks, 1965.

Corning, Peter A. *The Synergism Hypothesis: A Theory of Progressive Evolution,* esp. pp. 63–124. New York: McGraw-Hill, 1983.

Courtillot, Vincent E. "A Volcanic Eruption." *Scientific American* (October 1990) 263(4): 85–92.

Crick, Francis H. C. *Life Itself: Its Origin and Nature.* New York: Simon and Schuster, 1982.

———. *What Mad Pursuit: A Personal View of Scientific Discovery.* New York: Basic Books, 1988.

Cuénot, Claude. *Teilhard de Chardin: A Biographical Study.* Baltimore, Md.: Helicon, 1965.

Currie, Philip. "Long-Distance Dinosaurs." *Nature* (June 1989):60–65.

Darlington, C. D. *Darwin's Place in History.* Oxford: Basil Blackwell, 1960.
———. "The Origin of Darwinism." *Scientific American* 200(5):60–66.
Darnell, Regna, ed. *Readings in the History of Anthropology,* esp. pp. 32–45, 218–44, 407–39. New York: Harper & Row, 1974.
Darwin, Charles. *The Descent of Man, and Selection in Relation to Sex* (1871). New York: Modern Library, 1936.
———. *The Illustrated Origin of Species.* Edited by Richard E. F. Leakey. New York: Hill and Wang, 1979.
———. *On the Origin of Species by Means of Natural Selection, or the Preservation of Favored Races in the Struggle for Life* (1859). New York: Modern Library, 1936.
———. *The Voyage of the Beagle.* Edited by Leonard Engel. Garden City, N.Y.: Anchor Books, 1962.
Darwin, Francis, ed. *The Life and Letters of Charles Darwin.* 2 vols. New York: Basic Books, 1959.
Davie, Maurice R. *William Graham Sumner.* New York: Thomas Y. Crowell, 1963.
de Beer, Gavin. *Charles Darwin: A Scientific Biography.* Garden City, N.Y.: Anchor Books, 1965.
de Terra, Helmut. *Memories of Teilhard de Chardin.* New York: Harper & Row, 1964.
Deely, John N., and Raymond J. Nogar. *The Problem of Evolution: A Study of the Philosophical Repercussions of Evolutionary Science.* New York: Appleton-Century-Crofts, 1973.
DeGrood, David H. *Haeckel's Theory of the Unity of Nature.* Boston: Christopher Publishing House, 1965.
Denton, Michael. *Evolution: A Theory in Crisis.* Bethesda, Md.: Adler & Adler, 1986.
Dewey, John. *Experience and Nature* (1925). New York: Dover, 1958.
———. *The Influence of Darwin on Philosophy, and Other Essays in Contemporary Thought* (1910). Bloomington: Indiana University Press, 1965.
———. *On Experience, Nature, and Freedom: Representative Selections.* New York: Bobbs-Merrill, 1960.
Dick, Steven J. "The Origins of the Extraterrestrial Life Debate and Its Relation to the Scientific Revolution." *Journal of the History of Ideas* 41(1):3–27.
———. *Plurality of Worlds: The Origins of the Extraterrestrial Life Debate from Democritus to Kant.* Cambridge, England: Cambridge University Press, 1982.

Dickerson, Richard E. "Chemical Evolution and the Origin of Life." *Scientific American* 239(3):70–86, 242.

Diderot, Denis. *Interpretation of Nature: Selected Writings* (1754). 2d ed. New York: International Publishers, 1963.

Dillon, L. S. *Evolution: Concepts and Consequences*. Saint Louis, Miss.: C. V. Mosby, 1973.

Dobzhansky, Theodosius. *The Biology of Ultimate Concern*. New York: New American Library, 1967.

———. *Genetics and the Origin of Species*. New York: Columbia University Press, 1937.

Dodson, Edward O. *The Phenomenon of Man Revisited: A Biological Viewpoint on Teilhard de Chardin*. New York: Columbia University Press, 1984.

Dodson, Edward O., and Peter Dodson. *Evolution: Process and Product*. New York: D. Van Nostrand, 1976.

Draper, John William. *History of the Conflict Between Religion and Science* (1875). Brookfield, Vt.: Gregg International.

Durham, Frank, and Robert D. Purrington. *Frame of the Universe*. New York: Columbia University Press, 1983.

Eaton, T. H., Jr. *Evolution*. New York: W. W. Norton, 1970.

Eddington, Arthur. *The Expanding Universe*. New York: Cambridge University Press, 1987.

Edelman, Gerald M. *Neural Darwinism: The Theory of Neuronal Group Selection*. New York: Basic Books, 1987.

Edey, Maitland, and Donald C. Johanson. *Blueprints: Solving the Mystery of Evolution*. Boston: Little, Brown, 1989.

Einstein, Albert. *Essays in Humanism*. New York: Philosophical Library, 1983.

———. *Relativity: Special and General*. New York: Random House, 1961.

Eiseley, Loren C. "Alfred Russel Wallace." *Scientific American* 200(2):70–82, 84, 172.

———. "Charles Lyell." *Scientific American* 201(2):90–106, 168.

———. *Darwin's Century: Evolution and the Men who Discovered It*. Garden City, N.Y.: Anchor Books, 1961.

Eldredge, Niles. *Life Pulse: Episodes From the Story of the Fossil Record*. New York: Facts On File Publications, 1987.

———. *The Monkey Business: A Scientist Looks at Creationism*. New York: Pocket Books, 1982.

———. *Time Frames: The Rethinking of Darwinian Evolution and the Theory of Punctuated Equilibria*. New York: Simon and Schuster, 1985.

Eldredge, Niles, and Stephen Jay Gould. "Rates of Evolution Revisited." *Paleobiology* 2(2):174–79.

Engels, Friedrich. *Dialectics of Nature* (1927). New York: International Publishers, 1963.

———. *Ludwig Feuerbach and the Outcome of Classical German Philosophy* (1888). New York: International Publishers, 1967.

———. *The Origin of the Family, Private Property, and the State: In the Light of Researches of Lewis H. Morgan* (1891). New York: International Publishers, 1964.

———. *The Part Played by Labour in the Transition from Ape to Man* (1876). Moscow: Foreign Languages, 1953.

Fagan, Brian M. *People of the Earth: An Introduction to World Prehistory.* 5th ed. Boston: Little & Brown, 1986.

Farber, Marvin. *Basic Issues of Philosophy: Experience, Reality, and Human Values.* New York: Harper Torchbooks, 1968.

———. *Naturalism and Subjectivism.* Albany: State University of New York Press, 1968.

———. "On Subjectivism and the World-Problem." *Philosophy and Phenomenological Research* 34(1):34–41.

Farley, John. *The Spontaneous Generation Controversy from Descartes to Oparin.* Baltimore, Md.: Johns Hopkins University Press, 1979.

Feinberg, Gerald, and Robert Shapiro. *Life Beyond Earth.* New York: William Morrow, 1980.

Ferris, James P. "The Chemistry of Life's Origin." *Chemical and Engineering News* 62(35) (August 27, 1984): 22–35.

Ferris, Timothy. *Coming of Age in the Milky Way.* New York: William Morrow, 1988.

Feuerbach, Ludwig. *The Essence of Christianity* (1841). New York: Harper Torchbooks, 1957.

Fisher, R. A. *The Genetical Theory of Natural Selection.* New York: Dover, 1958.

Fiske, John. *Darwinism and Other Essays,* rev. ed. New York: Houghton Mifflin, 1885.

———. *The Destiny of Man Viewed in the Light of His Origin.* Boston: Houghton Mifflin, 1884.

———. *Outlines of Cosmic Philosophy Based on the Doctrine of Evolution, with Criticism on the Positive Philosophy.* 4 vols. Boston: Houghton Mifflin, 1874.

———. *Through Nature to God.* New York: Houghton Mifflin, 1889.

Fix, William R. *The Bone Peddlers: Selling Evolution.* New York: Macmillan, 1984.

Fletcher, Joseph. *The Ethics of Genetic Control: Ending Reproductive Roulette.* Buffalo, N.Y.: Prometheus Books, 1988.

Flew, Antony G. N. *Evolutionary Ethics*. New York: St. Martin's Press, 1967.

Foglino, Annette. "Space: Is Anyone Out There?" *Life* (July 1989):48–53, 57.

Folsome, Clair Edwin. *The Origin of Life: A Warm Little Pond*. San Francisco: W. H. Freeman, 1979.

Fossey, Dian. *Gorillas in the Mist*. Boston: Houghton Mifflin, 1983.

Fothergill, Philip G. *Historical Aspects of Organic Evolution*. London: Hollis and Carter, 1952.

Fox, Sidney W., and Klaus Dose. *Molecular Evolution and the Origin of Life*. San Francisco: W. H. Freeman, 1972.

Fraser, Nicholas. "At the Feet of the Dinosaurs." *New Scientist* 120(1633): 39–42.

Freeman, Derek. "The Evolutionary Theories of Charles Darwin and Herbert Spencer." *Current Anthropology* 15(3):211–37.

Frieden, Earl. "The Chemical Elements of Life." *Scientific American* 227(1): 52–60.

Futuyma, Douglas. *Science on Trial: The Case for Evolution*. New York: Pantheon Books, 1982.

Galdikas, Birute. "Orangutans: Indonesia's 'People in the Forest.' " *National Geographic* 148(4):444–73.

Gallant, Roy A. *Charles Darwin: The Making of a Scientist*. Garden City, N.Y.: Doubleday, 1972.

Garbarino, Merwyn S. *Sociocultural Theory in Anthropology: A Short History*, esp. pp. 20–24, 27–35, 87–91. Prospect Heights, Ill.: Waveland Press, 1977.

Gay, Peter. *The Godless Jew: Freud, Atheism, and the Making of Psychoanalysis*. New Haven, Conn.: Yale University Press, 1987.

Gee, Henry. "Dinosaur Finds from China." *Nature* 6 (July 1989):22.

Geffré, Claude, and Jean-Pierre Jossua, eds. *Nietzsche and Christianity*. New York: Harper & Row, 1981.

George, W. B. *Biologist Philosopher: A Study of the Life and Writings of Alfred Russel Wallace*. New York: Abelard-Schuman, 1964.

Gerver, Israel. *Lester Frank Ward*. New York: Thomas Y. Crowell, 1963.

Ghiselin, Michael T. *The Triumph of the Darwinian Method*. Chicago: University of Chicago Press, 1984.

Gillespie, Neal C. *Charles Darwin and the Problem of Creation*. Chicago: University of Chicago Press, 1979.

Gillispie, Charles Coulston. *Genesis and Geology: A Study in the Relations of Scientific Thought, Natural Theology, and Social Opinion in Great Britain, 1790–1850*. New York: Harper Torchbooks, 1959.

Glass, Bentley, Owsei Temkin, and William L. Straus, Jr., eds. *Forerunners of Darwin: 1745–1859.* Baltimore, Md.: Johns Hopkins Press, 1968.

Godfrey, Laurie Rhode, ed. *Scientists Confront Creationism.* New York: W. W. Norton, 1983.

———. *What Darwin Began: Modern Darwinism and Non-Darwinian Perspectives on Evolution.* Boston: Allyn and Bacon, 1985.

Goldschmidt, Richard. *The Material Basis of Evolution.* New Haven, Conn.: Yale University Press, 1982.

Goodall, Jane. *The Chimpanzees of Gombe: Patterns and Behavior.* Cambridge, Mass.: Belknap Press of Harvard University Press, 1986.

———. *In the Shadow of Man.* Boston: Houghton Mifflin, 1983.

Gore, Rick. "Extinctions." *National Geographic* 175(6):662–99.

Gosse, Edmund. *Father and Son: A Study of Two Temperaments* (1907). New York: W. W. Norton, 1963.

Gosse, Philip Henry. *Creation (Omphalos): An Attempt to Untie the Geological Knot.* London: John Van Voorst, 1857.

Goudge, T. A. *The Ascent of Life: A Philosophical Study of the Theory of Evolution.* Toronto: University of Toronto Press, 1961.

Gould, Stephen Jay. "Adam's Navel." *Natural History* 93(6):6, 8,10, 14.

———. "An Asteroid to Die For." *Discover* 10(10):60–65.

———. *The Complete Works of Vladimir Kovalevsky: Original Anthology.* Salem, N.H.: Ayer, 1980.

———. *Ever Since Darwin: Reflections in Natural History.* New York: W. W. Norton, 1979.

———. *The Flamingo's Smile: Reflections in Natural History.* New York: W. W. Norton, 1985.

———. "Full of Hot Air." *Natural History* (October 1989): 28, 30, 32–34, 36, 38.

———. *Hen's Teeth and Horse's Toes: Further Reflections in Natural History.* New York: W. W. Norton, 1983.

———. "In Praise of Charles Darwin." *Proteus: A Journal of Ideas* 6(2): 1–4.

———. *The Mismeasure of Man.* New York: W. W. Norton, 1981.

———. *Ontogeny and Phylogeny.* Cambridge, Mass.: Belknap Press of Harvard University Press, 1977.

———. *The Panda's Thumb: More Reflections in Natural History.* New York: W. W. Norton, 1980.

———. *Time's Arrow, Time's Cycle: Myth and Metaphor in the Discovery of Geological Time.* Cambridge, Mass.: Harvard University Press, 1987.

———. *An Urchin in the Storm: Essays about Books and Ideas.* New York: W. W. Norton, 1988.

Gould, Stephen Jay. *Wonderful Life: The Burgess Shale and the Nature of History*. New York: W. W. Norton, 1989.

Gould, Stephen Jay, and Niles Eldredge. "Punctuated Equilibria: The Tempo and Mode of Evolution Reconsidered." *Paleobiology* 3(2):115–51.

Graham, Loren R. *Science, Philosophy, and Human Behavior in the Soviet Union*. New York: Columbia University Press, 1987.

Grant, Verne. *The Evolutionary Process: A Critical Review of Evolutionary Theory*. New York: Columbia University Press, 1985.

Gray, Asa. *Darwiniana* (1876). Cambridge, Mass.: Harvard University Press, 1960.

Greene, John C. *Darwin and the Modern World View*. Baton Rouge: Louisiana State University Press, 1981.

———. *The Death of Adam: Evolution and Its Impact on Western Thought*. Ames: Iowa State University Press, 1981.

Grene, Marjorie. "Aristotle and Modern Biology." *Journal of the History of Ideas* 33(3):395–424.

———. "'Changing Concepts of Darwinian Evolution." *The Monist* 64(2):195–213.

Gribben, John, and Mary Gribben. *The One Percent Advantage: The Sociobiology of Being Human*. New York: Basil Blackwell, 1988.

Gribbin, John. *In Search of the Double Helix: Quantum Physics and Life*. New York: Bantam Books, 1987.

———. *The Omega Point*. New York: Bantam Books, 1988.

Gruber, Howard E., and Paul H. Barrett. *Darwin on Man: A Psychological Study of Scientific Creativity*. New York: E. P. Dutton, 1974.

———. *Darwin's Early Unpublished Notebooks*. New York: E. P. Dutton, 1974.

Grünbaum, Adolf. "The Pseudo-Problem of Creation in Physical Cosmology." *Philosophy of Science* 56 (3):373–94.

Guthrie, W. K. C. *The Greek Philosophers from Thales to Aristotle*. New York: Harper Torchbooks, 1960.

Haber, F. C. *Age of the World: Moses to Darwin*. Baltimore, Md.: Johns Hopkins Press, 1959.

Haeckel, Ernst. *Freedom in Science and Teaching*. London: C. Kegan Paul, 1879.

———. *Monism as Connecting Religion and Science*. London: Adam and Charles Black, 1895.

———. *The Riddle of the Universe at the Close of the Nineteenth Century* (1900). New York: Harper & Brothers, 1905.

Hahlweg, Kai, and Clifford A. Hooker, eds. *Issues in Evolutionary Epistemology*. Albany: State University of New York Press, 1989.

Haldane, J. B. S. *The Causes of Evolution*. Ithaca, N.Y.: Cornell University Press, 1966.

Haller, John S., Jr. *Outcasts from Evolution: Scientific Attitudes of Racial Inferiority, 1859–1900*. New York: McGraw Hill, 1975.

Hamilton, T. H. *Process and Pattern in Evolution*. New York: Macmillan, 1967.

Hamrum, Charles L., ed. *Darwin's Legacy*. San Francisco: Harper & Row, 1983.

Handwerker, W. Paul. "The Origins and Evolution of Culture." *American Anthropologist* 91(2):313–26.

Harris, Marvin. *Cultural Materialism: The Struggle for a Science of Culture*. New York: Vintage Books, 1980.

————. *The Rise of Anthropological Theory: A History of Theories of Culture*. New York: Thomas Y. Crowell, 1970.

Harrison, James. "Erasmus Darwin's View of Evolution." *Journal of the History of Ideas* 32(2):247–64.

Hawking, Stephen W. *A Brief History of Time: From the Big Bang to Black Holes*. New York: Bantam Books, 1988.

Hecht, Jeff. "Evolving Theories of Old Extinctions." *New Scientist* (12 November 1988):28–30.

Henig, Robin Marantz. "Exobiologists Continue to Search for Life on Other Planets." *BioScience* 30(1):9–12.

Heyler, Daniel, and Cecile M. Poplin. "The Fossils of Montcean-les-Mines." *Scientific American* 259(3):104–10.

Himmelfarb, Gertrude. *Darwin and the Darwinian Revolution*. New York: W. W. Norton, 1968.

Ho, A. W., and S. W. Fox, eds. *Evolutionary Processes and Metaphors*. Chichester: John Wiley & Sons, 1988.

Hofstadter, Richard. *Social Darwinism in American Thought*, rev. ed. Boston: Beacon Press, 1971.

Holland, Clifford G. "First Canadian Critics of Darwin." *Queen's Quarterly* 88(1):100–106.

————. "The Sage of Ottawa: William Dawson LeSueur." *Canadian Literature* 96:167–81.

Horowitz, Norman H. *To Utopia and Back: The Search for Life in the Solar System*. New York: W. H. Freeman, 1986.

Hoyle, Fred, and N. Chandra Wickramasinghe. *Evolution from Space: A Theory of Cosmic Creationism*. New York: Simon and Schuster, 1981.

————. *Lifecloud: The Origin of Life in the Universe*. New York: Harper & Row, 1978.

Hsü, Kenneth J. *The Great Dying: Cosmic Catastrophe, Dinosaurs, and the Theory of Evolution*. New York: Ballantine Books, 1986.

Hudson, William Henry. *Herbert Spencer*. New York: Dodge, 1908.

Hull, David L. *Darwin and His Critics*. Cambridge, Mass.: Harvard University Press, 1973.

———. *The Metaphysics of Evolution*. Albany: State University of New York Press, 1989.

Huxley, Julian S. *Evolution: The Modern Synthesis*. New York: John Wiley, 1963.

———. *Religion without Revelation*. New York: Mentor Books, 1957.

Huxley, Julian S., and H. B. D. Kettlewell. *Charles Darwin and His World*. New York: Viking Press, 1966.

Huxley, Julian S., et al. *A Book that Shook the World: Anniversary Essays on Charles Darwin's "Origin of Species."* Pittsburgh, Pa.: University of Pittsburgh Press, 1961.

Huxley, Thomas H. *Evidence as to Man's Place in Nature* (1863). Ann Arbor: University of Michigan Press, 1959.

———. *On the Origin of Species, Or the Causes of the Phenomena of Organic Nature* (1863). Ann Arbor: University of Michigan Press, 1968.

Irvine, William. *Apes, Angels, and Victorians: The Story of Darwin, Huxley, and Evolution*. New York: McGraw-Hill, 1972.

———. *Thomas Henry Huxley*. London: Longmans/Green, 1960.

James, William. *Essays in Pragmatism*. Edited by Alburey Castell. New York: Hafner, 1948.

———. *Pragmatism and Four Essays from "The Meaning of Truth."* New York: Meridian Books, 1955.

———. *The Will to Believe and Other Essays in Popular Philosophy and Human Immortality*. New York: Dover, 1956.

Jensen, U. J., and R. Harré, eds. *The Philosophy of Evolution*. New York: St. Martin's Press, 1981.

Johanson, Donald C., and James Shreeve. *Lucy's Child: The Discovery of a Human Ancestor*. New York: William Morrow, 1989.

Johanson, Donald C., and Maitland A. Edey. *Lucy: The Beginnings of Humankind*. New York: Simon and Schuster, 1981.

Johanson, Donald C., and Timothy D. White. "A Systematic Assessment of Early African Hominids." *Science* 202 (4378):321–30.

Jones, H. S. *Life on Other Worlds*. New York: New American Library, 1960.

Jones, F. Wood. *Arboreal Man*. New York: Longmans/Green, 1916.

Judson, Horace Freeland. *The Eighth Day of Creation: Makers of the Revolution in Biology*. New York: Simon and Schuster, 1980.

Kamshilov, M. M. *Evolution of the Biosphere*. Moscow: Mir, 1976.

Kant, Immanuel. *Universal Natural History and Theory of the Heavens* (1755). Ann Arbor: University of Michigan Press, 1969.

Kaplan, David, and Robert A. Manners. *Cultural Theory*, esp. pp. 38–55. Prospect Heights, Ill.: Waveland Press, 1972.

Karp, Walter. *Charles Darwin and the Origin of Species*. New York: Harper & Row, 1968.

Kaye, Howard L. *The Social Meaning of Modern Biology: From Social Darwinism to Sociobiology*. New Haven, Conn.: Yale University Press, 1986.

Kerkut, G. A. *Implications of Evolution*. New York: Pergamon Press, 1965.

King, Thomas M. "The Milieux Teilhard Left Behind." *America* 152(12): 249–53.

Kirk, G. S., and J. E. Raven. *The Presocratic Philosophers: A Critical History with a Selection of Texts*. Cambridge, England: Cambridge University Press, 1966.

Kitcher, Philip. *Abusing Science: The Case Against Creationism*. Cambridge, Mass.: MIT Press, 1982.

———. *Vaulting Ambition: Sociobiology and the Quest for Human Nature*. Cambridge, Mass.: MIT Press, 1985.

Klein, Richard G. *The Human Career: Human Biological and Cultural Origins*. Chicago: University of Chicago Press, 1989.

Knezević, Bozidar. *History, the Anatomy of Time: The Final Phase of Sunlight*. New York: Philosophical Library, 1980.

Kobler, John. "The Priest Who Haunts the Catholic World." *The Saturday Evening Post* (12 October 1963):42–51.

Kohn, David, ed. *The Darwinian Heritage*. Princeton, N.J.: Princeton University Press, 1985.

Krishtalka, Leonard. *Dinosaur Plots, and Other Intrigues in Natural History*. New York: Avon Books, 1990.

Kroeber, A. L. "The Superorganic." *American Anthropologist* n.s. 19:163–213.

Kropotkin, Peter A. *Ethics: Origin and Development* (1922). New York: Tudor, 1924.

———. *Mutual Aid: A Factor of Evolution* (1902). Boston: Extending Horizons Books, 1914. Also includes Thomas H. Huxley's "The Struggle for Existence in Human Society" (1888).

Kuhn, Thomas S. *The Structure of Scientific Revolutions*. 2d ed. Chicago: University of Chicago Press, 1970.

Kurtz, Paul. *Eupraxophy: Living Without Religion*. Buffalo, N.Y.: Prometheus Books, 1989.

———. *The Secular Humanist Declaration*, esp. pp. 21–22. Buffalo, N.Y.: Prometheus Books, 1980.

Kutter, G. Siegfried. *The Universe and Life: Origins and Evolution*. Boston: Jones and Bartlett, 1987.

Lacey, Cedric. "Galactic Evolution: Starbursts, Quasars and All That." *Nature* 340(6236):675–76.

Lamarck, J. B. *Zoological Philosophy: An Exposition with Regard to the Natural History of Animals* (1809). New York: Hafner, 1963.

Lamont, Corliss. *The Philosophy of Humanism*. 6th ed. New York: Frederick Ungar, 1982.

Langness, L. L. *The Study of Culture*, esp. pp. 12–39, 100–108, 121–28, 140–47. Novato: Chandler & Sharp, 1985.

Laszlo, Ervin. *Evolution: The Grand Synthesis*. Boston: Shambhala, 1987.

———. "Evolution in Nature, Development in Society." *Dialectics and Humanism: The Polish Philosophical Quarterly* (Spring; 1989) 16(2): 159–64.

Laudan, Rachel. *From Mineralogy to Geology: The Foundations of a Science, 1650–1830*. Chicago: University of Chicago Press, 1987.

Leakey, Louis S. B. *By the Evidence: Memoirs, 1932–1951*. New York: Harcourt Brace Jovanovich, 1974.

———. *White African: An Early Autobiography*. Cambridge, Mass.: Schenkman, 1966.

Leakey, Mary D. *Disclosing the Past*. New York: Doubleday, 1984.

Leakey, Richard E. F. *The Illustrated Origin of Species*. New York: Hill and Wang, 1979.

———. *One Life*. Salem, Mass.: Salem House, 1984.

Leakey, Richard E. F., and Alan Walker. "*Homo Erectus* Unearthed." *National Geographic* 168(5):624–29.

Lederberg, J. "Exobiology: Approaches to Life Beyond the Earth." *Science* 131(3412):1503–8.

Lehrman, Nathaniel S. "Human Sociobiology: Wilson's Fallacy." *The Humanist* 41(4):39–42, 58.

LeSueur, William Dawson. *Evolution and the Positive Aspects of Modern Thought*. Ottawa: A. S. Woodburn, 1884.

Levins, Richard, and Richard Lewontin. *The Dialectical Biologist*. Cambridge, Mass.: Harvard University Press, 1985.

Levinton, Jeffrey. *Genetics, Paleontology, and Macroevolution*. Cambridge, Mass.: Cambridge University Press, 1988.

Lewin, Roger. *Bones of Contention: Controversies in the Search for Human Origins*. New York: Simon and Schuster, 1987.

Lewontin, Richard C. "Evolution/Creation Debate: A Time for Truth." *BioScience* 31(8):559.

Lewontin, Richard C., Steven Rose, and Leon J. Kamin. *Not in Our Genes: Biology, Ideology, and Human Nature.* New York: Pantheon Books, 1984.

Linden, Mareta. *Untersuchungen zum Anthropologiebegriff des 18. Jahrhunderts.* Frankfurt/M. (BRD): Peter Lang GmbH, 1976.

Loewenberg, B. J. *Darwin, Wallace, and the Theory of Natural Selection.* Cambridge, Mass.: Arlington Books, 1959.

Lovejoy, Arthur O. *The Great Chain of Being: A Study of the History of an Idea* (1936). New York: Harper Torchbooks, 1965.

Lovejoy, C. Owen. "Evolution of Human Walking." *Scientific American* 259(5):118–25.

Lovelock, James E. *The Ages of Gaia: A Biography of Our Living Earth.* New York: W. W. Norton, 1988.

———. *Gaia: A New Look at Life on Earth.* New York: Oxford University Press, 1988.

Lowenstein, Jerold, and Adriene Zihlman. "The Invisible Ape." *New Scientist* (3 December 1988):156–59.

Lucretius. *De Rerum Natura.* Rev. 2d ed. Cambridge, Mass.: Harvard University Press, 1982.

Luijpen, William A., and Henry J. Koren. *Religion and Atheism.* Atlantic Highlands, N.J.: Humanities Press, 1982.

Lukas, Mary, and Ellen Lukas. *Teilhard: A Biography.* New York: McGraw-Hill, 1981.

Lumsden, Charles J., and Edward O. Wilson. *Promethean Fire: Reflections on the Origin of Mind.* Cambridge, Mass.: Harvard University Press, 1983.

Maddox, John. "Down with the Big Bang." *Nature* 340 (6233) (10 August 1989): 425.

Malthus, Thomas Robert. *An Essay on the Principle of Population.* Edited by Philip Appleman. New York: W. W. Norton, 1976.

Manners, Robert, and David Kaplan, eds. *Theory in Anthropology: A Sourcebook,* esp. pp. 15–31, 229–68. Chicago: Aldine, 1968.

Margulis, Lynn. *Early Life.* Boston: Jones and Bartlett, 1984.

Margulis, Lynn, and Karlene V. Schwartz. *Five Kingdoms: An Illustrated Guide to the Phyla of Life on Earth.* 2d ed. New York: W. H. Freeman, 1988.

Margulis, L., and D. Sagan. *Origins of Sex: Three Billion Years of Genetic Recombination.* New Haven, Conn.: Yale University Press, 1986.

Marshack, Alexander. "An Ice Age Ancestor?" *National Geographic* 174(4): 478–81.

Martindale, Don. *The Nature and Types of Sociological Theory.* 2nd ed. Prospect Heights, Ill.: Waveland Press, 1988.

Mayr, Ernst. *The Growth of Biological Thought: Diversity, Evolution, and Inheritance*. Cambridge, Mass.: Belknap Press of Harvard University Press, 1982.

McAlester, A. Lee. *The History of Life*. 2d ed. Englewood Cliffs, N.J.: Prentice-Hall, 1977.

McDermott, John J., ed. *The Writings of William James*. New York: Random House, 1967.

McGowan, Chris. *In the Beginning . . . : A Scientist Shows Why the Creationists Are Wrong*. Buffalo, N.Y.: Prometheus Books, 1984.

McKeon, Richard, ed. *The Basic Works of Aristotle*. New York: Random House, 1966.

McKibben, Bill. *The End of Nature*. New York: Random House, 1989.

McKillop, A. B. *A Critical Spirit: The Thoughts of William Dawson LeSueur*. Toronto: McClelland and Stewart, 1977.

————. *A Disciplined Intelligence: Critical Inquiry and Canadian Thought in the Victorian Era*. Toronto: McGill-Queens University Press, 1989.

McMullin, Ernan, ed. *Evolution and Creation*. Notre Dame, Ind.: University of Notre Dame Press, 1985.

Mead, Margaret. *Continuities in Cultural Evolution*. New Haven, Conn.: Yale University Press, 1964.

Meadows, Jack. *The Great Scientists*, esp. pp. 9–28, 149–68. New York: Oxford University Press, 1987.

Medawar, Peter Brian. *Induction and Intuition in Scientific Thought*. Vol. 75, esp. pp. 10–11. Philadelphia: American Philosophical Society, 1969.

————. *The Limits of Science*. Scranton, Pa.: Harper & Row, 1984.

Melchert, Norman Paul. *Realism, Materialism, and the Mind: The Philosophy of Roy Wood Sellars*. Springfield, Ill.: Charles C. Thomas, 1968.

Metchnikoff, Élie. *The Nature of Man: Studies in Optimistic Philosophy*. New York: G. P. Putnam's Sons, 1903.

Metz, Johannes Baptist, ed. *Evolving World and Theology*. Mahwah, N.J.: Paulist Press, 1967.

Mikulak, Maxim W. "Darwinism, Soviet Genetics, and Marxism-Leninism." *Journal of the History of Ideas* 31(3):359–76.

Milhauser, M. *Just Before Darwin: Robert Chambers and the Vestiges*. Middletown, Conn.: Wesleyan University Press, 1959.

Miller, Stanley L., and Leslie E. Orgel. *The Origins of Life on Earth*. Englewood Cliffs, N.J.: Prentice-Hall, 1974.

Montagu, Ashley. "As If Living and Loving Were One." *Free Inquiry* 7 (3) (Summer 1987):40–42.

————. *Darwin: Competition and Cooperation*. New York: Henry Schuman, 1952.

————. *Immortality, Religion, and Morals*. New York: Hawthorne Books, 1971.

Montagu, Ashley. ed. *Science and Creationism*. New York: Oxford University Press, 1984.

Moody, Paul Amos. *Introduction to Evolution*. 3d ed. New York: Harper & Row, 1970.

Moog, F. "Alfred Russel Wallace: Evolution's Forgotten Man." *The American Biology Teacher* 22(7):414–18.

Moorehead, Alan. *Darwin and the Beagle*. New York: Harper & Row, 1969.

Morgan, C. Lloyd. *The Emergence of Novelty*. London: Williams & Norgate, 1933.

————. *Emergent Evolution*. New York: Henry Holt, 1927.

Morris, Richard. *Time's Arrows: Scientific Attitudes toward Time*. New York: Simon & Schuster, 1985.

Morris, S. Conway. "Burgess Shale Faunas and the Cambrian Explosion." *Science* (20 October, 1989):339–46.

Morrison, David, and Clark R. Chapman. "The New Castastrophism." *Skeptical Inquirer* 14(2) (Winter 1990) :141–52.

Mortensen, Viggo, and Robert C. Sorensen, eds. *Free Will and Determinism*. Philadelphia: Coronet Books, 1987.

Mowat, Farley. *Woman in the Mists: The Story of Dian Fossey and the Mountain Gorillas of Africa*. New York: Warner Books, 1987.

Muller, Richard. *Nemesis: The Death Star*. New York: Weidenfeld & Nicolson, 1988.

Munitz, Milton K. *Space, Time and Creation: Philosophical Aspects of Scientific Cosmology*. New York: Collier Books, 1961.

————, ed. *Theories of the Universe: From Babylonian Myth to Modern Science*. New York: Free Press, 1965.

Napier, J. R., and P. H. Napier. *The Natural History of the Primates*. Cambridge, Mass.: MIT Press, 1985.

Naroll, Raoul. *The Moral Order: An Introduction to the Human Situation*. Beverly Hills, Calif.: Sage, 1983.

Nelkin, Dorothy. *The Creation Controversy: Science or Scripture in the Schools*. New York: W. W. Norton, 1982.

Newell, Norman D. *Creation and Evolution: Myth or Reality?* New York: Cambridge University Press, 1982.

Nietzsche, Friedrich. *The Portable Nietzsche*. New York: Viking Press, 1968.

Novik, Ilya. *Society and Nature: Socio-Ecological Problems*. Moscow: Progress Publishers, 1981.

Olby, R. C. *Charles Darwin*. London: Oxford University Press, 1967.

Oparin, A. I. *Genesis and Evolutionary Development of Life*. New York: Academic Press, 1968.

Oparin, A. I. *The Origin of Life*. 2d ed. New York: Dover, 1965.

Orgel, Leslie E. *The Origins of Life*. New York: John Wiley & Sons, 1973.

Oro, J., et al., eds. *Cosmochemical Evolution and the Origins of Life*. The Netherlands: D. Reidel, 1974.

Osborn, J. F. *From the Greeks to Darwin*. 2d ed. New York: Scribner's, 1929.

Ospovat, Dov. *The Development of Darwin's Theory: Natural History, Natural Theology, and Natural Selection, 1838–1859*. Cambridge, Mass.: Cambridge University Press, 1981.

Osterbrock, Donald E., and Peter H. Raven, eds. *Origins and Extinctions*. New Haven, Conn.: Yale University Press, 1988.

Ostrom, John H. "A New Look at Dinosaurs." *National Geographic* 154(2): 152–85.

Pagels, Heinz R. *The Cosmic Code: Quantum Physics as the Language of Nature*. New York: Bantam Books, 1983.

———. *Perfect Symmetry: The Search for the Beginning of Time*. New York: Bantam Books, 1986.

Paradis, James, and George C. Williams. *Evolution and Ethics*. Lawrenceville, N.J.: Princeton University Press, 1989.

Pastner, Stephen, and William Haviland, eds. *Confronting the Creationists*. Northeastern Anthropological Association, Occasional Proceedings No. 1, 1982.

Paterson, Dale. *The Deluge and the Ark: A Journey into Primate Worlds*. Boston: Houghton Mifflin, 1989.

Patterson, Colin. *Evolution*. British Museum (Natural History), 1978.

Paul, Gregory S. "Giant Meteor Impacts and Great Eruptions: Dinosaur Killers?" *BioScience* 39(3):162–72.

Peacocke, Arthur, and Svend Andersen, eds. *Evolution and Creation: A European Perspective*. Aarhus, Denmark: Aarhus University Press, 1987.

Peirce, Charles S. *Essays in the Philosophy of Science*, esp. pp. 105–25. Edited by Vincent Tomas. New York: Bobbs-Merrill, 1957.

———. "Evolutionary Love." *The Monist* 3 (1893):176–200.

Perdue, William D. *Sociological Theory: Explanation, Paradigm, and Ideology*, esp. pp. 47–68. Palo Alto, Calif.: Mayfield, 1986.

Perry, Donald R. *Life Above the Jungle Floor*. New York: Simon and Schuster, 1986.

Popper, Karl R. *Objective Knowledge: An Evolutionary Approach*, esp. pp. 256–80. Oxford: Clarendon Press, 1972.

Prenant, M. *Biology and Marxism*. London: Lawrence and Wishart, 1943.

Pringle, J. W. S., ed. *Essays on Philosophical Evolution*. New York: Macmillan, 1965.

Provine, William B. *Sewall Wright and Evolutionary Biology*. Chicago: University of Chicago Press, 1989.

Putnam, John J. "The Search for Modern Humans." *National Geographic* 174(4):438–77.

Racle, Fred A. *Introduction to Evolution*. Englewood Cliffs, N.J.: Prentice-Hall, 1979.

Rahner, Karl. *Hominisation: The Evolutionary Origin of Man as a Theological Problem*. New York: Herder and Herder, 1965.

Ralling, Christopher, ed. *The Voyage of Charles Darwin*. New York: Mayflower Books, 1979.

Rampiro, Michael. "Dinosaurs, Comets and Volcanoes." *New Scientist* (18 February 1989):54–58.

Raupe, David M. *The Nemesis Affair: A Story of the Death of Dinosaurs and the Ways of Science*. New York: W. W. Norton, 1966.

Reese, Ronald Land, Steven M. Evertt, and Edwin D. Craun. "The Chronology of Archbishop James Ussher." *Sky and Telescope* 62(5):404–5.

Reese, Hubert. *Atoms of Silence: An Exploration of Cosmic Evolution*. Cambridge, Mass.: MIT Press, 1984.

Reichmann, James B. *Philosophy of the Person*, esp. pp. 257–72. Chicago: Loyola University Press, 1985.

Richards, Robert J. *Darwin and the Emergence of Evolutionary Theories of the Mind and Behavior*. Chicago: University of Chicago Press, 1988.

Riddle, Oscar. *The Unleashing of Evolutionary Thought*. New York: Vantage Press, 1954.

Riepe, Dale, ed. *Phenomenology and Natural Existence: Essays in Honor of Marvin Farber*. Albany: State University of New York Press, 1973.

Rifkin, Jeremy. *Entropy: A New World View*. New York: Bantam Books, 1981.

Rigaud, Jean-Phillippe. "Art Treasures from the Ice Age: Lascaux Cave." *National Geographic* 174(4):482–99.

Ritzer, George. *Sociological Theory*, esp. pp. 168–71, 215–17. 2d ed. New York: Alfred A. Knopf, 1988.

Rogers, James Allen. "Darwinism and Social Darwinism." *Journal of the History of Ideas* 33(2):265–80.

Rohr, Janelle, ed. *Science & Religion: Opposing Viewpoints*. St. Paul, Minn.: Greenhaven Press, 1988.

Ross, H. H. *Understanding Evolution*. Englewood Cliffs, N.J.: Prentice-Hall, 1966.

Roth, Robert J. "The Importance of Matter." *America* 109(25):792–94.

Rumney, Jay. *Herbert Spencer's Sociology*. New York: Atherton Press, 1965.

Ruse, Michael. *Biology & Philosophy*. Vol. 1. Boston: D. Reidel, 1986.

Ruse, Michael. *But Is It Science? The Philosophical Question in the Creation/ Evolution Controversy*. Buffalo, N.Y.: Prometheus Books, 1988.

————. *The Darwinian Paradigm: Essays on Its History, Philosophy, and Religious Implications*. New York: Routledge, 1989.

————. *The Darwinian Revolution: Science Red in Tooth and Claw*. Chicago: University of Chicago Press, 1979.

————. *Darwinism Defended: A Guide to the Evolution Controversies*. Reading, Pa.: Addison-Wesley, 1982.

————. "Darwin's Theory: An Exercise in Science." *New Scientist* 90(1259):828–30.

————. *Taking Darwin Seriously*. New York: Basil Blackwell, 1987.

Sagan, Carl. *Cosmos*. New York: Random House, 1980.

Sagan, Carl, and I. S. Shklovskii. *Intelligent Life in the Universe*. San Francisco: Holden-Day, 1966.

Sahlins, Marshall D. "The Origin of Society." *Scientific American* 203(3): 76–87.

Sahlins, Marshall D., and Elman R. Service, eds. *Evolution and Culture*. Ann Arbor: University of Michigan Press, 1961.

Sauer, Carl O. *Agriculture, Origins and Dispersals: The Domestication of Animals and Foodstuffs*. 2d ed. Cambridge, Mass.: MIT Press, 1969.

Savage, Jay M. *Evolution*, 2d ed. New York: Holt, Rinehart and Winston, 1969.

Schmitz-Moormann, Karl. *Das Weltbild Teilhard de Chardins*. Köln und Opladen: Westdeutscher Verlag, 1966.

————. *Die Erbsünde: Übergolte Vorstellung bleibender Glaube*. Olten und Freiburg im Breisgau: Walter-Verlag, 1969.

————. "Einleitung." In *Teilhard de Chardin in der Diskussion*, pp. 1–19. Darmstadt: Wissenschaftliche Buchgesellschaft, 1986.

————. "Ein zu früh geborener Prophet? Weg und Werk des Teilhard de Chardin (1881–1955)." In *Gegenentwürfe: 24 Lebensläufe für eine andere Theologie*, pp. 283–95. Edited by Hermann Häring and Karl-Josef Kuschel. München: R. Piper, 1988.

————. "Ohne den Apfel leben?" *Die Sache mit dem Apfel*, pp. 151–68. Edited by Joachim Illies. Verlag Herder KG Freiburg im Breisgau, 1972.

————. "On the Evolution of Human Freedom." *Zygon* 22(4):443–58.

————. "On the Need to Rewrite the Story of Creation." In *The Desire to Be Human*, pp. 292–301. Edited by Leo Zonneveld and Robert Muller. Netherlands: Mirananda/Wassenaar, 1983.

————. "Pierre Teilhard de Chardin—100 Years: Is There Anything He Has to Tell Us?" *The Teilhard Review* 16(1, 2):78–83.

Schmitz-Moorman, Karl. "The Stephen Jay Gould Hoax and the Piltdown Conspiracy." *The Teilhard Review* 16(3):7–15.

———. "Theological Methods in an Evolving World." *The Teilhard Review* 9(3):66–72.

Schmitz-Moormann, Nicole, and Karl Schmitz-Moormann, eds. *Pierre Teilhard de Chardin: Scientific Works.* 11 vols. Olten und Freiburg im Breisgau: Walter-Verlag, 1971.

Schoenwald, Richard L., ed. *Nineteenth-Century Thought: The Discovery of Change.* Englewood Cliffs, N.J.: Prentice-Hall, 1965.

Schopenhauer, Arthur. *The World as Will and Representation* (1818, 1844). 2 vols. New York: Dover, 1966.

Schopf, J. William. "The Evolution of the Earliest Cells." *Scientific American* 239(3):110–12, 114, 116–20, 126, 128–34, 137–38.

Schopf, J. William, ed. *Earth's Earliest Biosphere: Its Origin and Evolution.* Princeton, N.J.: Princeton University Press, 1938.

Schubert-Soldern, Rainer. *Mechanism and Vitalism: Philosophical Aspects of Biology.* Notre Dame, Ind.: University of Notre Dame Press, 1962.

Schwartz, Jeffrey H. *The Red Ape.* Boston: Houghton Mifflin, 1987.

Schwartzbach, Martin. *Alfred Wegener: The Father of Continental Drift.* Madison, Wis.: Science Tech, 1986.

Sellars, Roy Wood. *Evolutionary Naturalism.* Chicago: Open Court, 1969.

———. *The Philosophy of Physical Realism.* New York: Macmillan, 1932.

———. *The Principles and Problems of Philosophy.* New York: Macmillan, 1926.

———, ed. *Philosophy for the Future: The Quest of Modern Materialism.* New York: Macmillan, 1949.

Sevastyanov, V., A. Ursul, and Yu Shkolenko. *The Universe and Civilization.* Moscow: Progress Publishers, 1981.

Shapiro, Robert. *Origins: A Skeptic's Guide to the Creation of Life on Earth.* New York: Bantam Books, 1987.

Sibley, Jack R., and Peter A. Y. Gunter, eds. *Process Philosophy: Basic Writings.* Washington, D.C.: University Press of America, 1978.

Silk, Joseph. *The Big Bang: The Creation and Evolution of the Universe.* San Francisco: W. H. Freeman, 1980.

Simon, Walter M. "Herbert Spencer and the 'Social Organism'. " *Journal of the History of Ideas* 21(2):294–99.

Simons, Elwyn L. "Human Origins." *Science* (22 September 1989):1343–50.

Simpson, George Gaylord. *The Meaning of Evolution.* New York: New American Library, 1959.

———. *Tempo and Mode in Evolution.* New York: Columbia University Press, 1944.

Simpson, George Gaylord. *This View of Life: The World of an Evolutionist*. New York: Harcourt, Brace and World, 1964.

Sims, R. W., J. H. Price, and P. E. S. Whalley, eds. *Evolution, Time and Space: The Emergence of the Biosphere*. New York: Academic Press, 1983.

Smith, George H. *Atheism: The Case Against God*. Buffalo, N.Y.: Prometheus Books, 1979.

Smith, John Maynard. "Evolution: Generating Novelty by Symbiosis." *Nature* (28 September 1989):284–85.

Smuts, Jan Christiaan. *Holism and Evolution*. New York: Viking Press, 1961.

Sober, Elliott. *The Nature of Selection: Evolutionary Theory in Philosophical Focus*. Cambridge, Mass.: MIT Press, 1985.

———, ed. *Conceptual Issues in Evolutionary Biology: An Anthology*. Cambridge, Mass.: MIT Press, 1984.

Soleri, Paolo. *The Omega Seed: An Eschatological Hypothesis*. Garden City, N.Y.: Anchor Press/Doubleday, 1981.

Sorokin, Pitirim A. "Sociocultural Dynamics and Evolutionism." In *Twentieth Century Sociology*, pp. 96–101. Edited by Georges Gurvitch and Wilbert E. Moore. Freeport: Books for Libraries Press, 1971.

Speaight, Robert. *Teilhard de Chardin: A Biography*. St. James's Place, London: Collins, 1967.

Spencer, Herbert. *First Principles* (1862). New York: De Witt Revolving Fund, 1958.

———. *Synthetic Philosophy*. 10 vols. New York: D. Appleton, 1862–1893.

Stanley, Steven M. *The New Evolutionary Timetable: Fossils, Genes, and the Origin of Species*. New York: Basic Books, 1981.

Stansfield, William D. *The Science of Evolution*. New York: Macmillan, 1977.

Stapleton, William Olaf. *Last and First Men* (1931). Los Angeles: J. P. Tarcher, 1988.

———. *To the End of Time*. New York: Funk & Wagnalls, 1953.

Stebbins, G. Ledyard. *Processes of Organic Evolution*. Englewood Cliffs, N.J.: Prentice-Hall, 1966.

Steedman, David W., and S. Zousmer. *Galapagoes: Discovery on Darwin's Islands*. Washington, D.C.: Smithsonian Institution, 1988.

Strickberger, Monroe W. "Evolution and Religion." *BioScience* 23(7):417–21.

———. *Evolution*. Boston: Jones and Bartlett, 1990.

Susman, Randall L. *The Pygmy Chimpanzee*. New York: Plenum Press, 1984.

Tax, Sol, ed. *Evolution after Darwin*. 3 vols. Chicago: University of Chicago Press, 1960.

Taylor, Gordon Rattray. *The Great Evolution Mystery*. New York: Harper & Row, 1983.

Teilhard de Chardin, Pierre. "The Antiquity and World Expansion of Human Culture." In *Man's Role in Changing the Face of the Earth*. Edited by William L. Thomas, Jr. Chicago: University of Chicago Press, 1956.

———. *The Divine Milieu*, rev. ed. New York: Harper Torchbooks, 1968.

———. *Man's Place in Nature: The Human Zoological Group*. New York: Harper & Row, 1966.

———. *The Phenomenon of Man*, 2d ed. New York: Harper Torchbooks, 1965.

———. *Scientific Works*. 11 vols. Edited by Nicole and Karl Schmitz-Moormann. Olten und Freiburg im Breisgau: Walter-Verlag, 1971.

Thaxton, Charles B., Walter L. Bradley, and Roger L. Olsen. *The Mystery of Life's Origin: Reassessing Current Theories*. New York: Philosophical Library, 1984.

Timasheff, Nicholas S. *Sociological Theory: Its Nature and Growth*. New York: Random House, 1967.

Toulmin, Stephen. "Historical Inference in Science: Geology as a Model for Cosmology." *The Monist* 47(1):142–58.

Trachtman, Paul. "The Search for Life's Origins and a First 'Synthetic Cell'." *Smithsonian* 15(3):42–51.

Tracy, David, and Nichols Lash, eds. *Cosmology and Theology*. New York: Seabury Press, 1983.

Turner, Jonathan H. *Herbert Spencer: A Renewed Appreciation*. Beverly Hills, Calif.: Sage, 1985.

Tylor, Edward Burnett. *Anthropology: An Introduction to the Study of Man and Civilization* (1881). Ann Arbor: University of Michigan Press, 1960.

———. *Primitive Culture: Researches into the Development of Mythology, Philosophy, Religion, Language, Art and Custom* (1871). 2 vols. New York: Harper Torchbooks, 1958.

———. *Researches into the Early History of Mankind and the Development of Civilization* (1865). Chicago: Phoenix Books, 1964.

Tyndall, John. *Fragments of Science*. 2 vols. Brookfield, Vt.: Gower, 1892.

Van Cleave, Kent B. *Evolutionary Foundations for Philosophy*. Phoenix, Ariz.: Kinetic Visions, 1989.

Van Melsen, Andrew G. *Evolution and Philosophy*. Pittsburgh, Pa.: Duquesne University Press, 1965.

Van Ness, Peter H. "Nietzsche on Solitude: The Spiritual Discipline of the Godless." *Philosophy Today* 32(4):346–58.

Villiers, Alan. "In the Wake of Darwin's Beagle." *National Geographic* 136(4):449–95.

Volpe, E. Peter. *Understanding Evolution*. 5th ed. Dubuque, Ia.: William C. Brown, 1985.

Von Schilcher, Florian, and Neil Tennant. *Philosophy, Evolution and Human Nature*. Boston: Routledge & Kegan Paul, 1984.

Vorzimmer, Peter. "Darwin, Malthus, and the Theory of Natural Selection." *Journal of the History of Ideas* 30(4):527–42.

Waddington, C. H. *The Ethical Animal*. Chicago: University of Chicago Press, 1960, esp. pp. 65–71, 84–174.

Wald, George. "The Origin of Life." *Scientific American* 191(2):44–53.

Wallace, Alfred Russel. *Contributions to the Theory of Natural Selection: A Series of Essays*. New York: Macmillan, 1870.

———. *Darwinism: An Exposition on the Theory of Natural Selection with Some of Its Applications*. London: Macmillan, 1899.

———. *The World of Life: A Manifestation of Creative Power, Directed Mind and Ultimate Purpose* (1910). New York: Moffat/Yard, 1916.

Ward, Lester Frank. *Glimpses of the Cosmos*. 6 vols. New York: G. P. Putnam's Sons, 1913–1918.

———. "Mind as a Social Factor." In *Darwinism and the American Intellectual: A Book of Readings*, pp. 149–61. Edited by R. J. Wilson. Homewood, Ill.: Dorsey Press, 1967.

Warren, Preston. "Roy Wood Sellars (1889–1973)." *Philosophy and Phenomenological Research* 34(2):300–301.

Wartofsky, Marx W. *Feuerbach*. Cambridge, Mass.: Cambridge University Press, 1977.

Watson, James D. *The Double Helix: A Personal Account of the Discovery of the Structure of DNA*. New York: Mentor Books, 1977.

———. *Molecular Biology of the Gene*, 2d ed. New York: W. A. Benjamin, 1970.

Weaver, Kenneth F. "The Search for Our Ancestors." *National Geographic* 168(5):560–623.

Wegener, Alfred. *The Origin of Continents and Oceans* (1915, 1920, 1922, 1929). New York: Dover, 1966.

Weinberg, Steven. *The First Three Minutes: A Modern View of the Origin of the Universe*. New York: Bantam Books, 1979.

Weiner, Jonathan. *Planet Earth*. New York: Bantam Books, 1986.

Weismann, August. *The Evolutionary Theory*. London: Edward Arnold, 1904.

Wells, H. G., Julian Huxley, and G. P. Wells. *The Science of Life*. New York: Doran, 1930.

Wheelwright, Philip, ed. *The Presocratics*. New York: Odyssey Press, 1966.

White, Andrew Dickson. *A History of the Warfare of Science with Theology in Christendom* (1896). 2 vols. Magnolia, Mass.: Peter Smith.

White, Leslie A. *The Evolution of Culture: The Development of Civilization to the Fall of Rome*. New York: McGraw-Hill, 1959.

Whitehead, Alfred North. *Process and Reality: An Essay in Cosmology* (1929). New York: Free Press, 1979.

——. *Science and the Modern World*. New York: Macmillan, 1967.

Whorf, Benjamin. *Language, Thought, and Reality*. New York: Wiley, 1956.

Wichler, G. *Charles Darwin: The Founder of the Theory of Evolution and Natural Selection*. New York: Permagon Press, 1961.

Wiener, Philip P., ed. *Charles S. Peirce: Selected Writings (Values in a Universe of Chance)*. New York: Dover, 1966.

Wiener, Philip P. *Evolution and the Founders of Pragmatism*. New York: Harper Torchbooks, 1965.

Wilders, N. M. *An Introduction to Teilhard de Chardin*. New York: Harper & Row, 1968.

Wilson, Edward O. *On Human Nature*. Cambridge, Mass.: Harvard University Press, 1978.

——. *Sociobiology: The New Synthesis*. Cambridge, Mass.: Belknap Press of Harvard University Press, 1975 (see also the abridged edition: Cambridge, Mass.: Belknap Press of Harvard University Press, 1980).

——. "Threats to Biodiversity." *Scientific American* 261(3):108–12, 114, 116.

Wilson, Leonard G., ed. *Sir Charles Lyell's Scientific Journals on the Species Question*. New Haven, Conn.: Yale University Press, 1970.

Wilson, R. J., ed. *Darwinism and the American Intellectual: A Book of Readings*. Homewood, Ill.: Dorsey, 1967.

Winson, Jonathan. *Brain and Psyche: The Biology of the Unconscious*. Garden City, N.Y.: Anchor Press, 1985.

Wolfe, Elaine Claire Daughetee. "Acceptance of the Theory of Evolution in America: Louis Agassiz vs. Asa Gray." *The American Biology Teacher* 37(4):244–47.

Wylie, Philip. "Cultural Evolution: The Fatal Fallacy." *BioScience* 21(13): 729–31.

Zeller, Eduard. *Outlines of the History of Greek Philosophy*. New York: Meridian Books, 1965.

Zetterberg, J. Peter, ed. *Evolution Versus Creationism: The Public Education Controversy*. Phoenix, Ariz.: Oryx Press, 1983.

Zimmerman, P. A., J. W. Klotz, W. H. Rusch, and R. F. Surburg. *Darwin, Evolution, and Creation*. St. Louis, Miss.: Concordia, 1959.

Zirkle, C. *Evolution, Marxian Biology and the Social Sciences*. Philadelphia: University of Philadelphia Press, 1959.

Zonneveld, Leo, ed. *Humanity's Quest for Unity*. Netherlands: Mirananda/
Wassenaar, 1985.
Zonneveld, Leo, and Robert Muller, eds. *The Desire to Be Human*. Nether-
lands: Mirananda/Wassenaar, 1983.

Index